ELECTRONICS
FOR TODAY AND TOMORROW

second edition

TOM DUNCAN

JOHN MURRAY

Also by Tom Duncan

Advanced Physics: Fourth Edition
Adventures with Digital Electronics
Adventures with Electronics
Adventures with Microelectronics
Basic Skills: Electronics
GCSE Physics: Third Edition
Physics for the Caribbean
   (with Deniz Önaç)
Physics for Today and Tomorrow
Science for Today and Tomorrow
   (with M A Atherton and D G Mackean)
Success in Electronics

© Tom Duncan 1985, 1997

First published 1985
by John Murray (Publishers) Ltd
50 Albemarle Street London W1X 4BD

Reprinted 1985, 1986, 1987, 1989, 1991, 1993, 1994

Second edition 1997

Reprinted 1998, 2000, 2001

Typeset in 11/13.5 Goudy by Wearset, Boldon, Tyne and Wear.
Layouts by Eric Drewery.
Illustrations by Wearset, Boldon, Tyne and Wear.
Cover design by John Townson/Creation.
Cover photo © The Telegraph Colour Library.
Printed and bound in Great Britain by Henry Ling Limited, Dorchester

A catalogue entry for this title may be obtained from the British Library.

ISBN 0 7195 7413 7

PREFACE TO THE SECOND EDITION

The rapid growth in the importance of electronics in many fields has been mirrored by the development of numerous courses in physics, electronics and technology for post-GCSE and equivalent overseas students. This book is intended to cover the content of these courses.

This second edition has been extensively revised to match the requirements of present-day syllabuses and to ensure the text and illustrations are up to date. There are two new chapters, *67 More Boolean algebra* and *78 Electronics and society*, and chapters *89 Telephone system*, *91 Other telephone services*, *93 Digital computers* and *94 Computer peripherals* have been rewritten to cover the rapid changes occurring in these areas.

In addition, as well as updating certain items, e.g. resistance and capacitance codes, some new sections have been added, e.g. on *stepper motors*, MOSFETs and *modems*, and others no longer required by syllabuses have been deleted.

Checklists of specific objectives have been added to the beginning of each chapter to aid study and revision, and more *Progress questions* at the end of groups of related topics are included from recent examination papers.

The book will meet the needs of students taking:

(i) *Advanced Supplementary (AS) Electronics* courses intended primarily, but not exclusively, for those studying other science-based A or AS subjects,

(ii) *A-level Electronics* core and options as offered by several examining boards, either as a separate subject or as a module of an A-level Physics course, and

(iii) *other post-GCSE Electronics* courses.

The organization and treatment of the 98 chapters is such that the order in which they are studied may be varied, within limits, according to individual situations and preferences.

Thanks remain due to the examining bodies listed below for permission to use questions from their papers in the first edition, and further thanks are due to the Northern Examinations and Assessment Board (N.) for permission to use additional questions in this edition. All answers given are the sole responsibility of the author.

Northern Examinations and Assessment Board (N.)
Oxford and Cambridge Schools Examination Board (O. and C.)
The Associated Examining Board (A.E.B.)
University of Cambridge Local Examinations Syndicate (C.)
University of London Examinations and Assessment Council (L.)
University of Oxford Delegacy of Local Examinations (O.L.E.)

Once again I am indebted to my wife for producing the typescript and to Jane Roth for her meticulous editing.

T.D.

CONTENTS

ANALOGUE ELECTRONICS: TRANSISTORS

ANALOGUE ELECTRONICS: OP AMPS

DIGITAL ELECTRONICS: LOGIC CIRCUITS

DIGITAL ELECTRONICS: MULTIVIBRATORS

INFORMATION, ELECTRONICS AND SOCIETY

AUDIO SYSTEMS

RADIO AND TELEVISION

TELEPHONY

COMPUTERS AND MICROPROCESSORS

PHOTO ACKNOWLEDGEMENTS

Thanks are due to the following for permission to reproduce copyright photographs:

p.20 *l* & *br* RS Components, *tr* Gaye Allen*; **p.21** RS Components; **p.22** *l* & *r* RS Components; **p.24** RS Components; **p.25** *t* Maplin Electronics plc, *b* RS Components; **p.36** *tl, tr, bl* RS Components; *br* Maplin Electronics plc; **p.37** RS Components; **p.38** RS Components; **p.42** RS Components; **p.44** RS Components; **p.46** RS Components; **p.53** Maplin Electronics plc; **p.60** RS Components; **p.62** RS Components; **p.65** RS Components; **p.76** Astrid & Hans Frieder Michler/Science Photo Library; **p.77** Brierley Photo Library; **p.78** *tl* David Parker/Seagate Microelectronics Ltd/Science Photo Library, *bl* Ray Ellis/Science Photo Library, *r* Takeshi Takahara/Science Photo Library; **p.82** RS Components; **p.86** RS Components; **p.90** Philip Harris; **p.91** Unilab; **p.93** Unilab; **p.95** Unilab; **p.96** Philip Harris; **p.132** Unilab; **p.145** Unilab; **p.165** Unilab; **p.168** Unilab; **p.171** Unilab; **p.182** *tl* Royal Shrewsbury Hospital/Simon Fraser/Science Photo Library, *tr* Racal Group Services Limited, *bl* University of St. Andrews, *br* Explorer/Robert Harding Picture Library; **p.185** Science Museum, London/Science & Society Picture Library; **p.191** Philips Consumer Electronics; **p.205** *t* Intelsat, *b* Barnaby's Picture Library; **p.207** BT Archives; **p.208** BT Archives; **p209** BICC plc; **p.210** *l* Cable & Wireless plc, *r* STC Components Ltd; **p.211** *l* BT Archives, *r* Roger Foley/Science Photo Library; **p.212** BT Archives; **p.213** *tl* Zefa, *tr* Aston Communications, *cr* BT Corporate Picture Library; *br* Robert Reichert/Robert Harding Picture Library; **p.218** *tl* Zefa, *tr* Lawrence Migdale/Science Photo Library, *bl* Philippe Plailly/Science Photo Library, *br* Larry Mulvehill/Science Photo Library; **p.220** Compaq Presario 8700 Series; **p.221** VR Solutions Limited; **p.225** © Patrick Llewelyn-Davies. Courtesy of Hi-Grade Computers plc; **p.227** Unilab; **p.230** STC/A. Sternberg/Science Photo Library; **p.231** Mondex UK Ltd; **p.234** *tl* Locktronics Ltd, *tr* Unilab, *bl* RS Components, *br* Gaye Allen*; **p.235** RS Components.

*Every effort has been made to contact these copyright holders, who gave their permission for reproduction in the previous edition; the publishers apologise for any omissions and will be pleased to rectify this at the earliest opportunity.

BASIC ELECTRICITY

1 ELECTRIC CURRENT

CHECKLIST

After studying this chapter you should be able to:
- state that electric current is a flow of electrons, measured in amperes by an ammeter,
- use the equation $Q = It$,
- recognize circuit symbols for wires, cells, batteries, switches, lamps and ammeters, and
- state and use Kirchhoff's first law for series and parallel circuits.

ATOMS AND ELECTRIC CHARGES

An atom consists of a nucleus containing *protons*, which have a positive (+) electric charge, surrounded by an equal number of *electrons* with a negative (−) charge. The charge on a proton is the same size as that on an electron, making the whole atom electrically neutral. The nucleus also contains uncharged *neutrons*. Fig. 1.1 is a simplified 'picture' of an atom.

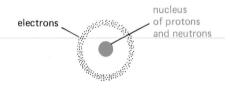

electrons — nucleus of protons and neutrons

Fig. 1.1

An atom can lose one or more electrons. If it does it becomes positively charged, because it then has more protons than electrons, and is called a *positive ion*. If it gains one or more electrons it becomes a *negative ion*, Fig. 1.2.

normal atom   positive ion   negative ion

Fig. 1.2

Electric charges exert forces on one another.

> *Like charges repel, opposite charges attract.*

ELECTRIC CURRENT: WHAT IS IT?

An electric current is produced when electric charges (electrons or ions) move in a definite direction.

In *metals*, the outer electrons are held loosely by their atoms and are free to move around the fixed positive

metal ions making up the regular lattice (i.e. crystalline) structure of the metal. This free electron motion is normally haphazard, with as many electrons moving in one direction as in the opposite direction at any time, Fig. 1.3a. Overall there is no net flow of charge in any direction, i.e. no current.

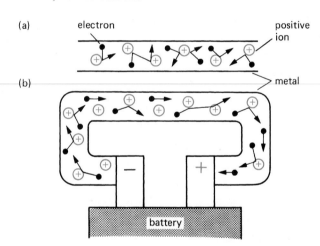

Fig. 1.3

When the metal is part of a circuit connected to a battery, the battery acts as an electron pump. It forces the free electrons to drift through the metal, in the direction from its negative (−) terminal towards its positive (+) terminal and then, in effect, through the battery itself, Fig. 1.3b.

> The flow of electrons in one direction is an *electric current* which can reveal itself by making the metal warmer and by deflecting a nearby magnetic compass.

The drift speed of the free electrons through the metal may, surprisingly, be less than 1 mm per second due to their frequent collisions with the fixed positive metal ions. But they *all* start drifting in the same direction as soon as the battery is connected, just as all the links in a bicycle chain move at the instant the pedals are pushed. If the circuit is broken the current stops but the haphazard motion of the free electrons goes on.

In some *liquids* and *gases*, under certain conditions, there are free positive and negative ions which can move when a battery is connected. The current produced consists of positive ions moving through the liquid or gas towards the − terminal of the battery and negative ions moving towards the + terminal, as shown in Fig. 1.4.

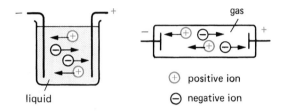

Fig. 1.4

CONDUCTORS AND INSULATORS

Electrical conductors are materials containing electric charges that are free to move. They allow current to pass through them easily. The best conductors are the metals silver, copper and gold.

In electrical insulators such as polythene, Perspex and PVC (polyvinyl chloride), all electrons are bound firmly to their atoms, making charge flow difficult. They can however be charged by rubbing. For example, when polythene is rubbed with a cloth, electrons are transferred from the cloth to the polythene. The cloth is left with a positive charge and the polythene becomes negatively charged, Fig. 1.5. The charges produced cannot move on the insulator, i.e. they are static electric charges.

Fig. 1.5

DIRECTION OF CURRENT

Before the electron was discovered it was thought that a current consisted of positive charges moving from the battery's + terminal round the circuit to its − terminal. We now know that this is partly true for conduction by liquids (electrolytes) and gases; it is not true for metals.

The original choice has been kept because it does not matter which direction is chosen provided we always keep

to the same one. Also, the laws of electricity were drawn up assuming it was true.

The conventional direction of current, called the *conventional current*, is the direction in which positive charges would flow. It is opposite to the direction of electron flow, Fig. 1.6.

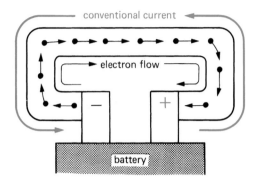

Fig. 1.6

UNITS OF CHARGE AND CURRENT

A huge number of charges drift past each point in a circuit per second. The quantity of charge carried by about six million million million (6.3×10^{18} to be more precise) electrons is called 1 *coulomb* (C).

When 1 coulomb passes each point in a circuit every second, the current is 1 *ampere* (A). That is,

1 ampere = 1 coulomb per second   (1A = 1 C/s)

If 2 C pass in 1 s, the current is 2 C/s = 2 A. If 6 C pass in 2 s, the current is 6 C/2 s = 3 A. In general if Q coulombs pass in t seconds, the current I in amperes is given by:

$$I = \frac{Q}{t} \quad \text{or} \quad Q = It$$

Two smaller units of currents are:

1 milliampere (mA) = 1/1000 A = $1/10^3$ A = 10^{-3} A
1 microampere (μA) = 1/1 000 000 A = $1/10^6$ A = 10^{-6} A

Electric current is measured by an *ammeter*.

CIRCUITS AND DIAGRAMS

Currents require complete conducting paths (circuits). Wires of copper are used to connect batteries, lamps, switches, etc. in a circuit. If the wires are covered with insulation, e.g. a plastic such as PVC, the insulation must be removed from the ends before connecting up. Some common symbols (signs) used in circuit diagrams are shown in Fig. 1.7.

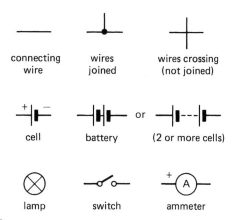

Fig. 1.7

KIRCHHOFF'S FIRST LAW

To measure a current, the circuit has to be broken and an ammeter connected in the gap with its + terminal going to the + terminal of the battery, as shown in Fig. 1.8a, b.

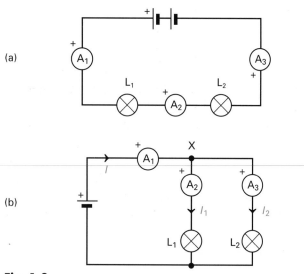

Fig. 1.8

Series circuit In Fig. 1.8a the two lamps L_1 and L_2 are in series, i.e. one after the other. The readings on the ammeters A_1, A_2 and A_3 will be equal and show that the current is the *same* in all parts of the circuit.

Parallel circuit In Fig. 1.8b L_1 and L_2 are in parallel, i.e. side by side, and there are alternative paths for the main current I which splits into I_1 and I_2. The readings on A_1, A_2 and A_3 show that:

$$I = I_1 + I_2$$

That is, the main current equals the sum of the branch currents. For example, if $I_1 = 0.2$ A and $I_2 = 0.3$ A then $I = 0.5$ A.

First law In neither of the above circuits is current used up. This is a consequence of the fact that electric charge is always conserved and, stated in terms of currents, this is summed up by Kirchhoff's first law:

The sum of the currents entering a junction (such as X in Fig. 1.8b) equals the sum of the currents leaving it.

QUESTIONS

1. Why are (i) metals, (ii) some liquids and gases, conductors?
2. If the current through a floodlamp is 5 A, what charge passes in (i) 1 s, (ii) 10 s, (iii) 5 minutes?
3. What is the current in a circuit if the charge passing each point is (i) 10 C in 2 s, (ii) 20 C in 40 s, (iii) 240 C in 2 minutes?
4. **a)** Express the following in mA: (i) 1 A, (ii) 0.5 A, (iii) 0.02 A.
   b) Express the following in μA: (i) 2 mA, (ii) 0.4 mA, (iii) 0.005 mA.

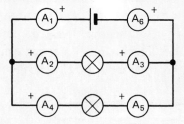

Fig. 1.9

5. The two lamps in Fig. 1.9 are the same. If A_1 reads 1 A, what do A_2, A_3, A_4, A_5 and A_6 read?

2 E.M.F., P.D. AND VOLTAGE

CHECKLIST

After studying this chapter you should be able to:
- state that potential difference (p.d.) or voltage causes current and is measured in volts by a voltmeter, and

- state Kirchhoff's second law for series and parallel circuits and use the equation $E = V + v$.

ELECTROMOTIVE FORCE OF A BATTERY

The battery in Fig. 2.1 supplies the electrical force and energy which drives the free electrons round the circuit as a current. As they drift along they give up most of their energy as heat and light in the lamp. It is useful to imagine that as each coulomb leaves the battery, it receives a fixed amount of electrical energy which depends on the battery.

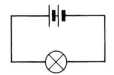

Fig. 2.1

The *electromotive force* (e.m.f.) E of a battery is defined to be 1 *volt* if it gives 1 joule of electrical energy to each coulomb of charge passing through it. That is,

1 volt = 1 joule per coulomb   (1 V = 1 J/C)

A 6 V battery gives 6 J of energy to each coulomb passing through it.

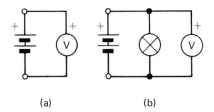

Fig. 2.2   (a)   (b)

The e.m.f. of a battery can be measured by connecting a *voltmeter* across it, Fig. 2.2a, with its + terminal (often coloured red) going to the + of the battery.

POTENTIAL DIFFERENCE ACROSS A DEVICE

The lamp in the circuit of Fig. 2.2b changes most of the electrical energy carried by the free electrons into heat and light. A negligible amount is lost as heat in the copper connecting wires.

The *potential difference* (p.d.) across a device (e.g. a lamp) in a circuit is 1 *volt* if it changes 1 joule of electrical energy into other forms of energy (e.g. heat and light) when 1 coulomb passes through it.

The p.d. across a device is 2 V if it changes 2 J when 1 C passes through it. In general, if W joules are changed when Q coulombs pass, the p.d. V in volts across the device is:

$$V = \frac{W}{Q}$$

A voltmeter is also used to measure p.d., being connected *across* the device as in Fig. 2.2b.

E.m.fs and p.ds are usually called *voltages*, since both are measured in volts.

The flow of electric charge in a circuit can be compared with the flow of water in a pipe. A pressure difference is required to make water flow. To move electric charge we consider that a p.d. is needed and whenever there is current between two points in a circuit there must be a p.d. across them.

TERMINAL P.D. OF A BATTERY

When a battery drives current round a circuit some of the electrical energy carried by the charge is needed to get the current through the battery itself. This energy is changed to heat in the battery. There is therefore less energy available to drive each coulomb round the rest of the circuit.

The *terminal p.d.* V of a battery when it is driving current, i.e. on closed circuit, is less than the e.m.f. E of the battery because some energy and volts v are 'lost'. On open circuit when the battery is not driving current, the *terminal p.d. equals the e.m.f.* We can say for each coulomb, assuming energy is conserved:

energy supplied = energy changed + energy 'lost'
by battery   in devices   inside battery

or,   e.m.f. = terminal p.d. + 'lost' volts

That is,

$$E = V + v$$

A voltmeter connected across a battery on open circuit only measures the e.m.f. if it does not require the battery to drive current through it.

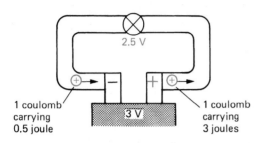

Fig. 2.3

In Fig. 2.3, if the e.m.f. of the battery is 3 V and considering conventional current, we can think of each coulomb as:

(i) leaving the + terminal of the battery with 3 J of electrical energy,

(ii) changing 2.5 J of electrical energy into heat and light in the lamp,

(iii) arriving at the − terminal of the battery with 0.5 J of electrical energy which it requires to get through the battery, and

(iv) picking up another 3 J of electrical energy as it leaves the + terminal.

In this example, the useful (terminal) p.d. available for driving current through the lamp is 2.5 V and the 'lost' volts is 0.5 V, making a total of 3 V, which is the e.m.f. of the battery.

In general,

e.m.f. = sum of p.ds across devices + 'lost' volts

KIRCHHOFF'S SECOND LAW

What happens to e.m.f. and p.d. round a circuit is summed up by the Second Law and follows from the fact that energy is always conserved.

The sum of all the p.ds round a circuit equals the e.m.f. of the battery.

CELLS AND BATTERIES

A battery consists of two or more *cells* joined in series, i.e. + of one to − of next. This gives a greater e.m.f. than one cell alone. In Fig 2.4a the two 1.5 V cells have a total e.m.f. of 3 V.

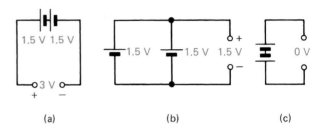

Fig. 2.4

If two 1.5 V cells are connected in parallel, Fig. 2.4b, the e.m.f. is still 1.5 V but they behave like one larger cell with more energy and so last longer.

The two 1.5 V cells in Fig. 2.4c are in opposition and their combined e.m.f. is zero.

POTENTIAL AT A POINT

Although it is usually the p.d. between two points in a circuit we have to consider, there are times when it is a help to deal with the *potential at a point*. To do this we choose one point in the circuit as having zero potential, i.e. 0 V. The potentials of all other points are stated with reference to it; the potential at any point is then the p.d. between the point and 0 V.

For example, if we take point C in Fig. 2.5a (in effect the − terminal of the battery) as our zero, the potential at B is +2.5 V and at A (the + terminal of the battery) it is +6 V. Alternatively, taking A as being at 0 V, then B is at −3.5 V and C at −6 V.

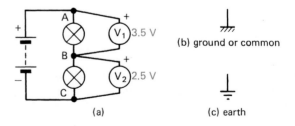

Fig. 2.5

In electronic circuits the point chosen as being at 0 V is called *ground* or *common*, or 'earth' if it is connected to earth (e.g. via the earthing pin on a 3-pin plug). The signs used are given in Fig. 2.5b, c.

QUESTIONS

1. If the e.m.f. of a battery is 12 V, how many joules of electrical energy are given to a charge passing through of   (i) 1 C,   (ii) 3 C?

2. The p.d. across a lamp is 6 V. What does this mean?

3. In the circuits of Fig. 2.6, the voltmeter V requires almost zero current to take a reading. What is
   a) the e.m.f. of the cell,
   b) its terminal p.d. when driving current through the lamp L,
   c) the 'lost' volts?

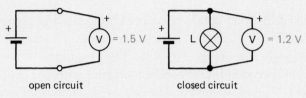

open circuit   closed circuit

Fig. 2.6

4. What are the e.m.fs of the batteries of 1.5 V cells connected as in Fig. 2.7a, b?

Fig. 2.7

5. Three voltmeters V, V_1 and V_2 are connected as in Fig. 2.8. Copy and complete the table of voltmeter readings (in volts) shown, which were obtained with three different batteries.

V	V_1	V_2
	12	6
6	4	
12		4

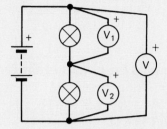
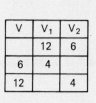

Fig. 2.8

6. In Fig. 2.9, L_1 and L_2 are identical lamps. Assuming the 'lost' volts in the 6 V battery is zero, what is the potential at each of points A, B and C if 'ground' (0 V) is taken as (i) C, (ii) A, (iii) B?

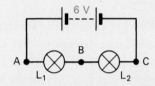

Fig. 2.9

7. a) Express the following in mV: (i) 1 V, (ii) 0.7 V,   (iii) 0.02 V.
   b) Express the following in V: (i) 1600 mV, (ii) 400 mV,   (iii) 50 mV.

3 RESISTANCE AND OHM'S LAW

CHECKLIST

After studying this chapter you should be able to:
• state that resistance is measured in ohms,
• state and use the equation $R = V/I$ in its three forms using different units,
• recognize the symbol for a resistor,
• recognize I–V graphs for ohmic and non-ohmic conductors,

• calculate the effective resistance of a number of resistors in series,
• calculate the effective resistance of two resistors in parallel, and
• explain the effect of temperature on the resistance of metals and semiconductors.

RESISTANCE

Conductors oppose current to a certain extent, some more than others. The *resistance R* of a conductor in *ohms* (Ω: pronounced 'omega') is defined by:

$$R = \frac{V}{I}$$

where V is the p.d. across the conductor in *volts* and I is the resulting current in *amperes*. When $V = 6\,V$ and $I = 2\,A$, $R = 6/2 = 3\,\Omega$: but if $I = 1\,A$ then $R = 6/1 = 6\,\Omega$.

Larger units of resistance are:

$$1\ \textit{kilohm}\ (k\Omega) = 1000\ \Omega = 10^3\ \Omega$$
$$1\ \textit{megohm}\ (M\Omega) = 1000\ k\Omega = 10^6\ \Omega$$

In electronics I is often in mA (or μA) and V in volts; this gives R in $k\Omega$ (or $M\Omega$).

Conductors especially made to have resistance are called *resistors* (—▭—).

The equation $R = V/I$ can also be written $V = IR$ and $I = V/R$. All are used in calculations and the triangle in Fig. 3.1 is a memory aid. If you cover the quantity you want with your finger, e.g. I, it equals what you still see, i.e. V/R.

Fig. 3.1

If there is current through a resistor, the *potential at one end must be greater than at the other*, i.e. a p.d. exists across it. In Fig. 3.2 the potential at X is 9 V if S is open and 0 V if it is closed. This fact is the basis of electronic switching circuits (p. 69).

Fig. 3.2

OHM'S LAW

For metals, carbon and some alloys, V/I is constant whatever the value of V, if their temperature does not change. Since $R = V/I$ it follows that the resistance of such conductors is constant for different p.ds. We can write:

$$\frac{V}{I} = \text{constant} \quad \text{or} \quad I \propto V$$

This is Ohm's law which states that *the current through a conductor is directly proportional to the p.d. across it if the temperature is constant.* Doubling or trebling V therefore doubles or trebles I.

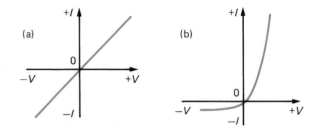

Fig. 3.3

Ohmic or *linear* conductors obey Ohm's law and a graph of I against V, called a *characteristic curve*, is a straight line through the origin, Fig. 3.3a. The resistance of a non-ohmic or non-linear conductor varies with the p.d. and its I–V graph is curved. The one in Fig. 3.3b is for a semiconductor diode (p. 58).

RESISTOR NETWORKS

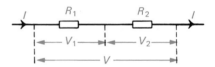

Fig. 3.4

Series Two resistors in series are shown in Fig. 3.4. Note two points:

(i) I is the same in R_1 and R_2, and
(ii) $V = V_1 + V_2$, where $V_1 = IR_1$ and $V_2 = IR_2$, i.e. $V_1/V_2 = R_1/R_2$.

It can be shown that the combined resistance R is:

$$R = R_1 + R_2$$

Parallel Two resistors in parallel are shown in Fig. 3.5. Note two points:

(i) $I = I_1 + I_2$ and $I_1 > I_2$ if $R_1 < R_2$ and vice versa, and
(ii) V is the same across R_1 and R_2, i.e. $V = I_1R_1 = I_2R_2$.

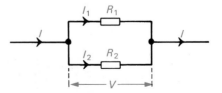

Fig. 3.5

It can be shown that if the combined resistance is R then:

$$\frac{1}{R} = \frac{1}{R_1} + \frac{1}{R_2} \quad \text{or} \quad R = \frac{R_1 \times R_2}{R_1 + R_2}$$

If $R_1 = R_2$ then $R = R_1^2/2R_1 = R_1/2$.

Example For the network in Fig. 3.6 calculate **(a)** the combined resistance R_p of 3 Ω and 6 Ω in parallel, **(b)** the total resistance R of the network, **(c)** I, **(d)** V_p and V_3, and **(e)** I_1 and I_2.

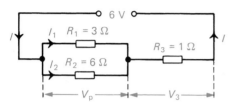

Fig. 3.6

(a) $\dfrac{1}{R_p} = \dfrac{1}{3} + \dfrac{1}{6} = \dfrac{2}{6} + \dfrac{1}{6} = \dfrac{3}{6} \quad \therefore R_p = \dfrac{6}{3} = 2\ \Omega$

Note that R_p is less than either resistance.

(b) $R = R_p + R_3 = 2 + 1 = 3\ \Omega$

(c) The p.d. V across the network is 6 V.

$\therefore I = \dfrac{V}{R} = \dfrac{6}{3} = 2\ \text{A}$

(d) $V_p = IR_p = 2 \times 2 = 4\ \text{V}$

But $V = V_p + V_3 \quad \therefore V_3 = V - V_p = 6 - 4 = 2\ \text{V}$

(e) $I_1 = \dfrac{V_p}{R_1} = \dfrac{4}{3}\ \text{A}$

But $I = I_1 + I_2 \quad \therefore I_2 = I - I_1 = 2 - \dfrac{4}{3} = \dfrac{2}{3}\ \text{A}$

RESISTANCE AND TEMPERATURE

If the temperature of a conductor rises, due for example to the heat produced by the current in it, its resistance increases if it is a metal but decreases if it is a pure semiconductor (p. 56) or carbon.

The change is given by the *temperature coefficient of resistance* α, defined by:

$$\alpha = \frac{R_\theta - R_0}{R_0 \times \theta}$$

where R_θ and R_0 are the resistances at θ and 0 °C. For metals α is positive, for semiconductors and carbon it is negative.

To find α for say copper, the resistance of a coil of thin copper wire is measured at different temperatures. A graph of resistance against temperature is plotted. Between 0 and 100 °C it is a straight line, Fig. 3.7, and α is calculated from it.

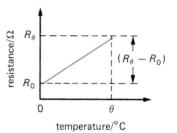

Fig. 3.7

The different behaviour of metals and semiconductors is due to the fact that in metals the positive ions vibrate more vigorously if the temperature rises and make it more difficult for the free electrons to flow. In pure semiconductors this also happens but more charge carriers are 'freed' as well, thereby, in effect, decreasing the resistance.

RESISTIVITY

The resistance R of a conductor is directly proportional to its length l, i.e. $R \propto l$, and inversely proportional to its cross-sectional area A, i.e. $R \propto 1/A$. Combining these two statements we get $R \propto l/A$ or

$$R = \rho \frac{l}{A}$$

where ρ (rho) is a constant called the *resistivity* of the material. If $l = 1$ m and $A = 1$ m^2 then $\rho = R$. That is, the resistivity equals the resistance of a 1 m length of cross-sectional area 1 m^2. Knowing ρ (in ohm metre) of a material, the resistance of different-sized samples can be calculated.

VARIABLE RESISTORS

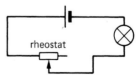

Fig. 3.8

A variable resistor has a connection at each end and one which can slide along it. When used as a *rheostat* to control the current in a circuit, connections are made to one end only and to the sliding contact, Fig. 3.8.

QUESTIONS

1. a) If $V = 9$ V and $I = 5$ mA, find R.
   b) Find V when $I = 0.5$ mA and $R = 10$ kΩ.
   c) What is R if $I = 3$ μA when $V = 6$ V?

2. What is the resistance in kilohms of the network in Fig. 3.9 between (i) P and Q, (ii) Q and R, (iii) P and R?

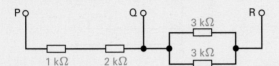

Fig. 3.9

3. a) Calculate the total effective resistance between P and Q in the network of Fig. 3.10.
   b) 6 V is now applied between P and Q. Calculate the voltage between R and S. (A.E.B.)

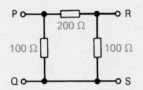

Fig. 3.10

4. Calculate I, I_1 and I_2 in the network of Fig. 3.11.

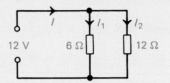

Fig. 3.11

5. In Fig. 3.12 what are the potentials at A and B when S is (i) open, (ii) closed?

Fig. 3.12

4 METERS AND MEASUREMENT

CHECKLIST

After studying this chapter you should be able to:
- state that an ammeter has a low resistance and is connected in series,
- state that a voltmeter has a high resistance and is connected in parallel,
- perform calculations using shunts and multipliers,
- state that the sensitivity of a voltmeter is measured in ohms per volt, and
- measure resistance.

MOVING-COIL METER

Most ammeters and voltmeters are moving-coil microammeters (galvanometers) adapted to measure a range of currents and voltages.

A moving-coil meter consists of a coil of wire pivoted so that it can turn between the poles of a U-shaped magnet, Fig. 4.1. The coil has a pointer fixed to it which moves over a scale. When current passes through the coil, it becomes a magnet, turns and winds up a spiral hair spring until the force caused by the current balances the resisting force of the wound-up spring. The greater the current, the more the coil turns and the larger the pointer deflection. The scale is linear, i.e. the divisions are evenly spaced.

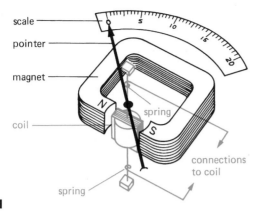

Fig. 4.1

Two important properties of the meter are the *resistance* of the coil and the *current to give a full scale deflection* (f.s.d.). If, for example, the coil resistance is $1000\,\Omega$ and the f.s.d. current is $100\,\mu A$ (0.1 mA), a p.d. $= 1000 \times 0.1 = 100$ mV will cause an f.s.d. The meter is both a microammeter reading up to $100\,\mu A$ and a millivoltmeter reading up to 100 mV.

Connecting a meter into a circuit to make a measurement should cause minimum disturbance to the conditions which existed before.

AMMETERS AND SHUNTS

An ammeter is inserted in *series* in a circuit. It must have a *low* resistance compared with the rest of the circuit, otherwise it changes the current to be measured.

To convert the previous microammeter to an ammeter reading up to 1 A, a low value resistor called a *shunt* is connected in *parallel* with the meter. It must have a value S which, when the current is 1 A, allows only $100\,\mu A$ (0.0001 A) to go through the meter and provides a bypass itself for the remaining 0.9999 A.

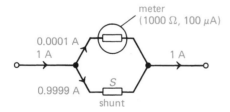

Fig. 4.2

From Fig. 4.2 we have:

p.d. across meter = p.d. across shunt

$\therefore 0.0001 \times 1000 = 0.9999 \times S$ (from $V = IR$)

$$\therefore S = \frac{0.0001 \times 1000}{0.9999} \approx 0.1\,\Omega$$

The resistance of the ammeter (meter + shunt) is $0.1\,\Omega$.

VOLTMETERS AND MULTIPLIERS

A voltmeter is connected in *parallel* with the component, e.g. a resistor, across which the p.d. is to be measured. It should have a *high* resistance compared with the resistor (at least 10 times greater), otherwise the total resistance of the whole circuit is reduced by the 'loading' effect of the voltmeter and the required p.d. changes.

To convert the same microammeter to a voltmeter reading $0 - 1$ V, a high value resistor, called a *multiplier*, is connected in *series* with the meter. It must have a value M which, when a p.d. of 1 V is applied across multiplier and meter together, restricts the current to $100\,\mu A$ (0.0001 A) and an f.s.d. is recorded.

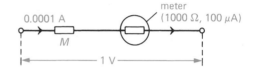

Fig. 4.3

From Fig. 4.3 we have:

p.d. across multiplier and meter at f.s.d.
$$= 0.0001 (M + 1000) = 1 \text{ (from } V = IR)$$

$\therefore M + 1000 = 1/0.0001 = 10\,000$

$\therefore M = 9000\,\Omega$

The resistance of the voltmeter (meter + multiplier) is $10\,000\,\Omega$.

SENSITIVITY OF A VOLTMETER

Definition The *sensitivity* (or quality) of a voltmeter is stated in ohms per volt (Ω/V). The larger it is, the smaller is the current taken and the less is the disturbance to the circuit. It is calculated from:

$$\text{sensitivity} = \frac{\text{resistance of meter + multiplier}}{\text{p.d. required to give f.s.d.}}$$

For example, for the voltmeter above,

$$\text{sensitivity} = 10\,000\,\Omega/1\,V = 10\,000\,\Omega/V$$

To extend its range to 10 V, a multiplier of $99\,000\,\Omega$ would be required to make the resistance of the voltmeter (meter + multiplier) $100\,000\,\Omega$ and so keep the full scale current at $100\,\mu A$. The sensitivity is still $10\,000\,\Omega/V$. To read up to 100 V this voltmeter would need to have a resistance of $1\,M\Omega$.

Example In Fig. 4.4, $\widehat{V_1}$ and $\widehat{V_2}$ are two voltmeters of sensitivity $10\,k\Omega/V$ used on their 10 V range. The resistance of each is therefore $100\,k\Omega$.

(V_1) gives a fairly accurate reading of the p.d. (4.5 V) which existed across R_2 (10 kΩ) before (V_1) was connected; its resistance is 10 times greater than R_2.

(V_2) gives a misleading reading of the p.d. across R_4 (100 kΩ) because its resistance is the same as R_4. The combined resistance of R_4 and (V_2) in parallel is 50 kΩ. The loading effect of (V_2) is large and it records the p.d. across R_4 as 3 V instead of 4.5 V.

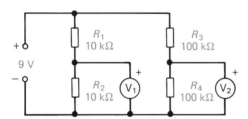

Fig. 4.4

Electronic voltmeters (p. 91) have much higher resistances, e.g. 10 MΩ, than moving-coil types.

MEASUREMENT OF RESISTANCE BY THE AMMETER–VOLTMETER METHOD

The resistance R of a conductor can be found from $R = V/I$ using the circuit of Fig. 4.5. The ammeter (A) gives the sum of the currents in R and the voltmeter (V), which can be taken as the current I through R if the resistance of (V) is much greater than that of R. If not, (V) is connected across (A) and R in series so that (A) records the exact current I in R and the reading on (V) nearly equals the p.d. V across R if the resistance of (A) is much lower than R.

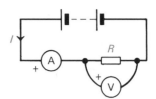

Fig. 4.5

QUESTIONS

1. A meter has a resistance of 50 Ω and an f.s.d. current of 1 mA. How can it be adapted to measure
   a) currents up to 1 A,
   b) p.ds up to 10 V?
2. If a voltmeter has a sensitivity of 1000 Ω/V, what is its resistance on (i) the 1 V range, (ii) the 5 V range? On which range will it give a more accurate measurement of the p.d. across a certain resistor? Why?
3. What is the f.s.d. current of a voltmeter with a sensitivity of (i) 2000 Ω/V, (ii) 20 000 Ω/V?
4. In Fig. 4.6 what is the reading on V if it has a resistance on the scale used of (i) 20 kΩ, (ii) 2 kΩ?

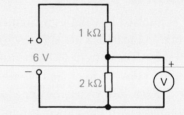

Fig. 4.6

5 POTENTIAL DIVIDER

CHECKLIST

After studying this chapter you should be able to:
• calculate the output voltage from a potential divider, and

• state the effect of the load on the output voltage.

A potential divider divides a voltage so that its output voltage is some fraction of the input voltage.

USING TWO FIXED RESISTORS

In the circuit of Fig. 5.1a the 1 kΩ and 2 kΩ resistors divide the 6 V input voltage into three (1 + 2) equal parts. One part (2 V) appears across the 1 kΩ and two

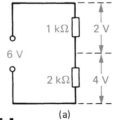 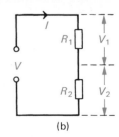

Fig. 5.1

parts (4 V) across the 2 kΩ resistor. The division is in the ratio of the two resistances, i.e. 1 kΩ/2 kΩ = 1/2 since the same current passes through each.

For the more general circuit of Fig. 5.1b we can say:

$$V_1 = IR_1 \tag{1}$$
$$V_2 = IR_2 \tag{2}$$
$$V = V_1 + V_2 = I(R_1 + R_2) \tag{3}$$

Dividing (1) by (3):

$$\frac{V_1}{V} = \frac{IR_1}{I(R_1 + R_2)} = \frac{R_1}{R_1 + R_2}$$

$$\therefore V_1 = \frac{R_1}{R_1 + R_2} \cdot V$$

Similarly from (2) and (3) we get:

$$V_2 = \frac{R_2}{R_1 + R_2} \cdot V$$

Note also from (1) and (2) that:

$$\frac{V_1}{V_2} = \frac{R_1}{R_2}$$

USING A VARIABLE RESISTOR

If all three connections on a variable resistor are used, it acts as a potential divider or potentiometer ('pot') in which the ratio R_1/R_2 is readily changed.

In Fig. 5.2 the resistance between A and B represents R_1 and that between B and C represents R_2. A continuously variable output voltage, from 0 to V, is available between X and Y, depending on the position of the sliding contact. It will be V/2 when $R_1 = R_2$, which, in a linear pot (p. 22), occurs when AB = BC.

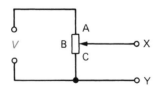

Fig. 5.2

EFFECT OF LOAD ON OUTPUT VOLTAGE

When a potential divider drives current through another circuit, the latter 'loads' the potential divider and the output voltage is less than the calculated value. The difference is small if the resistance R_L of the load is at least *ten times greater* than that of the part of the potential divider across which it is connected. In other words the current

drawn by the load should not exceed 1/10 of the current through the potential divider. (As we saw when considering voltmeters (p. 12) the same problem arises.)

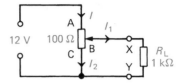

Fig. 5.3

For example, if the potential divider in Fig. 5.3 has a resistance of 100 Ω between A and C then I_1 through the 1 kΩ load will always be at least ten times smaller than I_2 through the potential divider. Therefore when B is midway between A and C, the output voltage between X and Y will be 6 V. If however R_L is 50 Ω, the output voltage is only 4 V (see question 3).

QUESTIONS

1. For the potential divider in Fig. 5.1b opposite, calculate V_2 when V, R_1 and R_2 have the values in this table.

	(a)	(b)	(c)	(d)
V	3 V	5 V	6 V	9 V
R_1	2 kΩ	30 kΩ	100 kΩ	100 Ω
R_2	1 kΩ	20 kΩ	50 kΩ	200 Ω

2. In Fig. 5.2, if BC = 2/3 AC and V = 9 V, what is the output voltage across XY?

3. What is the p.d. between X and Y in Fig. 5.4?

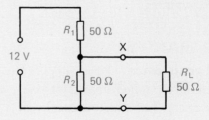

Fig. 5.4

6 ELECTRIC POWER

CHECKLIST

After studying this chapter you should be able to:
- state that power is measured in watts,
- use the equations $P = IV$, $P = I^2R$ and $P = V^2/R$,
- calculate the maximum safe current which can be passed through a resistor of a certain power rating, and
- state the maximum power theorem.

POWER

Expression for power   The power of a device is 1 *watt* (1 W) if it changes energy at the rate of 1 joule per second (1 J/s) from one form to another. Every electronic component has a maximum power rating which should not be exceeded. For example a $\frac{1}{2}$ W resistor can safely change $\frac{1}{2}$ J of electrical energy per second into heat without damage.

The *power P* of a component in *watts* is calculated from:

$$P = V \times I$$

where V is the p.d. across it in *volts* and I is the current through it in *amperes*. If it has resistance R in *ohms*, then since $V = IR$ we have:

$$P = (IR) \times I = I^2R$$

If $R = 10\,\Omega$ and $I = 1$ A, $P = 1^2 \times 10 = 10$ W but if $I = 2$ A then $P = 2^2 \times 10 = 40$ W. Doubling I quadruples P.

We can also write:

$$P = V \times \left(\frac{V}{R}\right) = \frac{V^2}{R}$$

Example   To find the maximum safe current which can be passed through, for instance, a 10 kΩ $\frac{1}{4}$ W resistor, we have:

$$R = 10\,\text{k}\Omega = 10^4\,\Omega \quad \text{and} \quad P = 0.25\,\text{W}$$

But $P = I^2R \quad \therefore I^2 = P/R$

$$\therefore \quad I^2 = 0.25/10^4 = 0.25 \times 10^2/(10^4 \times 10^2)$$

$$= 25/10^6$$

$$\therefore \quad I = \sqrt{25/10^6} = 5/10^3\,\text{A} = 5\,\text{mA}$$

POWER TRANSFER

Internal resistance   Power supplies have some resistance themselves, called *internal* or *source resistance*. It causes the energy loss which occurs inside a battery (i.e. the 'lost' volts, p. 5) when a current is driven through a load in a circuit. The greater the current the greater is the loss and the smaller is the terminal p.d. of the battery. Before we can calculate the maximum power a supply can deliver to a load of resistance R we need to know the e.m.f. E of the supply and its internal resistance r.

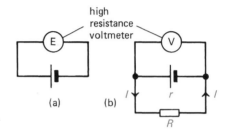

Fig. 6.1

In Fig. 6.1a the high resistance voltmeter measures E (since the cell is on open circuit) and in Fig. 6.1b it measures the terminal p.d. V which drives I through R, i.e. $V = IR$. The 'lost' volts $v = E - V = Ir$ since the current all round the circuit is I. But $E = V + v$ (p. 5).

$$\therefore \quad E = IR + Ir = I(R + r)$$

$$\therefore \quad I = E/(R + r)$$

$$\therefore \quad P = I^2R = E^2R/(R + r)^2$$

For example if $E = 1.5$ V, $V = 1.2$ V and $R = 4\,\Omega$, then from $V = IR$ we get $I = V/R = 1.2/4 = 0.3$ A. Also $v = E - V = Ir$, therefore $r = v/I = 0.3/0.3 = 1\,\Omega$ and $P = I^2R = 0.3^2 \times 4 = 0.36$ W.

Maximum power theorem   This states that the maximum power is delivered to a load *when its resistance equals the internal resistance of the power supply*, i.e. when $R = r$. Putting $r = R$ in the equation for P we get:

$$\begin{aligned}
\text{maximum power in } R &= E^2R/(R + R)^2 \\
&= E^2R/4R^2 \\
&= E^2/4R
\end{aligned}$$

In the above example maximum power transfer would occur when $R = r = 1\,\Omega$ and would be $(1.5)^2/4 = 0.56$ W. (The power wasted in the cell is $E^2/4r$, i.e. the same as that in R since $R = r$. The *efficiency* of the power transfer process is therefore 50%, only half of the total power available being supplied to R.)

QUESTIONS

1. Calculate the maximum safe current through a resistor of (i) $100\,\Omega$ 1 W, (ii) $100\,\Omega$ 4 W, (iii) $1.8\,k\Omega\,\frac{1}{2}$ W.

2. If the p.d. across a resistor is 9 V and it carries a current of 3 mA, what is
a) its resistance,
b) the power delivered to it?

3. a) In Fig. 6.2, V is a high resistance voltmeter. With the switch open it reads 4.0 V and when the switch is closed it falls to 3.2 V. Calculate the e.m.f. and internal resistance of the battery.
b) By what resistor must the $10\,\Omega$ resistor be replaced to give maximum power output? What will be the value of that power output? (L.)

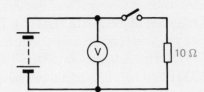

Fig. 6.2

7 ALTERNATING CURRENT

CHECKLIST

After studying this chapter you should be able to:
• state that in an alternating current the direction of electron flow reverses regularly,
• state the meaning of the terms frequency, period, peak value (amplitude) and peak-to-peak value, and indicate them on a sine-wave voltage–time graph,
• state that frequency is measured in hertz,
• use the equation $T = 1/f$,
• state what is meant by r.m.s. current,
• state and use the equation $r.m.s.\ value \approx 0.7 \times peak\ value$, and
• explain the term 'mark–space ratio' in connection with rectangular and square waves.

DIRECT AND ALTERNATING CURRENTS

A *direct current* (d.c.) flows in one direction only; batteries produce d.c. The *waveform* of a current is a graph whose shape shows how the current varies with time. Those in Fig. 7.1 are for steady and varying d.c.

An *alternating current* (a.c.) is one that continually changes direction; car and power station alternators produce a.c. Fig. 7.2 shows the simplest a.c. waveform—it has a sine wave or sinusoidal shape. The current rises from zero to a maximum in one direction (+), falls to zero again before becoming a maximum in the opposite direction (−) and then rises zero once more and so on. The circuit symbol for an a.c. power supply is ~.

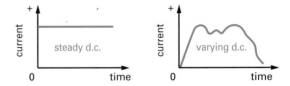

Fig. 7.1

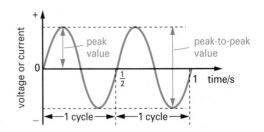

Fig. 7.2

Electric heaters and lamps work off either a.c. or d.c. but most electronic systems require d.c.

The pointer of a moving-coil meter is deflected one way by d.c.; a.c. makes it move to and fro about the scale zero if the direction changes are slow enough, otherwise no deflection occurs.

FREQUENCY OF A.C.

The *frequency f* of a.c. is the number of complete alternations or cycles made in 1 second and the unit is the *hertz* (Hz). The frequency of the a.c. in Fig. 7.2 is 2 Hz; that of the electricity mains supply in many countries (including the UK) is 50 Hz. Larger units are the kilohertz (1 kHz = 1000 Hz = 10^3 Hz) and the megahertz (1 MHz = 1000 kHz = 10^6 Hz).

The *period T* is the time for which one cycle of the a.c. lasts. If $f = 2$ Hz, there are 2 cycles per second, so $T = 1/2$ s. Smaller units are the millisecond (1 ms = 1/1000 s = 10^{-3} s and the microsecond (1 μs = 1/1000 ms = 10^{-6} s). In general:

$$T = 1/f$$

A useful conversion table is given below.

f	1 Hz	1 kHz	1 MHz
T	1 s	1 ms	1 μs

The *peak value* or *amplitude* of an a.c. is its maximum positive or negative value. For a sine wave it is half the peak-to-peak value.

Audio frequency (a.f.) currents have frequencies from 20 Hz or so to about 20 kHz. They produce an audible sound in a loudspeaker.

Radio frequency (r.f.) currents have frequencies above 20 kHz. They produce radio waves from an aerial.

ROOT-MEAN-SQUARE VALUES

Since the value of an alternating quantity changes, the problem arises of what value to take to measure it. The average value of a sine wave over a complete cycle is zero; the peak value might be used. However, the *root-mean-square* (r.m.s.) value is chosen because many calculations can then be done as they would for d.c.

The r.m.s. (or effective) value of an alternating current (or voltage) is the value of the steady direct current (or voltage) which would give the same heating effect.

For example, if the lamp in the circuit of Fig. 7.3 is lit first by a.c. (by moving the 2-way switch to the left) and its

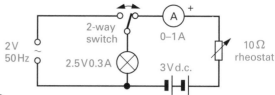

Fig. 7.3

brightness noted, then if 0.3 A d.c. produces the same brightness (when the switch is moved to the right and the rheostat adjusted), the r.m.s. value of the a.c. is 0.3 A. A lamp designed to be fully lit by a current of 0.3 A d.c. will also be fully lit by a.c. of r.m.s. value 0.3 A.

It can be shown that for a sine wave a.c.:

$$\text{r.m.s. value} = \frac{\text{peak value}}{\sqrt{2}} \approx 0.7 \times \text{peak value}$$

The r.m.s. voltage of the UK mains supply is 230 V; the peak value is much higher and is given by:

$$\text{peak value} \approx \frac{\text{r.m.s. value}}{0.7} = \frac{230}{0.7} \approx 330 \text{ V}$$

The value given for an alternating current or voltage is always assumed to be the r.m.s. one, unless stated otherwise. The power P of a device on an a.c. supply is given by the same expression as for d.c., i.e. $P = V \times I$ where V and I are r.m.s. values.

If the average value of a sine wave over *half* a cycle is required, it can be found from:

$$\text{r.m.s. value} \approx 1.1 \times \text{average value}$$

WAVEFORMS

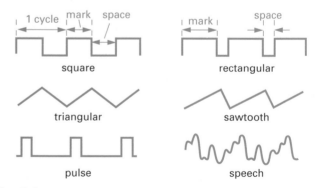

Fig. 7.4

Types Some of the many types of repetitive waveform occurring in electronics (apart from the sine wave) are shown in Fig. 7.4.

The term *mark–space ratio* is used in connection with rectangular and square waves. It is given by:

$$\text{mark–space ratio} = \frac{\text{mark time}}{\text{space time}}$$

For a square wave the ratio is 1 since the mark and space times are equal.

Harmonics All waveforms, except sine waves, can be shown to consist of:

(i) a *fundamental*, which is a sine wave having the same frequency f as the original waveform, and

(ii) *harmonics*, which are sine waves with frequencies that are multiples of the fundamental and with amplitudes that are usually smaller. The second harmonic has frequency $2f$, the third $3f$ and so on.

A sine wave is the only waveform that consists of one frequency (it produces a pure tone in the a.f. range). All others can be analysed into a number of sine waves of different frequencies and amplitudes.

Conversely, they can be built up by adding a number of sine waves together. For example, a triangular wave consists of the fundamental with an infinite number of *even* harmonics. A square wave consists of the fundamental and an infinite number of *odd* harmonics. Fig. 7.5 shows that an approximate square wave is obtained with frequencies of 1, 3, 5 and 7 Hz.

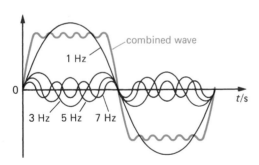

Fig. 7.5

Varying d.c. waveform Sometimes it is useful to consider that a varying d.c. consists of a steady d.c. plus an a.c., Fig. 7.6.

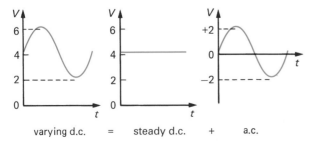

Fig. 7.6

1. The waveform of an alternating voltage is shown in Fig. 7.7. What is
   a) the period,
   b) the frequency,
   c) the peak voltage,
   d) the r.m.s. voltage?

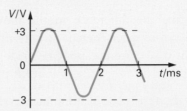

Fig. 7.7

2. An a.c. supply lights a lamp with the same brightness as does a 12 V battery. What is
   a) the r.m.s. voltage,
   b) the peak voltage,
   c) the power of the lamp,
   if it takes a current of 2 A on the a.c. supply?

3. What are the mark–space ratios of the waves in Fig. 7.8a, b?

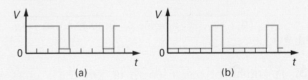

Fig. 7.8

4. What values of *steady* direct voltage and *peak* alternating voltage when added give the varying direct voltage of Fig. 7.9?

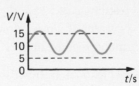

Fig. 7.9

8 PROGRESS QUESTIONS

1. a) An electric current in a metal wire can be considered as a drift of free electrons in one direction through the wire.

(i) Explain why the 'free electrons' are so called.

(ii) What causes the drift of free electrons in the wire?

(iii) Describe the difference in the movement of the free electrons when there is no current flowing.

b) Explain why current cannot be made to flow in an insulator. (*O.L.E. part qn.*)

2. How does the resistance of **(a)** a metal wire, and **(b)** a slice of pure (intrinsic) semiconductor, change as the temperature rises? Fig. 8.1 shows graphs of current against time for (i) a metal wire, and (ii) a slice of semiconductor across which a constant p.d. is applied. Explain the shape of each graph. (*L.*)

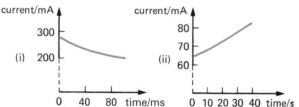

Fig. 8.1

3. A conductor has a length of 5 m and a cross-sectional area of 0.1 mm². If the resistivity of the material is 45×10^{-8} Ωm, calculate the resistance of the conductor.

What length of this conductor is needed to form a 1 Ω resistor? (*C.*)

4. A meter has a resistance of 5 Ω and gives a full scale deflection for a current of 2 mA. How can it be converted to a voltmeter reading up to 10 V? (*C.*)

5. The diagrams in Fig. 8.2a, b show two ways of measuring the p.d. across and the current through a resistor. State, giving your reasons, which arrangement is to be preferred if (i) $R = 2\,\Omega$ and $S = 4\,\Omega$, (ii) $R = 500$ kΩ and $S = 1$ MΩ. (*L.*)

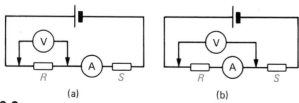

Fig. 8.2

6. An input voltage of 16 V is applied to the resistor chain in the circuit of Fig. 8.3. Calculate the voltage at terminal A when (i) switch S is open, (ii) switch S is closed. (*N.*)

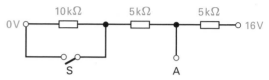

Fig. 8.3

7. The circuit in Fig. 8.4 is designed to allow a variable voltage to be applied across the 500 Ω resistor by using a 1 kΩ linear potentiometer. What is the value of the p.d. across the 500 Ω resistor when the potentiometer slider Y is (i) at the end X, (ii) at the mid-point of the potentiometer, (iii) at the end Z?

Comment on these values.

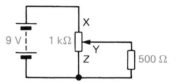

Fig. 8.4

8. In Fig. 8.5, V is a high resistance voltmeter and A is an ammeter of negligible resistance. When the switch S is open the voltmeter reads 6 V and when switch S is closed the voltmeter reads 5 V.

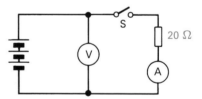

Fig. 8.5

a) Calculate

(i) the e.m.f. and the internal resistance of the battery,

(ii) the ammeter reading when switch S is closed,

(iii) the power consumption in the 20 Ω resistor with switch S closed.

b) The 20 Ω resistor is then replaced by another resistor R.

(i) State the value of R required to obtain maximum power from the battery.

(ii) Calculate the maximum power output. (*C.*)

9. For the circuit in Fig. 8.6, calculate
   a) the resistance of the circuit,
   b) the electric current through the 6 Ω resistor,
   c) the p.d. between the cell terminals.   (C.)

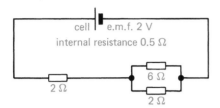

Fig. 8.6

10. If in the circuit of Fig. 8.7 the voltmeter has a resistance of 8 kΩ calculate the p.d. across the 2 kΩ resistor with and without the voltmeter, V, across it. Comment upon the results.   (L.)

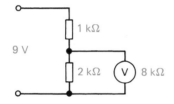

Fig. 8.7

11. a) Define the volt. Distinguish between the e.m.f. of a cell and the p.d. across its terminals.
   b) State Ohm's law and describe how it may be verified for a metallic conductor.

c) Two identical cells each of e.m.f. *E* are connected as shown in Fig. 8.8. What is the e.m.f. of the combination in each case?

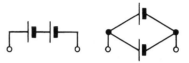

Fig. 8.8

   d) Two cells each of e.m.f. 1.8 V and internal resistance 0.5 Ω pass a current through a 2 Ω resistor. Calculate the p.d. across the resistor when the cells are (i) in series, (ii) in parallel.   (C.)

12. Calculate the r.m.s. voltage and frequency of the signal shown in Fig. 8.9.   (L.)

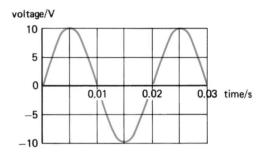

Fig. 8.9

COMPONENTS AND CIRCUITS

9 RESISTORS

CHECKLIST

After studying this chapter you should be able to:
- state the resistance of a resistor from the colour code or printed code,
- state the tolerance of a resistor from the colour code or printed code, and hence calculate the maximum and minimum resistances possible, and
- select an appropriate preferred value from the E12 series of resistors.

ABOUT RESISTORS

Basically resistors are used to limit the current in circuits. When choosing one, three factors need to be considered, apart from the value.

(i) The *tolerance*: exact values cannot be guaranteed by mass-production methods but this is not a disadvantage because in most electronic circuits the values of resistors are not critical. A resistor with a stated (called nominal) value of 100 Ω and a tolerance of $\pm10\%$ could have any value between 90 and 110 Ω.

(ii) The *power rating*: this is the maximum power which can be developed in a resistor without damage occurring by overheating. For most electronic circuits 0.25 or 0.5 W ratings are adequate. The greater the physical size of a resistor the greater is its rating.

(iii) The *stability*: this is the ability to keep the same value with changes of temperature and with age.

FIXED RESISTORS

Three types are shown in Fig. 9.1. The table shows their properties.

Since exact values of fixed resistors are unnecessary in most electronic circuits, only certain *preferred values* are made. The number of values depends on the tolerance. Fig. 9.2 shows the values required to give maximum coverage with minimum overlap for tolerances of $\pm5\%$ and $\pm10\%$. For example, a 22 Ω $\pm10\%$ resistor may have any value between $(22 + 2.2) \approx 24$ Ω and $(22 - 2.2) \approx 20$ Ω. The next higher value of 27 Ω (for $\pm10\%$ tolerance) more or less covers the range 24 Ω to 30 Ω.

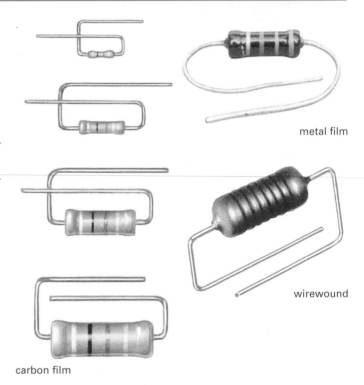

metal film

wirewound

carbon film

Fig. 9.1

Property \ Type	Carbon film	Metal film	Wirewound
Maximum value	10 MΩ	10 MΩ	4.7 kΩ
Tolerance	$\pm5\%$	$\pm1\%$	$\pm5\%$
Power rating	0.25–2 W	0.6 W	2.5 W
Stability	good	very good	very good
Use	general	accurate work	low values

Preferred values (multiple × 10) for tolerance			
±5% (E24 series)	±10% (E12 series)	±5% (E24 series)	±10% (E12 series)
1.0	1.0	3.3	3.3
1.1		3.6	
1.2	1.2	3.9	3.9
1.3		4.3	
1.5	1.5	4.7	4.7
1.6		5.1	
1.8	1.8	5.6	5.6
2.0		6.2	
2.2	2.2	6.8	6.8
2.4		7.5	
2.7	2.7	8.2	8.2
3.0		9.1	

Fig. 9.2

Resistors with ±10% tolerance belong to the E12 series (it has 12 basic values). Those with ±5% tolerance form the E24 series.

RESISTANCE CODES

Band colour codes   In this method the resistance value and tolerance are shown by either four or five coloured bands round the resistor, the latter giving the value more exactly. The way both systems work is shown by the examples in Fig. 9.3.

The *first* band to read is the one at the end of the resistor where the bands are closer together; its colour gives the first digit. The *second* band from that end gives the second digit. The *third* band in the *four band code* gives the multiplier (or the number of 0s to be added), but it gives the third digit (often 0, i.e. black) in the *five band code*. In the latter system the multiplier is given by the *fourth* band. The five band code tends to be used with higher precision resistors, e.g. tolerance ±1%.

In both systems the colour of the band on its own at the other end gives the tolerance. If this band is missing on the resistor, the tolerance is ±20%.

Printed code   The code is printed on the resistor and consists of letters and numbers. It is also used on circuit diagrams and on variable resistors. The examples in the table below show how it works. R means ×1, K means ×10³ and M means ×10⁶, and the position of the letter gives the decimal point.

Tolerances may be indicated by adding a letter at the end: F = ±1%,   G = ±2%,   J = ±5%,   K = ±10%, M = ±20%. For example 5K6K = 5.6 kΩ ± 10%.

Value	0.27 Ω	3.3 Ω	10 Ω	220 Ω
Printed code	R27	3R3	10R	220R
Value	1 kΩ	68 kΩ	100 kΩ	4.7 MΩ
Printed code	1K0	68K	100K	4M7

VARIABLE RESISTORS

Rotary type   A rotary variable resistor is shown in Fig. 9.4a, and its construction in Fig. 9.4b. Connections are made to each end of the track by terminal tags A and B. Tag C is connected to the sliding contact or wiper which varies the resistance between C and the end tags when it is rotated by the spindle. For power ratings up to 2 W, the track is made either of carbon or of cermet (ceramic and metal oxide); above 2 W it is wirewound.

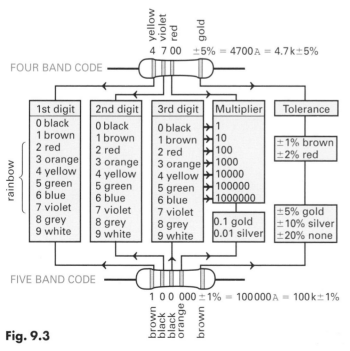

Fig. 9.3

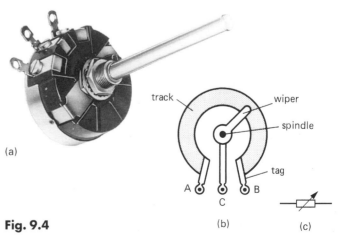

Fig. 9.4   (a)   (b)   (c)

If the track is 'linear' equal changes of resistance occur when the wiper is rotated through equal angles. In a 'log' track the resistance change for equal angular rotations is greater at the end of the track than at the start. Common values are 10 kΩ, 50 kΩ, 100 kΩ, 500 kΩ and 1 MΩ.

The symbol for a variable resistor of this type is shown in Fig. 9.4c.

Preset type   Two preset variable resistors are shown in Fig. 9.5a, b. These have carbon or cermet tracks and when adjustment is necessary, a screwdriver is used. Ratings vary from 0.25 to 1 W.

(a)   (b)

Fig. 9.5

QUESTIONS

1. a) What is the value and tolerance of R_1, R_2 and R_3 shown by the *four band* colour code in the table below?

Band	1	2	3	4
R_1	brown	black	red	silver
R_2	yellow	violet	orange	gold
R_3	green	blue	yellow	none

b) What is the value and tolerance of R_4, R_5 and R_6 shown by the *five band* colour code in the table below?

Band	1	2	3	4	5
R_4	brown	black	black	brown	brown
R_5	orange	orange	black	black	gold
R_6	green	brown	black	red	red

2. a) What is the *four band* colour code for the following: (i) 150 Ω ± 10%, (ii) 10 Ω ± 5%, (iii) 3.9 kΩ ± 10%, (iv) 10 kΩ ± 1%, (v) 330 kΩ ± 2%, (vi) 1 MΩ ± 10%?
b) What is the *five band* colour code for the following: (i) 160 Ω ± 2%, (ii) 2.4 kΩ ± 5%, (iii) 750 kΩ ± 1%?

3. What is the value and tolerance of a resistor marked (i) 2K2M, (ii) 270KJ, (iii) 1M0K, (iv) 15RF?

4. What is the printed code for the following: (i) 100 Ω ± 5%, (ii) 4.7 kΩ ± 2%, (iii) 100 kΩ ± 10%, (iv) 56 kΩ ± 20%?

5. What E12 preferred value would you use if you calculated that a circuit needed a resistor having a value of (i) 1.3 kΩ, (ii) 5.0 kΩ, (iii) 72 kΩ, (iv) 350 kΩ?

10 CAPACITORS

CHECKLIST

After studying this chapter you should be able to:
- state what a capacitor does and recognize its symbols,
- state that capacitance is measured in farads (usually microfarads),
- interpret the marking on a capacitor to determine its capacitance and voltage rating,
- state that some types of capacitor (electrolytic) are polarized and must be connected with the correct polarity,
- realize the difference in the use of non-polarized and polarized capacitors because of their different capacitance values,

- calculate the capacitance of capacitors in parallel and in series,
- state that a capacitor blocks d.c.,
- state that a capacitor passes a.c. and that the higher the frequency and the larger the capacitance, the smaller is the reactance X_C to the a.c.,
- use the equation $X_C = 1/(2\pi f C)$ for capacitive reactance,
- state that there is a phase shift in a capacitive circuit and that I leads V by 90°, and
- state some uses of capacitors.

ABOUT CAPACITORS

A capacitor stores electric charge. Basically it consists of two metal plates separated by an insulator called the *dielectric*.

Charging When connected to a battery, Fig. 10.1, the positive of the battery attracts electrons from plate X and the negative repels electrons to plate Y. Positive charge (deficit of electrons) builds up on X and an equal negative charge (excess of electrons) builds up on Y. During the charging, there is a brief flow of electrons round the circuit from X to Y. Charging stops when the p.d. between X and Y equals (and opposes) the e.m.f. of the battery. The process takes time, i.e. the response of a capacitor to a change of p.d. is not immediate.

If the connections to the battery are removed, the charge may take a long time to leak away from the capacitor unless a conductor is connected across it.

Capacitance The *capacitance* C of a capacitor measures its charge-storing ability. It is 1 *farad* (F) if it stores a charge of 1 coulomb when the p.d. across it is 1 volt. If the charge is 6 C when the p.d. is 2 V, then C = 6 C/2 V = 3 F. In general, for charge Q and p.d. V,

$$C = \frac{Q}{V} \quad \text{or} \quad Q = VC$$

Smaller more convenient units are:

$$1 \text{ microfarad } (\mu\text{F}) = 10^{-6} \text{ F}$$
$$1 \text{ nanofarad } (\text{nF}) = 10^{-9} \text{ F}$$
$$1 \text{ picofarad } (\text{pF}) = 10^{-12} \text{ F}$$

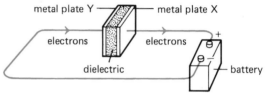

Fig. 10.1

C is large when the area of the plates is large, the plate separation is small and certain dielectrics are used.

Energy stored A charged capacitor stores electrical energy. For charge Q and p.d. V, the energy W stored is:

$$W = \tfrac{1}{2}QV = \tfrac{1}{2}CV^2 \quad (\text{since } Q = VC)$$

where W is in joules if Q is in coulombs and V in volts.

In a photographic flash unit a capacitor discharges through a lamp and its energy is changed to heat and light.

PRACTICAL CAPACITORS

When choosing a capacitor two factors need to be considered, apart from its value and tolerance.

(i) The *voltage rating*: this is the maximum voltage (d.c. or peak a.c.) it can withstand before the dielectric breaks down (it is often marked on it).
(ii) The *leakage current*: no dielectric is a perfect insulator but the loss of charge by leakage through it should be small.

Fixed capacitors Non-polarized types (—■|■—) can be connected either way round. Polarized types (+—[]■—) have

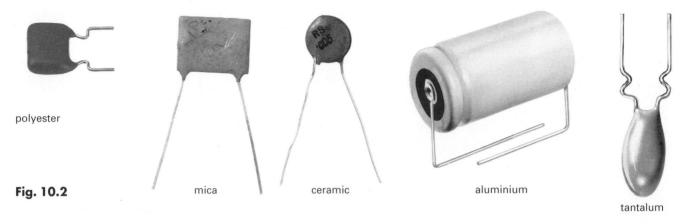

Fig. 10.2   polyester   mica   ceramic   aluminium   tantalum

	Type	Non-polarized			Polarized (electrolytic)	
Property		Polyester	Mica	Ceramic	Aluminium	Tantalum
Values		0.01–10µF	1 pF–0.01 µF	10 pF–1 µF	1–100 000 µF	0.1–100 µF
Tolerance		±5%	±1%	−25 to +50%	−10 to +50%	±20%
Leakage		small	small	small	large	small
Use		general	high frequency	decoupling	low frequency	low voltage

a positive and a negative terminal and must be connected so that there is d.c. through them in the correct direction (to maintain the dielectric by electrolytic action).

Five fixed capacitors which use different dielectrics are shown in Fig. 10.2. The table shows their properties, including the different value ranges of the various types.

The 'Swiss roll' method of construction used for polarized capacitors is shown in Fig. 10.3. Some other types are also made in this way, but a very thin metal film is deposited on each side of a flexible strip of the dielectric to act as the plates. In mica and ceramic capacitors the plates consist of a deposit of silver on a thin sheet of mica or ceramic.

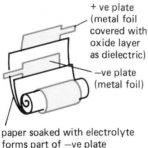

+ ve plate (metal foil covered with oxide layer as dielectric)

−ve plate (metal foil)

paper soaked with electrolyte forms part of −ve plate

Fig. 10.3

Capacitance codes   As for resistors, only certain *preferred values* of capacitors are made, i.e. those of the E12 range (Chapter 9). Various ways are used to indicate capacitance values.

In one method, the value is marked on the capacitor in µF, nF or pF with the submultiple abbreviation being used to indicate any decimal point. For example, 2.2 nF is shown as 2n2 and 4.7 pF as 4p7.

In another method, a *three digit code* is used like the resistor colour code but the numbers are printed on the capacitor rather than encoded in colours. The first two digits are the first two numbers of the value and the third gives the number of 0s to be added to the first two digits to give the value in *picofarads*.

For example, a capacitor marked '103' has a value of 10 plus 3 zeros, which is 10 000 pF or 10 nF or 0.01 µF. The table below gives more examples.

Code	Value (pF)	Value (nF)	Value (µF)
101	100	0.1	0.0001
222	2200	2.2	0.0022
333	33 000	33	0.033
474	470 000	470	0.47

Tolerance values are shown using the same letters as in the resistor printed code (Chapter 9). For example, a capacitor marked '102K' has a value of 1000 pF ±5%.

Variable capacitors   These are used to tune radio receivers. Their value is varied (e.g. up to 500 pF) by altering the overlap between a fixed set of metal plates and a moving set, separated by a dielectric of air. Often two or more are 'ganged', Fig. 10.4.

Fig. 10.4

Fig. 10.5

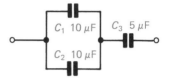

Small variable capacitors called *trimmers* or *presets* are used to make fine, infrequent changes to the capacitance of a circuit. Fig. 10.5 shows a type in which metal foils with a polypropylene dielectric are compressed more or less by a screw to change the value.

CAPACITOR NETWORKS

Parallel In Fig. 10.6a the p.d. across each capacitor is the same but the charges are different (unless $C_1 = C_2$). The total charge $Q = Q_1 + Q_2$; it therefore follows that the combined capacitance C is:

$$C = C_1 + C_2$$

Series In Fig. 10.6b each capacitor has the same charge but the p.ds are different (unless $C_1 = C_2$). Since the

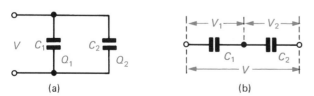

Fig. 10.6

total p.d. $V = V_1 + V_2$, it follows that the combined capacitance C is given by:

$$\frac{1}{C} = \frac{1}{C_1} + \frac{1}{C_2} \quad \text{or} \quad C = \frac{C_1 \times C_2}{C_1 + C_2}$$

Example What is the combined capacitance of the network in Fig. 10.7?

Fig. 10.7

The capacitance of C_1 and C_2 in parallel is $C_1 + C_2 = 10 + 10 = 20\ \mu F$.

This is in series with $C_3 = 5\ \mu F$ and their combined capacitance C is:

$$C = \frac{5 \times 20}{5 + 20} = \frac{100}{25} = 4\ \mu F$$

CAPACITORS IN A.C. CIRCUITS

Action In Fig. 10.8 the lamp lights only when S is in the a.c. position, i.e. C blocks d.c. but allows a.c. to pass. With d.c. there is a brief current which charges C. With a.c., as its direction changes each half cycle, C is charged, discharged, charged in the opposite direction and discharged again 50 times a second. No current actually passes through C but it *seems* to because electrons flow to and fro in the wires joining the plates to the a.c. supply.

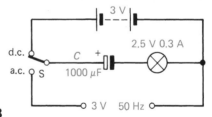

Fig. 10.8

Capacitive reactance The opposition of a capacitor to a.c. is measured by its *capacitive reactance* X_C, given by:

$$X_C = \frac{1}{2\pi f C}$$

where X_C is in *ohms* if f is in hertz and C in farads. This expression agrees with the facts that if C is large, electron flow is large and if f is high, electrons flow rapidly on and off the plates. Large C and large f thus lead to a large current, i.e. small opposition to the a.c.

For example, if $C = 1000 \, \mu F = 10^{-3} \, F$ and $f = 50 \, Hz$, then $X_C = 1/(2\pi \times 50 \times 10^{-3}) = 3.2 \, \Omega$. X_C would decrease if f increased (as Fig. 10.9 shows) or C increased.

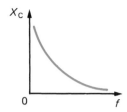

Fig. 10.9

X_C is to a capacitor in an a.c. circuit as R is to a resistor in a d.c. circuit, and if an r.m.s. voltage V is applied to a capacitor of reactance X_C, the r.m.s. current I is given by:

$$I = \frac{V}{X_C}$$

Phase shift For a resistor in an a.c. circuit the current and p.d. reach their peak values at the same time, i.e. they are in *phase*, Fig. 10.10.

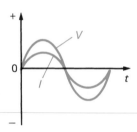

Fig. 10.10

For a capacitor, at the start of the cycle of applied p.d. there is a 'rush' of current I which falls to zero when the p.d. V across the capacitor equals the applied p.d., i.e. I is a maximum when $V = 0$ and vice versa, Fig. 10.11a. There is a phase shift of $\frac{1}{4}$ cycle (or 90°) between I and V with I ahead, i.e. it reaches its peak value first.

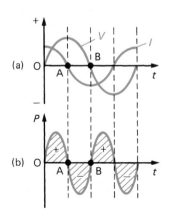

Fig. 10.11

Power The power P taken by a capacitor from an a.c. supply can be found at any instant by multiplying (taking $+$ and $-$ signs into account) the values of V and I at that instant since $P = I \times V$. This can be done using the V–t and I–t graphs of Fig. 10.11a (remembering that $+ \times + = +$, $+ \times - = -$ and $- \times - = +$). The P–t graph so obtained is shown in Fig. 10.11b. It is a sine wave like the V–t and I–t graphs but with twice the frequency. The average power taken by the capacitor over a cycle is zero since the graph is symmetrical about the t-axis. A capacitor is said to be a 'wattless' component.

To explain this behaviour we consider that during the first quarter-cycle OA (V and I are both positive, making P positive) power is drawn from the supply and energy is stored in the charged capacitor. In the second quarter-cycle AB (V is positive, I is negative so P is negative) the capacitor discharges and returns its stored energy to the supply.

COUPLING AND DECOUPLING

Capacitors are used to separate a.c. from d.c. For example, if the output from one circuit, X, contains d.c. and a.c. (e.g. is varying d.c.) but only the a.c. is wanted as the input to another circuit Y, the circuits can be *coupled* by a suitable capacitor offering low reactance at the frequencies involved, Fig. 10.12a.

Capacitors are also used to prevent a.c. passing through a circuit or component, i.e. it *decouples* them. In Fig. 10.12b C has a low reactance at the frequencies involved and decouples R by acting as a bypass for a.c. while d.c. goes through R.

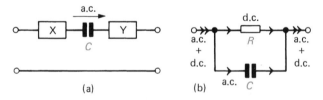

Fig. 10.12

USES OF CAPACITORS

  (i) To separate a.c. from d.c. by coupling or decoupling.
 (ii) To control current in an a.c. circuit (p. 25).
(iii) To smooth the output of a power supply by storing charge (p. 83).
(iv) To tune a radio receiver (p. 197).
 (v) To control the frequency of an oscillator (p. 35).
(vi) In time delay circuits (p. 70).

QUESTIONS

1. a) Calculate Q if V = 10 V and C = 100 000 µF.
b) What is C if Q = 12 µC and V = 6 V?
c) Find V if C = 10 µF and Q = 50 µC.

2. If 500 V is applied across a 2 µF capacitor, calculate
a) the charge,
b) the energy stored.

3. What is the combined capacitance of (i) 2.2 µF and 4.7 µF in parallel, (ii) two 100 µF capacitors in series?

4. Calculate the reactance of a 1 µF capacitor at frequencies of (i) 1 kHz, (ii) 1 MHz.

5. A constant voltage variable frequency a.c. supply is connected to a 2 µF capacitor.

a) Draw a graph to show how the current 'through' the capacitor varies with the frequency.
b) If the supply is 6 V r.m.s. and the frequency is 100 Hz, what is the current?

6. Why should a capacitor with a working voltage of 250 V not be used on a 230 V a.c. supply?

7. a) Express the following in nF: (i) 100 pF, (ii) 8200 pF, (iii) 10 000 pF.
b) Express the following in µF: (i) 330 nF, (ii) 1000 nF, (iii) 47 nF.

8. a) What is the value in pF of a capacitor with a three digit code of (i) 821, (ii) 102, (iii) 563, (iv) 104?
b) What is the three digit code for a capacitor of value: (i) 220 pF, (ii) 82 nF, (iii) 0.1 µF?

11 INDUCTORS

CHECKLIST

After studying this chapter you should be able to:
• state what an inductor is and recognize its symbols,
• state that inductance is measured in henrys,
• state what factors determine the inductance of an inductor,
• state that an inductor opposes a.c., and that the higher the frequency and the larger the inductance, the greater is the reactance X_L to the a.c.,
• use the equation $X_L = 2\pi fL$ for inductive reactance,
• state that there is a phase shift in an inductive circuit and that I lags on V by 90°, and
• state some uses of inductors.

ABOUT INDUCTORS

An inductor is a coil of wire with a core of air or a magnetic material. Four types with their symbols are shown in Fig. 11.1. Inductors have *inductance* (symbol L); they oppose changing currents as we will now see.

In the circuit of Fig. 11.2, if the rheostat is first adjusted so that the lamps are equally bright when S is closed, the resistance of the rheostat then equals the resistance (due to its coil) of the iron-cored inductor. When S is opened and closed again, the lamp in series with the inductor

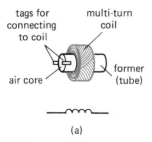
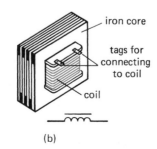
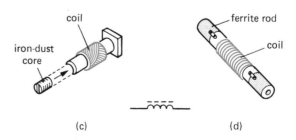

(a)   (b)   (c)   (d)

Fig. 11.1

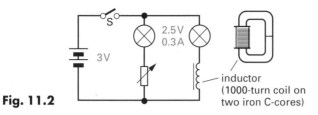

Fig. 11.2

inductor
(1000-turn coil on
two iron C-cores)

lights up a second or two *after* that in series with the rheostat. The inductor opposes the *rise* of the d.c. from zero to its steady value.

If the 3 V battery is replaced by a 3 V a.c. supply, the lamp in series with the inductor never lights, unless the inductance is reduced by removing the iron core.

To explain the behaviour of an inductor we need to know about electromagnetic induction.

ELECTROMAGNETIC INDUCTION

When a conductor is in a *changing* magnetic field, an e.m.f. is produced in it. This can be shown by pushing a magnet into a coil, one pole first, Fig. 11.3a, holding it still inside the coil and then withdrawing it, Fig. 11.3b. A centre-zero galvanometer in series with the coil shows that there is a current when the magnet is *moving* but not when it is at rest. The current is in opposite directions when the magnet enters and leaves the coil. The result is the same if the coil is moved towards and away from the stationary magnet.

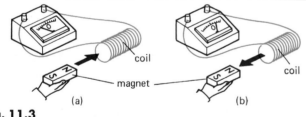

Fig. 11.3

Experiments show that the *induced e.m.f.* (which causes the current in the coil):

(i) increases when the rate at which the magnetic field changes also increases, e.g. it increases if the magnet is moved faster — this is *Faraday's law*;

(ii) always opposes the change causing it, e.g. in Fig. 11.3a the end of the coil nearest the magnet becomes a N pole (due to the induced current), in Fig. 11.3b it becomes a S pole — in both cases it tries to oppose the motion of the magnet — this is *Lenz's law*.

HOW INDUCTORS WORK

A changing magnetic field can be produced by an electric current if the value of the current alters.

In d.c. circuits When a direct current increases in a coil from zero to its steady value, the accompanying magnetic field builds up to its final shape. During the process the field is changing and induces an e.m.f. in the coil *itself* which opposes the change causing it, i.e. the rising current that is trying to establish the field. The opposition delays the rise of current, Fig. 11.4a.

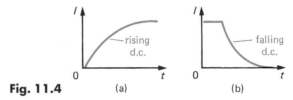

Fig. 11.4   (a)   (b)

When the current is switched off, the field collapses rapidly and induces a large e.m.f. in the coil which opposes the collapsing field caused by the falling current. It tries to keep the current flowing longer, so delaying its fall to zero, Fig. 11.4b. This effect causes sparking at switch contacts, the energy for which comes from that stored in the magnetic field round the coil.

In a.c. circuits Since an alternating current is changing all the time, the magnetic field it produces is also changing continuously. There is therefore always an e.m.f. in the coil and permanent opposition to the a.c. on this account.

INDUCTANCE

An inductor has an inductance of 1 *henry* (H) if a current changing in it at the rate of 1 ampere per second induces an e.m.f. of 1 volt. If the induced e.m.f. is 2 V, the inductance is 2 H. The millihenry (1 mH = 10^{-3} H) and the microhenry (1 μH = 10^{-6} H) are more convenient sub-units.

In general, the inductance (also called the self-inductance) of a coil increases if:

(i)   its cross-sectional area is large and its length small,
(ii) it has a large number of turns, and
(iii) it has a core of magnetic material.

It can be shown that the energy W stored in the magnetic field of an inductor of inductance L carrying a current I is:

$$W = \tfrac{1}{2}LI^2$$

where W is in joules if L is in henrys and I in amperes.

INDUCTORS IN A.C. CIRCUITS

Inductive reactance In a d.c. circuit when the current is constant, the opposition of an inductor is due entirely to the resistance of the copper wire used to wind it. In an a.c. circuit the current is changing all the time and opposition arises not only from the resistance of the coil but also because of its inductance. The opposition due to the latter is called the *inductive reactance* X_L of the inductor and its value is calculated from:

$$X_L = 2\pi f L$$

where X_L is in ohms if f is in hertz and L in henrys. X_L increases if either f or L increase, since in both cases the e.m.f. induced by the changing current (and magnetic field) would be greater. Fig. 11.5 shows the linear relationship between X_L and f for a given L; compare it with Fig. 10.9 for X_C and f. As an example, if $f = 50$ Hz and $L = 10$ H, then $X_L = 2\pi \times 50 \times 10 = 3.1$ kΩ; if $f = 500$ Hz, $X_L = 31$ kΩ.

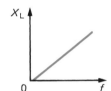

Fig. 11.5

If an r.m.s. voltage V is applied to an inductor of reactance X_L, the r.m.s. current I is given by:

$$I = \frac{V}{X_L}$$

Phase shift The current 'through' a capacitor in an a.c. circuit *leads* the p.d. across it (see Fig. 10.11a). In an inductor the current I *lags* behind the p.d. V by $\frac{1}{4}$ cycle (or 90°), Fig. 11.6. The phase shift arises from the fact that when the current starts to flow, although it is small at first, it is increasing at its fastest rate, therefore so too is the build-up of the magnetic field. As a result the e.m.f. induced in the inductor has its maximum value and opposes the applied p.d.

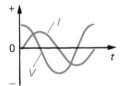

Fig. 11.6

Power Like a perfect capacitor, a perfect inductor, i.e. one with zero resistance, is a 'wattless' component. Its power against time (P–t) graph is obtained from Fig. 11.6 (as for a capacitor) and is the same as Fig. 10.11b.

Again the average power taken from the a.c. supply over a cycle is zero. In this case the energy drawn from the supply on the first quarter-cycle is stored in the mag-

netic field of the inductor. During the second quarter-cycle, the current and magnetic field decrease and the e.m.f. induced in the inductor makes it act as a generator returning the energy stored to the supply.

In practice an inductor has resistance and some energy is taken from the supply on this account and is not replaced.

CURRENT CONTROL BY AN INDUCTOR IN A.C. CIRCUITS

If the current in an a.c. circuit is to have a certain value, the necessary 'opposition' can be provided by a resistor but it wastes electrical energy as heat. The opposition of an inductor to a.c. is due to its reactance X_L and its resistance R. Although both are measured in ohms they cannot be added directly (because of phase shifts) to give the combined opposition. This is called the *impedance Z* of the inductor. It is also measured in ohms and is given by:

$$Z = \sqrt{R^2 + X_L^2}$$

For example, if the resistance R of an inductor coil = 50 Ω and the reactance X_L at a certain frequency = 120 Ω, then the impedance $Z = \sqrt{50^2 + 120^2} = \sqrt{2500 + 14\,400} = \sqrt{16\,900} = 130$ Ω at that frequency. The r.m.s. current I is found from:

$$I = \frac{V}{Z} \quad \left(\text{compare } I = \frac{V}{R} \text{ for d.c.}\right)$$

where V is the r.m.s. voltage.

If an inductor is used to control current in an a.c. circuit instead of a resistor, only part of the required opposition is resistive and so less heat is produced since the reactance is 'wattless'.

Current control in an a.c. circuit is also possible using a capacitor, the opposition being its reactance X_C. Usually, however, it is only used for high frequency currents. At low frequencies C would be inconveniently large, giving low values of X_C.

USES OF INDUCTORS

Air-cored types These have small inductances (e.g. 1 mH) and are used at high frequencies, either in radio tuning circuits above 2 MHz (p. 197) or as r.f. 'chokes' to stop radio frequency currents taking certain paths in a circuit. Their reactance is large to radio frequencies but small to low frequencies, enabling them to separate r.f. from a.f. signals.

Iron-cored types In a current-carrying coil with an iron core, the core becomes magnetized and the strength of this magnetic field is several hundred times greater than that due to the coil alone. For a typical iron-cored inductor $L = 10$ H.

Iron cores are laminated, that is, they consist of flat sheets which are coated thinly with an insulator. The laminations thus offer a high resistance to currents, called *eddy currents*, that would otherwise be produced in the core and cause energy losses as heat. E- and I-shaped laminations are shown in Fig. 11.7.

Iron-cored inductors are used in relays.

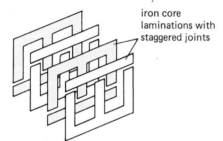

iron core laminations with staggered joints

Fig. 11.7

Iron-dust and ferrite types   These are used at high frequencies. The core of the iron-dust type is in the form of a powder, coated with an insulator and pressed to give a magnetic core of high resistance which reduces eddy current losses. Ferrite cores are made from non-conducting, magnetic materials.

Both types are used in radio tuning circuits up to about 2 MHz. They enable a small coil to give the required inductance, which may be varied by screwing the core in or out of the coil, Fig. 11.1c. The aerial of a radio receiver is often a coil on a ferrite rod, Fig. 11.1d.

QUESTIONS

1. What property do all inductors have as well as inductance?
2. At a certain frequency of a.c. a resistor, a capacitor and an inductor each offer the same 'opposition' to the a.c. How does the opposition of each change (if it does) when the frequency of the a.c. increases?
3. What is the inductance of an inductor in which a current changing at the rate of 1 A s^{-1} induces an e.m.f. of 100 mV?
4. Calculate the inductive reactance of
   a) a 15 H inductor at 100 Hz,
   b) a 1 mH r.f. choke at 1 MHz.
5. When 9 V is applied to an inductor the current through it is 3 A if the supply is d.c. and 0.1 A if it is a.c. with a frequency of 50 Hz. What is
   a) the resistance R,
   b) the impedance Z,
   c) the reactance X_L (ignore R compared with Z),
   d) the inductance L of the inductor?

12 *CR* AND *LR* CIRCUITS

CHECKLIST

After studying this chapter you should be able to:
- draw and interpret, in terms of the time constant, voltage–time graphs for a capacitor charging and discharging through a resistor,
- state that the time constant in seconds is given by $t = CR$ where C is in farads and R in ohms,
- state that in CR seconds the p.d. across a capacitor has risen to 0.63 of its final value if charging and has fallen by 0.63 of its initial value if discharging,
- state that after $5CR$ seconds a capacitor is fully charged if charging and fully discharged if discharging,
- draw voltage–time graphs to show the effect of CR-coupled circuits on square waves, and
- describe the rise and fall of current in an LR circuit in terms of the time constant.

CAPACITOR CHARGING IN A *CR* CIRCUIT

In the circuit of Fig. 12.1, when S is in position 1, C charges through R from the supply. The microammeter measures the charging current I and the voltmeters record the p.ds V_C and V_R across C and R respectively at different times t.

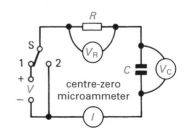

Fig. 12.1

centre-zero microammeter

Graphs like those in Fig. 12.2a and b can be plotted from the results and show that:

(i) I has its maximum value at the start and decreases more and more slowly to zero as C charges up;

(ii) V_C rises rapidly from zero and slowly approaches the supply voltage V which it equals when C is fully charged; and

(iii) V_R behaves like I.

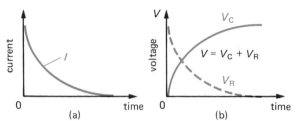

Fig. 12.2

All three graphs are *exponential* curves, I and V_R being 'decaying' ones and V_C a 'growing' one. In general we can say that, at any time,

$$I = \frac{V - V_C}{R}, \quad V_C = V - V_R \quad \text{and} \quad V_R = IR$$

When charging starts, $V_C = 0$ ∴ $I = V/R$ and $V_R = V$.

When charging stops, $V_C = V$ ∴ $I = 0$ and $V_R = 0$.

CAPACITOR DISCHARGING IN A *CR* CIRCUIT

In Fig. 12.1, when S is moved from position 1 to position 2, C discharges through R. If graphs of I, V_C, and V_R are plotted as before, they are again exponential curves, like those in Figs. 12.3a and b.

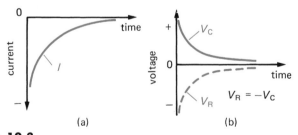

Fig. 12.3

They show that:

(i) I, the discharge current, has its maximum value at the start but is in the opposite direction to the charging current (as is V_R); and

(ii) V_C and V_R fall as C discharges and are equal and opposite at all times.

In general we can say that, at any time,

$$I = \frac{V_C}{R} \quad \text{and} \quad V_R = IR$$

TIME CONSTANT OF A *CR* CIRCUIT

The charging and discharging of a capacitor through a resistor do not occur instantaneously. The *time constant* is a useful measure of how long these processes take in a particular CR circuit.

Charging If a capacitor of capacitance C is charged at a constant rate through a resistor of resistance R by a *steady current I*, it would be fully charged with charge Q and p.d. V after a time t where $t = CR$ seconds if C is in farads and R in ohms. To prove this we use:

$$Q = It = \frac{V}{R} \times t \quad \text{(from } V = IR\text{)}$$

also $$Q = VC$$

therefore $$\frac{V}{R} \times t = VC$$

$$t = CR$$

In fact, as we have seen, the charging current decreases exponentially with time, Fig. 12.2a, and it can be shown that:

(i) in CR seconds, called the *time constant*, the p.d. V_C across the capacitor rises to only 0.63 of its final value; and

(ii) in 5CR seconds the capacitor is, in effect, fully charged and V_C has its final value.

For example, if $C = 500\,\mu\text{F} = 500 \times 10^{-6}\,\text{F}$ and $R = 100\,\text{k}\Omega = 10^5\,\Omega$, then $CR = 500 \times 10^{-6} \times 10^5 = 50$ s. Hence, if the charging p.d. $V = 9$ V then after 50 s, $V_C = 0.63 \times 9 \approx 2/3 \times 9 = 6$ V. After 100 s, V_C rises by 2/3 of the p.d. remaining across C after 50 s, i.e. by 2/3 of $(9 - 6) = 2/3 \times 3 = 2$ V, so making $V_C = 6 + 2 = 8$ V. After 250 s, $V_C \approx 9$ V. See Fig. 12.4.

Fig. 12.4

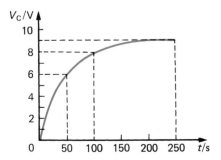

Discharging The time constant CR is useful here too and is the time for V_C to fall by 0.63 ($\approx 2/3$) of its value at the start of discharging. For example, considering the discharge of C in the example above, we can say that after 50 s, V_C will fall by 2/3 of 9 V, i.e. by 6 V, from 9 V to 3 V. In the next 50 s it falls by 2/3 of 3 V, i.e. by 2 V, making $V_C = 1$ V. After about 250 s (5CR), $V_C = 0$.

Note The values given for V_C in the examples above are based on the assumptions that the leakage current through the capacitor is negligible and that the capacitor has its stated value.

WORKED EXAMPLE

In the circuit of Fig. 12.1, $C = 10\ \mu F$, $R = 10\ k\Omega$ and $V = 10$ V. If S is switched to position 1, what will be **(a)** the maximum current through R, **(b)** the maximum charge on C?

a) Maximum current I occurs at the instant S is closed, i.e. before the p.d. builds up across C and opposes the voltage V of the supply. It is given by:

$$I = V/R = 10\ V/10\ k\Omega = 1\ mA$$

b) Maximum charge Q occurs when the charging current has fallen to zero and the p.d. across C equals V, i.e.

$$Q = VC = 10\ V \times 10\ \mu F = 100\ \mu C$$

CAPACITORS AND P.D. CHANGES

A capacitor tends to hold constant the charge and so also the p.d. between its plates. A rise or fall of potential on one plate creates a rise or fall of potential on the other. Hence an a.c. voltage which changes potential rapidly is passed on from one plate to the other. A d.c. voltage on the other hand gives the capacitor time to alter its charge to suit the new p.d. and is blocked. The time needed to adjust to different p.ds depends on the time constant of the circuit.

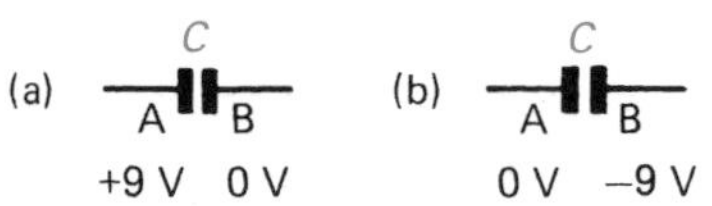

Fig. 12.5

Suppose C in Fig. 12.5a is charged to a p.d. of 9 V with plate A having a potential of $+9$ V and plate B being at 0 V. If the potential of plate A suddenly falls to 0 V, the p.d. cannot change instantaneously since charging and discharging take time. Therefore the potential of plate B must also drop by 9 V, to -9 V, to maintain the p.d. of 9 V across it, Fig. 12.5b.

CR-COUPLED CIRCUITS

Parts of a circuit are often coupled (joined) by a capacitor and resistor. The output of the circuit may then not have the same waveform as the input, depending on the time constant of the coupling. The effect can be studied with an input of a square wave since this contains many frequencies (p. 17).

Capacitor coupling The coupling circuit is given in Fig. 12.6a and the square-wave input of period T in Fig. 12.6b. When the input rises from 0 to V (i.e. AB in Fig. 12.6b), the output p.d. V_R, which is taken from across R, rises immediately to V because the p.d. across C cannot change suddenly, i.e. $V_C = 0$. C starts to charge through R, V_C rises and V_R falls and the rate at which they do so depends on the time constant CR.

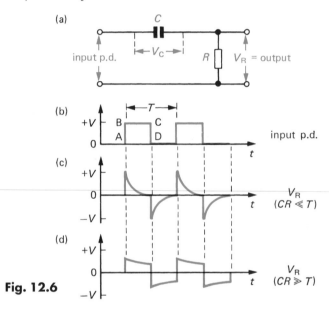

Fig. 12.6

If CR is much *smaller* than T, C has time to charge and discharge before the input reverses again. During charge, V_R varies as in Fig. 12.2b and during discharge (CD in Fig. 12.6b), it varies as in Fig. 12.3b. The complete V_R waveform is shown in Fig. 12.6c and is very different from the input, i.e. distortion has occurred.

If CR is much *greater* than T, C charges and discharges slowly and V_R more or less follows the input. Its waveform settles down to that in Fig. 12.6d, i.e. very little distortion occurs. This coupling circuit is used to connect the stages of an audio amplifier, the time constant being about 10 times greater than the period of the input waveform (e.g. 1/100 s for a 1 kHz input).

Also note that V_R is an alternating p.d. whereas the input p.d. is direct. If we think of the input as being steady d.c. plus a.c., then only the a.c. passes through C as the output.

Resistor coupling In this case, Fig. 12.7a, the output is the p.d. V_C across C and the waveforms are as in Fig. 12.7b, c and d. When CR is much greater than T, V_C does not change much and the input p.d. changes are smoothed in the output. This kind of coupling is used for smoothing power supplies (p. 83) and for detection in radio receivers (p. 198).

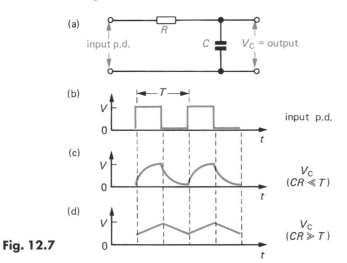

Fig. 12.7

For each case c and d in Figs. 12.6 and 12.7, the graphs show that $V_C + V_R$ = input p.d. at every instant.

LR CIRCUIT

The time constant can also be used as a measure of the time taken by the current to rise or fall in a circuit containing inductance L and resistance R, Fig. 12.8. It is given in seconds by:

$$t = \frac{L}{R}$$

if L is in henrys and R in ohms.

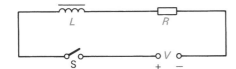

Fig. 12.8

When S is closed the current I rises to 0.63 (about 2/3) of its final steady value in L/R seconds. In $5L/R$ seconds I will have its final value of V/R, which does not depend on L, Fig. 12.9.

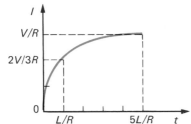

Fig. 12.9

When S is opened, the resistance of the circuit becomes very large, making L/R very small and causing the rapidly collapsing magnetic field to induce a large 'back' e.m.f. in L.

QUESTIONS

1. What p.d. is reached across a 100 µF capacitor when it is charged by a constant current of 20 µA for 1 minute?
2. What is the time constant of a circuit in which (i) $C = 1$ µF and $R = 1$ MΩ, (ii) $C = 100$ µF and $R = 50$ kΩ?
3. In the circuit of Fig. 12.10, C starts to charge up through R when S is closed. What is the p.d. (approximately) across
   a) the capacitor after (i) 1 s, (ii) 2 s, (iii) 5 s,
   b) the resistor after (i) 1 s, (ii) 2 s, (iii) 5 s?

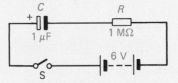

Fig. 12.10

4. In the circuit of Fig. 12.11, C starts to discharge through R when S is closed. What is the p.d. (approximately) across
   a) the capacitor after (i) 5 s, (ii) 10 s, (iii) 25 s,
   b) the resistor after (i) 5 s, (ii) 10 s, (iii) 25 s?

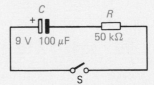

Fig. 12.11

5. A 1.5 V cell is connected to a 1000 µF capacitor in series with a 150 Ω resistor.
   a) What is the maximum current which flows through the resistor during charging?
   b) What is the maximum charge on the capacitor?
   c) How long does the capacitor take to charge to 1.0 V?
6. A capacitor is charged to 6 V so that one plate, A, is at +6 V and the other plate, B, at 0 V. What is the potential of plate B if plate A is suddenly connected to (i) 0 V, (ii) +12 V, (iii) −6 V? (Assume in each case that initially A is at +6 V and B at 0 V.)

13 *LCR* CIRCUITS

LCR SERIES CIRCUIT

Impedance   If an a.c. supply of r.m.s. voltage V and frequency f is applied to a series circuit having components of inductance L, capacitance C and total resistance R, Fig. 13.1, each offers some opposition to the current. The total opposition is the *impedance* Z of the circuit. It can be shown that:

$$Z = \sqrt{R^2 + (X_L - X_C)^2} \quad (1)$$

where $X_L = 2\pi fL$ and $X_C = 1/(2\pi fC)$.

The r.m.s. current I in the circuit is given by:

$$I = \frac{V}{Z}$$

Example   If $L = 2.0$ H, $C = 10$ μF, $R = 100\ \Omega$, $V = 24$ V and $f = 50$ Hz, calculate X_L, X_C, Z and I.

We have:

$$X_L = 2\pi \times 50 \times 2.0 = 200\pi \approx 630\ \Omega$$

$$X_C = 1/(2\pi \times 50 \times 10 \times 10^{-6}) \approx 320\ \Omega$$

$$\therefore Z = \sqrt{100^2 + (630 - 320)^2} \approx 330\ \Omega$$

Hence $I = \dfrac{V}{Z} = \dfrac{24}{330} = 0.073$ A $= 73$ mA

RESONANT CIRCUITS

A resonant circuit consists of a capacitor and inductor in series or in parallel and is used for frequency selection.

Series   In the circuit of Fig. 13.2a, when the frequency of the a.c. has a certain value f_0, called the *resonant frequency*, $X_L = X_C$. From equation (1) it follows that Z then has a minimum value, equal to R which in this case is the small resistance of the inductor coil. The *current* I is a maximum at resonance, given by V/R.

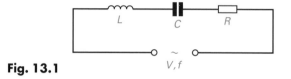

Fig. 13.1

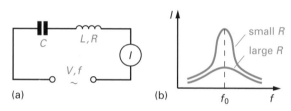

(a)   (b)

Fig. 13.2

The expression for f_0 is obtained from $X_L = X_C$, that is,

$$2\pi f_0 L = \frac{1}{2\pi f_0 C}$$

or $$4\pi^2 f_0^2 LC = 1$$

$$f_0 = \frac{1}{2\pi\sqrt{LC}}$$

where f_0 is in Hz if L is in henrys and C in farads.

The graph of I against f for the circuit is shown in Fig. 13.2b. The greatest response occurs at f_0 but the selectivity, i.e. the sharpness of the peak, falls off as R increases.

A circuit with a sharp peak is said to have a high Q-factor (Q for 'quality'). The higher the Q, the smaller the range of frequencies selected, i.e. the smaller the *bandwidth* (Chapter 77).

Parallel   For the circuit in Fig. 13.3a, f_0 is given by the same expression but Z is a maximum and I a minimum at resonance. The *voltage* developed across the circuit at resonance is thus a maximum (for the same I at other frequencies it would be smaller due to Z being smaller). Fig. 13.3b shows the response curve, this time of Z against f.

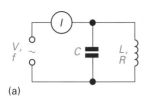

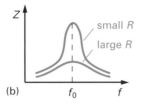

(a)   (b)

Fig. 13.3

Uses Tuning in a radio receiver (p. 197) is done using resonant circuits in which C (or L) is varied until f_0 equals the frequency of the wanted signal.

Note As in all cases of resonance, a 'driving force' (here the applied p.d. V) is coupled to a 'driven system' (the LCR circuit) which responds to the correct driving frequency.

OSCILLATORY CIRCUIT

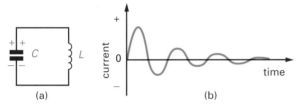

(a)   (b)

Fig. 13.4

If a charged capacitor C discharges through an inductor L in a circuit of low resistance, Fig. 13.4a, an a.c. of constant frequency and decreasing amplitude, called a damped oscillation, is produced which eventually dies off, Fig. 13.4b. Its frequency f, known as the *natural frequency* of the circuit, has the same value as the resonant frequency, that is,

$$f = \frac{1}{2\pi\sqrt{LC}}$$

The oscillations occur because the energy stored initially in C is transferred to the magnetic field of L which, when C has discharged, collapses and tries to keep the current going. C then charges in the opposite direction. This is repeated but energy is gradually lost due to the resistance of L.

The LC circuit is the basis of some oscillators (p. 112) for producing a.c. of constant amplitude, i.e. undamped oscillations.

QUESTIONS

1. Calculate Z if $R = 6\ \Omega$, $X_L = 58\ \Omega$ and $X_C = 50\ \Omega$.

2. What is the resonant frequency if $L = 1$ mH and $C = 1$ nF?

14 TRANSFORMERS

CHECKLIST

After studying this chapter you should be able to:
- state what a transformer does and recognize its symbols,
- recall the transformer turns ratio equation $V_s/V_p = n_s/n_p$ and use it to solve simple problems on step-up and step-down transformers, and

- name three types of transformer.

ABOUT TRANSFORMERS

A transformer changes (transforms) an alternating p.d. from one value to another. It consists of two coils, called the *primary* and the *secondary*, which are not connected electrically. The windings are either one on top of the other or are side-by-side on an iron, iron-dust or air core. Transformer symbols are given in Fig. 14.1.

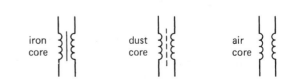

Fig. 14.1

A transformer works by electromagnetic induction. An alternating p.d. is applied to the primary and produces a changing magnetic field which passes through the secondary, thereby inducing an alternating p.d. in it. A practical arrangement for ensuring as much as possible of the magnetic field links the secondary is shown in the iron-cored transformer of Fig. 14.2; the secondary is wound on top of the primary.

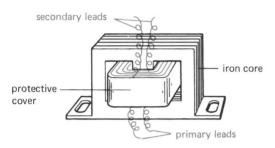

secondary leads

iron core

protective cover

primary leads

Fig. 14.2

TRANSFORMER EQUATIONS

Turns ratio   It can be shown that if a transformer is 100% efficient at transferring electrical energy from primary to secondary (many are nearly so), then:

$$\frac{\text{secondary p.d.}}{\text{primary p.d.}} = \frac{\text{turns on secondary}}{\text{turns on primary}}$$

In symbols,

$$\frac{V_s}{V_p} = \frac{n_s}{n_p}$$

If n_s is twice n_p, the transformer is a *step-up* one and V_s will be twice V_p. In a *step-down* transformer there are fewer turns on the secondary than on the primary and V_s is less than V_p. The ratio n_s/n_p is the *turns ratio*.

Power   If the p.d. is stepped-up by a transformer the current is stepped-down in proportion and vice versa. This must be so if we assume that all the electrical energy given to the primary appears in the secondary. Hence:

$$V_p \times I_p = V_s \times I_s$$

where I_p and I_s are the primary and secondary currents. A transformer (unlike an amplifier, p. 98) gives no power gain.

WORKED EXAMPLE

If the transformer in Fig. 14.3 is 100% efficient, calculate **(a)** the p.d. V_s across the 4 Ω resistor and **(b)** the primary current I_p.

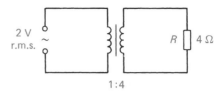

2 V r.m.s.   R   4 Ω

1:4

Fig. 14.3

a) $V_s/V_p = n_s/n_p = 4/1$
   $\therefore \ V_s = 4 \times V_p = 4 \times 2 = 8\ \text{V}$

b) $I_s = V_s/4 = 8/4 = 2\ \text{A}$
But $I_p \times V_p = I_s \times V_s$
   $\therefore \ I_p \times 2 = 2 \times 8$
   $\therefore \ I_p = 8\ \text{A}$

TYPES OF TRANSFORMER

Mains   A mains transformer is shown in Fig. 14.4a. The primary is connected to the a.c. mains supply (230 V 50 Hz in the UK). The secondary may be step-up or step-down, or there may be one or more of each. They have laminated iron cores and are used in power supplies.

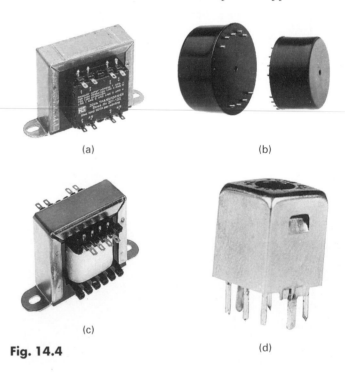

(a)   (b)

(c)

(d)

Fig. 14.4

Step-down *toroidal* types, Fig. 14.4b, have become popular due to their smaller size and weight and are now replacing the laminated type. They have virtually no external magnetic field and a screen between primary and secondary windings gives safety and electrostatic screening.

Audio frequency   An audio frequency transformer is shown in Fig. 14.4c. This also has a laminated iron core and acts as a matching transformer to ensure maximum power transfer from, for example, an amplifier to a loudspeaker (p. 187).

Radio frequency   A radio frequency transformer is shown in Fig. 14.4d. This has an iron-dust core and forms part of the tuning circuit in a radio (p. 197), being surrounded by a small metal 'screening' can to stop radiation from it getting to other parts of the circuit.

QUESTION

1. A 12 V lamp is operated from a 240 V a.c. mains step-down transformer.
a) What is the turns ratio?
b) How many turns are on the primary if the secondary has 80 turns?
c) What is the primary current if the current in the lamp is 2 A?

15 SWITCHES

CHECKLIST

After studying this chapter you should be able to:
• recognize various types of switch and their symbols and be familiar with their use in circuits, e.g. SPST, SPDT, DPST, DPDT, push button.

ABOUT SWITCHES

In a mechanical switch, metal contacts have to be brought together or separated to make or break a circuit.

A switch is rated according to (i) the *maximum current* it can carry and (ii) its *working voltage*. These both depend on whether it is to be used in a.c. or d.c. circuits. For example, one with an a.c. rating of 250 V 1.5 A has a d.c. rating of 20 V 3 A. If these values are exceeded the life of the switch is shortened. This is due to overheating while it carries current, or to vaporization of the contacts due to sparking as a result of the current trying to keep flowing in the air gap when it is switched off. In general sparking lasts longer with d.c. than with a.c. because a.c. falls to zero twice per cycle.

Switches have different numbers of *poles* and *throws*. The poles (P) are the number of separate circuits the switch makes or breaks at the same time. The throws (T) are the number of positions to which each pole can be switched. The symbols for various types are given in Fig. 15.1. In a SPDT (single pole double throw) switch, for example, there are two positions for the switch (B or C) and only one circuit (that joined to A) is switched.

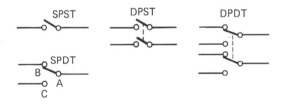

Fig. 15.1

TYPES OF SWITCH

Push-button   The push-button switch in Fig. 15.2a is a 'push-on, release-off' type; its symbol is given in Fig. 15.2b and that for the 'push-off, release-on' variety in Fig. 15.2c. 'Push-to-change-over' switches are also made; their symbol is shown in Fig. 15.2d.

Fig. 15.2

(a) 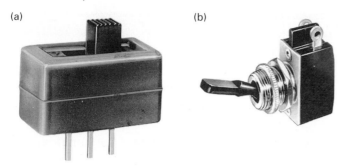 (b)  (c)

Fig. 15.3

Slide   The slide switch in Fig. 15.3a is a 'change-over' SPDT type.

Toggle   This type is often used on equipment as a power supply 'on-off' switch, either in the SPST form shown in Fig. 15.3b or as a SPDT, DPST or DPDT type.

Keyboard   The one shown in Fig. 15.3c is a SPST 'push-to-make' (momentary) type which can be mounted on a printed circuit board (p.c.b.).

Rotary wafer   One or more insulating plastic discs or wafers are mounted on a twelve-position spindle as shown in Fig. 15.4a. The wafers have metal contact strips on one or both sides and rotate between a similar number of fixed wafers with springy contact strips. The contacts on the wafers can be arranged to give switching that is 1 pole 12 throw, 2 pole 6 throw, 3 pole 4 throw, 4 pole 3 throw (as in Fig. 15.4b) or 6 pole 2 throw.

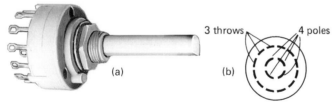

Fig. 15.4

TWO SWITCH CIRCUITS

Two circuits in which switches are used to perform logical tasks are given below.

Two-way control of staircase lighting   See Fig. 15.5. The light can be switched on or off either by the SPDT switch at the top or that at the bottom of the stairs.

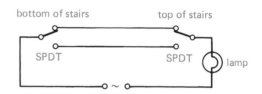

Fig. 15.5

Reversing polarity of supply to electric motor   See Fig. 15.6. When the polarity of the supply to the motor (a permanent magnet type, p. 52) is changed by the DPDT switch, its direction of rotation is reversed.

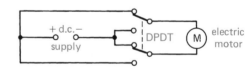

Fig. 15.6

16 PROGRESS QUESTIONS

1. A resistor may have four coloured bands on it (a, b, c and d in Fig. 16.1). What do each of these bands indicate? What is the maximum current (approximately) that can be conducted safely by a 1 kΩ ½ W resistor? *(O. and C.)*

Fig. 16.1

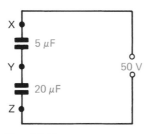

a b c d

2. A number of 1000 Ω resistors are available.
a) State the [four band] colour coding for a 1000 Ω resistor.
b) By means of diagrams, show how the resistors could be connected to make: (i) a combination of resistance 3000 Ω; (ii) a combination of resistance 500 Ω; (iii) a combination of resistance 1250 Ω.
c) Each of the 1000 Ω resistors will safely dissipate a maximum power of 0.5 W. Show how they could be connected to make a combination of resistance 1000 Ω and power rating 2 W. Explain. *(O.L.E.)*

3. A manufacturer's catalogue lists capacitors as 0.1 μF, 40 V d.c. What does this mean? What charge is stored by one such capacitor when it has a p.d. of 20 V across it? *(O. and C.)*

4. A 2 μF capacitor is charged to a p.d. of 100 V and immediately discharged through a 5 Ω resistor. Calculate **(a)** the charge taken by the capacitor, **(b)** the heat generated in the resistor. *(C.)*

5. For the circuit in Fig. 16.2 calculate **(a)** the capacitance between X and Z, **(b)** the charge on the 5 μF capacitor, **(c)** the p.d. between X and Y. *(C.)*

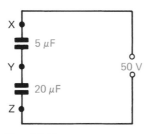

X
5 μF
Y
50 V
20 μF
Z

Fig. 16.2

6. If an a.c. signal of peak p.d. 5 V and frequency 1000 Hz is applied to the circuit in Fig. 16.3, calculate **(a)** the r.m.s. current through the 200 Ω resistor, **(b)** the reactance of the 8 μF capacitor, **(c)** the peak current through the 8 μF capacitor. *(C.)*

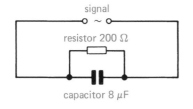

signal
resistor 200 Ω
capacitor 8 μF

Fig. 16.3

7. Draw a labelled diagram to show the structure of a low-frequency choke. What is meant by the *self-inductance* of a choke? On what factors does its value depend?

A choke (of negligible resistance) has a self-inductance of 0.5 H. Draw a graph to show how its reactance varies with frequency between 0 and 1000 Hz. This choke is connected across a 2 V peak-to-peak generator of negligible internal resistance. The generator frequency is 50 Hz. Draw a graph to show how the current and voltage in the circuit vary in one cycle, and explain their phase relationship to one another.

What is the r.m.s. current I in the circuit at 50 Hz? At what frequency will the r.m.s. current be $\frac{1}{2}I$? *(C.)*

8. Calculate, for the circuit of Fig. 16.4, **(a)** the maximum current, **(b)** the maximum charge which can be stored in the capacitor, **(c)** the time constant of the circuit.

Hence draw sketch maps of the current against time and the charge stored against time. Take $t = 0$ as the moment when the switch is closed. Label all axes. *(L.)*

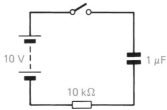

10 V
1 μF
10 kΩ

Fig. 16.4

9. For the circuit in Fig. 16.5 sketch a graph showing:
a) how the voltage V across the capacitor varies with time when the switch is moved from position 1 to position 2,
b) the maximum value of V,
c) the approximate value of V after a time equal to the time constant.

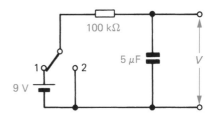

100 kΩ
5 μF
V
1
2
9 V

Fig. 16.5

10. In Fig. 16.6 three circuits are shown each with an input waveform. Copy the input waveforms and below each draw the output waveform V_0.

(*L. part qn.*)

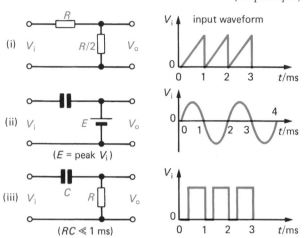

Fig. 16.6

11. a) What is meant by the self-inductance of a coil? Define the henry and explain what one means by a self-inductance of 0.1 H.

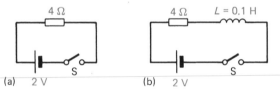

Fig. 16.7

b) Draw a graph over a time scale of 1 s of the current in the circuit shown in (i) Fig. 16.7a, (ii) Fig. 16.7b, when the position of switch S is altered every 0.25 s (that is S is on for 0.25 s then off for 0.25 s, etc.).

How would the current variation alter in the circuit in Fig. 16.7b if the value of the inductance were made larger? (C.)

12. Calculate the value of the capacitor C to make the circuit in Fig. 16.8 have a resonant frequency of 1 kHz. Explain briefly what is meant by the *impedance* of a circuit. (C.)

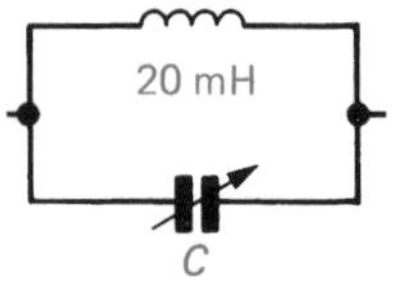

Fig. 16.8

13. What is the tuning range in MHz of a circuit containing a 1 μH inductor in parallel with a 100–500 pF variable capacitor?

14. The output of an oscillator is a sinusoidal voltage whose frequency can be varied but whose amplitude is constant and equal to 1 V r.m.s. The output is connected to a 1 μF capacitor in series with a 10 kΩ resistor and the frequency is varied from 1 Hz to 1 MHz.

What will be the *approximate* value of **(a)** the current flowing from the oscillator, and **(b)** its phase relative to the output voltage for a frequency of (i) 1 Hz, (ii) 1 MHz? (*O. and C.*)

15. A supplier's stocklist has the following entry:

Transformer: primary 240 V 50 Hz
          secondaries 20 V, 0.5 A; 20 V, 0.5 A

Explain carefully what the entry means. If the transformer is 100% efficient, what is the step-down ratio for one of the secondary windings? (*O. and C.*)

16. The circuit shown in Fig. 16.9 is used to investigate the discharge of an electrolytic capacitor of nominal value 2000 μF. Initially the switch S is closed and the digital voltmeter (DVM) indicates a potential difference of 10.00 V across the resistor. The switch is now opened.

a) Calculate the time for the reading on the DVM to fall to 3.70 V.

b) (i) What assumption concerning the capacitor did you make when carrying out the calculation in part **a)**? (ii) Would you expect the experimentally measured time to differ from the calculated time? Give reasons, excluding experimental errors, for your answer. (N.)

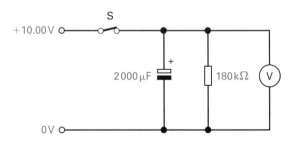

Fig. 16.9

TRANSDUCERS

17 MICROPHONES

CHECKLIST

After studying this chapter you should be able to:
- state what input and output transducers do,
- state what a microphone does and recognize its symbol,
- recall four important properties of a microphone, and
- recall the chief features of moving-coil, ribbon, capacitor, crystal, carbon and wireless microphones, giving impedances and output voltages where possible.

TRANSDUCERS

Transducers change energy from one form to another. Those in which electrical energy is the input or output will be considered; they enable electronic systems to communicate with the outside world, Fig. 17.1.

Fig. 17.1

A transducer must be matched to the system with which it is designed to work. Sometimes this means ensuring there is maximum voltage transfer between transducer and system, at other times we have to transfer maximum power. In all cases, as we will see later (p. 108), the *impedance* of the transducer is a critical factor (rather than its resistance since many transducers have inductance or capacitance and we are generally dealing with a.c.).

ABOUT MICROPHONES

A microphone changes sound into electrical energy: its symbol is given in Fig. 17.2.

Fig. 17.2

There are four important properties of a microphone.

(i) The *impedance*.

(ii) The *frequency response*. Ideally this should cover the audio frequency range from 20 Hz or so to about 20 kHz, and the sensitivity should be the same for the full range, i.e. sounds of the same intensity but different frequency should produce the same output.

(iii) The *directional sensitivity*. A microphone may be more sensitive, i.e. give a bigger output, for sound coming from one or more directions than from others. This is shown by polar diagrams like those in Fig. 17.3a, b, c for uni-, bi- and omni-directional microphones respectively. In these, the length of the line from the origin O to any point P on the diagram is a measure of the sensitivity in the direction OP.

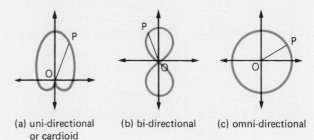

(a) uni-directional or cardioid   (b) bi-directional   (c) omni-directional

Fig. 17.3

(iv) The *output voltage*.

Microphones fall into three main groups: velocity-dependent, amplitude-dependent and carbon.

VELOCITY-DEPENDENT MICROPHONES

The output voltage of these depends on the *velocity* of vibration of the moving parts, which in turn depends on the sound received.

Moving-coil or dynamic type
This is a popular type because of its good quality reproduction, robustness and reasonable cost. One is shown in Fig. 17.4a. It consists of a small coil of many turns of fine wire wound on a tube (the former) which is attached to a light disc (the diaphragm) as in Fig. 17.4b.

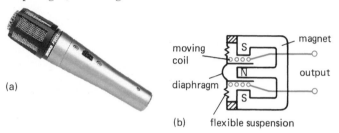

Fig. 17.4

When sound strikes the diaphragm, the coil moves in and out of the circular gap between the poles of a strong permanent magnet. Electromagnetic induction occurs and the alternating e.m.f. induced in the coil (typically 1 to 10 mV) has the same frequency as the sound.

A moving-coil microphone has an impedance of 200 to 600 Ω. Uni- and near-omni-directional models are available.

Ribbon type
The moving part is a thin corrugated ribbon of aluminium foil suspended between the poles of a powerful magnet, Fig. 17.5. Sound falling on either side of the ribbon makes it vibrate at the same frequency and a small e.m.f. is induced in it.

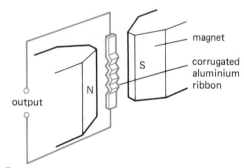

Fig. 17.5

Ribbon microphones are bi-directional and have a very low impedance (e.g. 0.2 Ω). Their performance is similar to that of the moving-coil type but they can be damaged by blasts of air.

AMPLITUDE-DEPENDENT MICROPHONES

In these the output depends on the amount by which the sound waves displace the diaphragm from its rest position, i.e. on the *amplitude* of the motion.

Capacitor (or condenser) type
This is used in broadcasting studios, for public address systems and for concerts where the highest quality is necessary. It consists of two capacitor plates, A and B in Fig. 17.6a. Plate B is fixed while the metal foil disc A acts as the movable diaphragm. Sound waves make the diaphragm vibrate and as its distance from B varies, the capacitance changes. A small battery in the microphone causes the resulting charging and discharging current to produce a varying p.d. across a resistor. This p.d. provides the input to a small amplifier in the stem of the microphone.

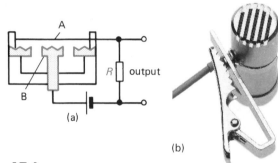

Fig. 17.6

Capacitor microphones, like the tie (clip-on) model in Fig. 17.6b, may be uni- or omni-directional and have impedances of the order of 1 kΩ.

Crystal type
The action of a crystal microphone depends on the *piezoelectric effect*. In this, an electric charge and so also a p.d. is developed across opposite faces of a slice of a crystal when it is bent, because of the displacement of ions. The effect is given by natural crystals such as Rochelle salt (sodium potassium tartrate) and quartz and by synthetic ceramic crystals of lead zirconium titanate. When sound waves set the diaphragm vibrating, the crystal bends to and fro, Fig. 17.7, and an alternating p.d. is produced between metal electrodes deposited on two of its faces.

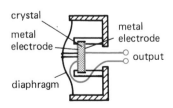

Fig. 17.7

The crystal microphone has a high output (e.g. 100 mV), is usually omni-directional and has a very high impedance (several megohms). Its performance is not as good as the previous types but it is inexpensive and much used in cassette recorders.

CARBON MICROPHONE

This is used in telephone handsets but it is not a high quality device. The principle of the latest type is shown in Fig. 17.8.

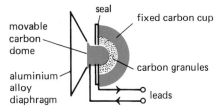

Fig. 17.8

Sound makes the diaphragm vibrate and vary the pressure on the carbon granules between the movable carbon dome, which is attached to the diaphragm, and the fixed carbon cup at the back. An increase of pressure squeezes the granules together and reduces their electrical resistance. Reduced pressure increases the resistance. *Direct current* through the microphone from a supply (of 50 V at the telephone exchange) varies accordingly to give a varying d.c. which can be regarded as a steady d.c. plus a.c. having the frequency of the sound.

Current variations represent information in a carbon microphone, whereas in other types, e.g. crystal, the representation is in the form of a varying voltage. In the first case a *current analogue* is produced, in the second it is a *voltage analogue*.

WIRELESS MICROPHONE

This type incorporates a small radio transmitter which sends the sound falling on the microphone as a frequency modulated (FM) v.h.f. signal (Chapter 84) up to a distance of about 100 metres. Its short aerial (e.g. 6 cm) transmits to any normal FM receiver and allows the performer/lecturer/public speaker greater freedom of movement since it has no long trailing cables attached. It is battery operated and usually tunable over the range 88 to 108 MHz.

QUESTIONS

1. What does a transducer do?
2. Name *four* types of microphone. Give *one* use for each.
3. What is the piezoelectric effect?

18 LOUDSPEAKERS, HEADPHONES AND EARPIECES

CHECKLIST

After studying this chapter you should be able to:
- state what a loudspeaker does and recognize its symbol,
- state three important properties of a loudspeaker,
- recall the chief features of moving-coil and crystal loudspeakers, giving frequency responses and impedances where possible, and
- recall the properties of moving-coil and magnetic headphones.

Loudspeakers, headphones and earpieces change electrical energy into sound. They each have three important properties.

(i) The *frequency response*. As with microphones, this should be uniform over the audio frequency range.
(ii) The *impedance*. This must be known for matching purposes.
(iii) The *power rating*. This should not be exceeded or damage may occur.

LOUDSPEAKERS

Moving-coil type Most loudspeakers in use today are of this type, shown in Fig. 18.1. Their construction, Fig. 18.2a, is basically the same as that of the moving-coil microphone, in fact a loudspeaker can be used as a low impedance microphone. The loudspeaker symbol is given in Fig. 18.2b.

Fig. 18.1

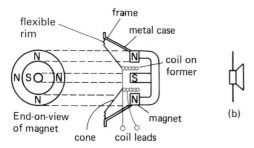

Fig. 18.2

When audio frequency a.c. (e.g. from an amplifier) passes through the coil of the speaker, it vibrates in or out between the poles of the magnet depending on the current direction. The paper, polypropylene or carbon fibre cone, which is fixed to the coil, also vibrates and passes the motion on to the large area of air touching it. A sound wave of the same frequency as the a.c. in the coil is thus produced. (The motion of the coil is due to the magnetic field caused by the current in it, interacting with the field of the magnet.)

The loudspeaker coil has inductive reactance X_L and resistance R and its impedance $Z = \sqrt{R^2 + X_L^2}$. The reactance X_L, and therefore Z, varies with the frequency of the a.c. but is usually given for 1 kHz. Common values are 4 Ω and 8 Ω.

Crystal type This type depends on the reverse piezo-electric effect (p. 42), i.e. an alternating p.d. applied across opposite faces of the crystal makes it vibrate and produce sound of the same frequency as the a.c. In speaker systems with two or more speakers it is used as a

tweeter to handle high (treble) frequencies since it rejects low (bass) frequencies, having a typical frequency response of 3.5 kHz to 35 kHz. It has a high impedance, of the order of 1 kΩ.

Efficiency The small loudspeakers used in radios and televisions are only 5 to 10% efficient at changing electrical energy into sound. For large speakers at low power levels the value is near 50%.

HEADPHONES

Moving-coil type High quality headphones, as used for listening to stereo, Fig. 18.3a, are moving-coil (dynamic) types with, in a typical case, a frequency response of 20 Hz to 20 kHz and an impedance of 8 Ω per earphone. Often each earphone is fitted with its own volume control. The headphone symbol is given in Fig. 18.3b.

Fig. 18.3

Magnetic type Headphones for general use, e.g. mono listening, work on a different principle; they have a higher impedance, e.g. 1 kΩ per earphone, and a smaller frequency response, e.g. 30 Hz to 15 kHz. Their action can be explained from Fig. 18.4. Current passes through the coils of an electromagnet and attracts the iron diaphragm more, or less, depending on the value of the current. As a result the diaphragm vibrates and produces sound.

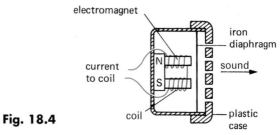

Fig. 18.4

TELEPHONE HANDSET RECEIVER

The construction of the latest telephone handset receiver, called the 'rocking armature' type, is shown in Fig. 18.5. (An armature is a piece of iron which is made to move by an electromagnet.)

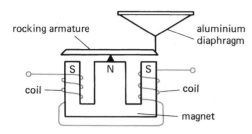

Fig. 18.5

The coils are wound in opposite directions on the two S poles of the magnet so that if the current goes round one in a clockwise direction, it goes round the other anti-clockwise. Current in one direction therefore makes one S pole stronger and the other weaker so causing the iron armature to rock on its pivot towards the stronger S pole. When the current reverses, the S pole which was the stronger before will now be the weaker and the armature rocks the other way. These movements of the armature are passed on to the aluminium diaphragm, making it vibrate and produce sound of the same frequency as the a.c. speech current in the coil.

EARPIECES

Earpieces like that in Fig. 18.6a are used in deaf aids and with transistor radios. The *magnetic* type works on the moving-coil principle and has an impedance of 8 Ω. The *crystal* type depends on the reverse piezoelectric effect and has an impedance of several megohms. The symbol for an earpiece is shown in Fig. 18.6b.

Fig. 18.6　　(a)　　　　　(b)

QUESTIONS

1. A manufacturer's catalogue gives the following information about a certain loudspeaker.
　　Frequency response: 25 Hz to 19 kHz
　　Impedance: 8 Ω
　　Power (r.m.s.): 5 W
Explain each statement.

2. a) The lowest frequency to which a certain loudspeaker, with a paper cone of diameter 12 cm responds is 25 Hz. For a similar 20 cm diameter speaker, the lowest frequency is 18 Hz. What conclusion can you draw from these figures?
b) Why are loudspeakers with an elliptical-shaped cone often used in television sets and transistor radios?

19 HEAT, LIGHT AND STRAIN SENSORS

CHECKLIST

After studying this chapter you should be able to:
- state how thermistors are affected by temperature and recognize the thermistor symbol,
- state how an LDR is affected by light intensity and recognize its symbol, and
- state how a strain gauge is affected electrically by mechanical strain.

THERMISTORS

Thermistors or *thermal resistors* are semiconductor devices whose use as transducers is due to the fact that their resistance changes markedly when their temperature changes. They are used for the measurement and control of tem-perature, being heated either externally or internally by the current they carry. Rod-, disc- and bead-shaped vari-eties are shown in Figs. 19.1a, b and c respectively, and the thermistor symbol is shown in d.

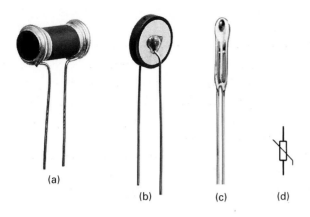

Fig. 19.1

There are two types of thermistor.

n.t.c. thermistors The letters *n.t.c.* stand for negative temperature coefficient, p. 9. These are the commonest type of thermistor. Their resistance *decreases* as the temperature increases as shown by the typical resistance–temperature graph of Fig. 19.2. They are made from oxides of nickel, manganese and other elements which, after heat treatment, have a ceramic-like appearance.

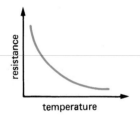

Fig. 19.2

The circuit for a simple electronic thermometer using a small bead thermistor is shown in Fig. 19.3a.

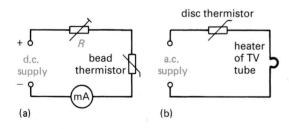

Fig. 19.3

Temperature changes cause the resistance of the thermistor to change and the current through the milliammeter (mA) changes. By adjusting the preset resistor *R* to give a suitable deflection at room temperature and then using several known temperatures, the meter can be calibrated to read °C directly.

In Fig. 19.3b the heater of the TV tube (p. 49) has a low resistance when cold and would, but for the disc thermistor, allow a dangerous current surge when the tube is switched on. As the current warms up the thermistor and the heater, the rise in heater resistance (a metal) is compensated by the fall in thermistor resistance (a semiconductor).

p.t.c. thermistors The letters *p.t.c.* stand for *p*ositive *t*emperature *c*oefficient. The resistance of this type of thermistor *increases* sharply above a certain temperature. They are used mainly to prevent damage in circuits which might experience a large temperature rise, e.g. one in which an electric motor becomes overloaded.

LIGHT-DEPENDENT RESISTOR (LDR)

This type of transducer changes light into electrical energy and is also called a photoconductive cell. Its action depends on the fact that the resistance of certain semiconductors, such as cadmium sulphide, decreases as the intensity of the light falling on them increases. The effect (also given by infrared and ultraviolet radiation) is due to light supplying the energy to set free charge carriers from atoms of the semiconductor, so increasing its conductivity, i.e. reducing its resistance.

A common LDR (the ORP12) is shown in Fig. 19.4a, b, and the LDR symbol is shown in c. There is a 'window' over the grid-like metal electrodes to allow light to fall on a thin layer of cadmium sulphide. Its resistance varies from about 10 MΩ in the dark to 1 kΩ or so in daylight.

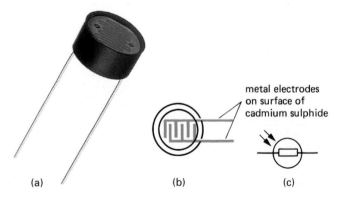

Fig. 19.4

LDRs have many uses, e.g. in photographic exposure meters, but their response to light changes is slow compared with a photodiode (p. 62), typically 100 milliseconds compared with 1 microsecond. This makes them unsuitable for fast counting operations.

STRAIN GAUGE

One device which engineers use to obtain information about the size and distribution of strains in structures such as buildings, bridges and aircraft is the electrical strain gauge. It converts mechanical strain into a resistance change in itself by using the fact that the resistance of a wire depends on its length and cross-sectional area.

One type of gauge consists of a very fine wire cemented to a piece of thin paper as in Fig. 19.5. In use it is securely attached with very strong adhesive to the component under test so that it experiences the same strain as the component. If, for example, an increase of length strain occurs, the gauge wire gets longer and thinner and on both counts its resistance increases. Thick leads connect the gauge to a resistance-measuring circuit and previous calibration of the gauge enables the strain to be measured directly.

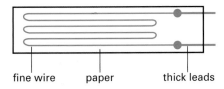

fine wire   paper   thick leads

Fig. 19.5

1. A lamp in series with a thermistor is just alight. Does it get brighter if the thermistor is cooled when the thermistor is (i) an n.t.c., (ii) a p.t.c., type?
2. A lamp in series with an LDR is just alight. What happens if the intensity of the light on the LDR (i) increases, (ii) decreases?
3. What is the advantage of using a parallel-folded wire for a strain gauge, Fig. 19.5?
4. The log–log graph in Fig. 19.6 shows the characteristic curve of an LDR. What is the resistance of the LDR when the light intensity is 400 lux?

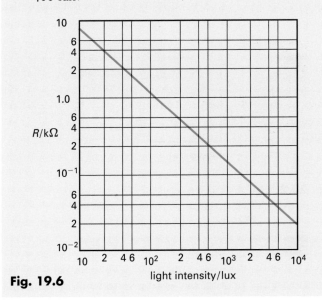

Fig. 19.6

20 DIGITAL DISPLAYS

CHECKLIST

After studying this chapter you should be able to:
- explain how seven-segment displays can display the digits 0 to 9 and the letters A to F, and
- compare LED, GDD, fluorescent and LCD displays.

Digital displays are transducers for changing electrical energy into light and are used as numerical indicators in calculators, digital watches, cash registers, weighing machines and other measuring instruments. They consist of seven segments arranged as a figure eight so that the digits 0 to 9 and sometimes the letters A to F can be displayed when different segments are energized, Fig. 20.1.

Fig. 20.1

Four types of digital display are in common use.

LIGHT-EMITTING DIODE (LED) DISPLAY

The segments are light-emitting diodes or LEDs (p. 61) made from the semiconductor gallium arsenide phosphide. When a typical current of 10 mA passes through the diode it emits red or green light, depending on its composition.

Each segment has an anode lead at which the current enters and a cathode lead at which it leaves. All seven anodes (or cathodes) are joined together to form a common anode (or cathode), Fig. 20.2. Such displays are often designed to work on a 5 V d.c. supply with a current-limiting resistor (e.g. 330 Ω) in series with each segment.

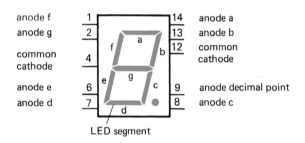

Fig. 20.2

GAS DISCHARGE DISPLAY (GDD)

Each segment of a GDD consists of a small glass tube containing mainly neon gas at low pressure which produces a bright orange glow when a current passes through it. Although about 170 V is required to start the display, the current taken by a segment may only be 0.2 mA or so.

FLUORESCENT VACUUM DISPLAY

The display is a blue–green colour, popular for calculators. The light is produced when electrons, emitted from an electrically heated filament (p. 49) are accelerated in an evacuated glass tube and strike a fluorescent screen. The filament needs, in a typical case, a current of 50 mA at 2 V a.c. or d.c. and an accelerating p.d. of 15 V is also required.

LIQUID CRYSTAL DISPLAY (LCD)

These are much used in digital watches where their very small current needs (typically 5 µA for all segments on) prolongs battery life. Liquid crystals are organic (carbon) compounds which exhibit both solid and liquid properties. A 'cell' with transparent electrodes on opposite faces, containing a thin layer of liquid crystal, Fig. 20.3a, and on which light falls, goes 'dark' when a p.d. is applied across the electrodes. The effect is due to molecular rearrangement within the liquid crystal.

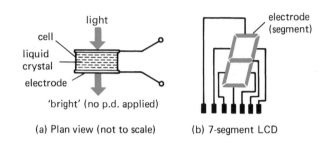

(a) Plan view (not to scale)   (b) 7-segment LCD

Fig. 20.3

Only the liquid crystal under those electrodes to which the p.d. is applied goes 'dark'. The display has a silvered background which reflects back incident light and it is against this continuously visible background (except in darkness when it has to be illuminated) that the numbers show up as dark segments, Fig. 20.3b.

Thousands of tiny LCDs are used to form the picture elements (pixels) of the screen in portable personal computers and in some pocket television receivers. (See Chapter 94.)

QUESTIONS

1. Explain the abbreviations (i) LED, (ii) GDD, (iii) LCD.
2. Give one advantage and one disadvantage of (i) LED, (ii) liquid crystal, displays.

21 CATHODE RAY TUBE (CRT)

CHECKLIST

After studying this chapter you should be able to:
• name the three main parts of a CRT,

• state what each part does, and
• name four devices in which a CRT is used.

A cathode ray tube changes electrical energy into light and is used in oscilloscopes (p. 93), television receivers (p. 201), computer monitors, radar and other electronic systems. It has three parts:

(i) an *electron gun* for producing a narrow beam of high-speed electrons (a cathode ray) in an evacuated glass tube;

(ii) a *beam deflection system* (electrostatic or magnetic); and

(iii) a *fluorescent screen* at the end of the tube which emits light where the beam falls on it.

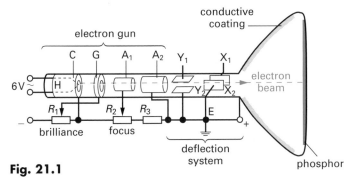

Fig. 21.1

A CRT with electrostatic deflection is shown in Fig. 21.1; a voltage divider provides appropriate voltages from a high voltage power supply.

ELECTRON GUN

The *heater* H is a tungsten wire inside, but electrically insulated from, the hollow cylindrical nickel *cathode* C. When current passes through H (usually from a 6 V a.c. supply) it raises the temperature of C to dull red heat, causing a special mixture on the tip of C to emit electrons copiously. The process is called *thermionic emission*.

The negatively charged electrons are accelerated towards the *anodes* A_1 and A_2 (metal discs or cylinders), which have positive potentials with respect to C, A_2 more so than A_1. The *grid* G, another hollow metal cylinder, has, relative to C, a negative potential which can be varied by R_1. The more negative it is, the fewer the number of electrons emerging from the hole in the centre of

G. The spot produced on the *screen* S by the beam, after it has shot through the central holes in A_1 and A_2, is then less bright. R_1 is the *brilliance* control.

Focusing of the beam to give a small luminous spot on S is achieved by changing the p.d. between A_1 and A_2, i.e. by altering the *focus* control R_2. Typical voltages for a small CRT are 1 kV for A_2, 200 to 300 V for A_1 and -50 V to 0 V for G.

There must be a return path for the electrons, from S to C, otherwise unwanted negative charge would build up on S. This does not happen because when struck by electrons, S emits a more or less equal number of *secondary* electrons. These are attracted to and collected by a conducting coating (e.g. of graphite) on the inside of the tube near S and which is connected to A_2. From there the circuit is completed through the power supply to C.

It is common practice to earth A_2 (and the conducting coating), thus preventing earthed objects (e.g. people) near S upsetting the electron beam. The other electrodes in the gun are then negative with respect to A_2 but the electrons are still accelerated through the same p.d. between C and A_2.

BEAM DEFLECTION SYSTEM

In electrostatic deflection (used in oscilloscopes) the beam from A_2 passes between two pairs of metal plates, Y_1Y_2 and X_1X_2, at right angles to each other. If one plate of each pair is at a positive potential with respect to the other, the beam moves towards it. The Y-plates, which are farther from the screen, are horizontal and deflect the beam vertically, i.e. along the y-axis. The X-plates are vertical and produce horizontal deflections, i.e. along the x-axis.

To minimize the defocusing effect which the voltages on the deflecting plates might have on the beam, one plate of each pair is at the same potential as A_2, i.e. earth. In practice X_2, Y_2 and A_2 are connected internally.

In magnetic deflection (used in TV receiver tubes whose need for a very large deflecting angle of 110° would require extremely high electrostatic deflecting voltages) two pairs of current-carrying coils are placed outside the tube at right angles to each other, Fig. 21.2. The magnetic field produced by one pair causes deflections in the y-direction and by the other pair in the x-direction.

FLUORESCENT SCREEN

The inside of the wide end of the tube is coated with a substance called a *phosphor* which emits light where it is hit by the electrons. The colour of the light emitted and the 'afterglow' or 'persistence', i.e. the time for which the emission is visible after the electron bombardment has stopped, depends on the phosphor.

Phosphors emitting bluish light are popular for oscilloscopes. For radar tubes long-persistence types producing orange light are common. For black-and-white TV tubes white phosphors of short persistence are used while for colour TV red, green and blue phosphors are employed.

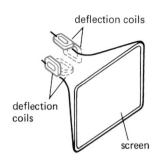

Fig. 21.2

QUESTIONS

1. State the job done by each of the three main parts of a CRT.

2. Explain why the electrons in the CRT of Fig. 21.1 (p. 49) travel with more or less constant speed between A_2 and S.

22 RELAYS AND REED SWITCHES

CHECKLIST

After studying this chapter you should be able to:
• state what a relay does and recognize its symbol, and

• state what a reed switch does.

RELAYS

A relay is a switch used to turn other circuits on or off. It is a transducer which changes an electrical signal to movement and then back to an electrical signal. It is useful if we want (i) a small current in one circuit to control another circuit containing a device such as a lamp or electric motor which needs a large current, or (ii) several different switch contacts to be operated simultaneously.

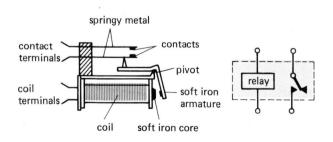

Fig. 22.1

The structure of a relay is shown in Fig. 22.1 with its symbol; the contacts may be normally open (as shown), normally closed or change-over. When the controlling current flows through the coil, the soft iron core is temporarily magnetized and attracts the iron armature. This rocks on its pivot and operates the contacts in the circuit being controlled. When the coil current stops, the armature is no longer attracted and the contacts return to their normal positions.

A relay with a nominal coil voltage of 12 V may, for example, work off any voltage between 9 V and 15 V. The coil resistance (e.g. 185 Ω) is also quoted as well as the current and voltage ratings of the contacts. The current needed to operate a relay is called the *pull-in* current, and the *drop-out* current is the smaller current in the coil when the relay just stops working. For the above relay, the pull-in current I is given by $I = V/R = 12 \text{ V}/185 \text{ Ω} = 0.065 \text{ A} = 65 \text{ mA}$.

RELAY CIRCUITS

Simple intruder alarm When light falls on the LDR in the circuit of Fig. 22.2a, the LDR resistance decreases, and the current through it and the relay coil is large enough to operate a relay. The normally open contacts close and the bell rings. If the light is removed the LDR current is too small to 'hold' the relay and the ringing stops.

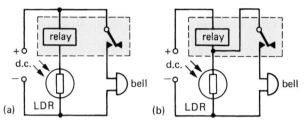

Fig. 22.2

Latched intruder alarm By connecting the relay contacts in series with the relay coil as in Fig. 22.2b, the bell keeps ringing when the LDR is no longer illuminated, i.e. the alarm is 'latched'.

REED SWITCHES

Relays operate relatively slowly so for fast-switching reed switches are used. One with normally open contacts is shown in Fig. 22.3. The reeds are thin strips of easily magnetized and demagnetized nickel-iron sealed in a small glass tube.

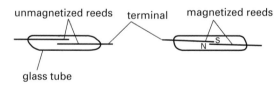

Fig. 22.3

The switch is operated either by passing a current through a coil surrounding it or by bringing a magnet near. In both cases the reeds become magnetized, attract and on touching complete the circuit connected to the terminals. The time for this may only be 1 millisecond and there is almost no *contact bounce*, i.e. no repeated making and breaking of the contacts when the reeds come together. When the current in the coil stops or the magnet is removed, the reeds separate.

QUESTIONS

1. What voltage is required to operate a relay having a coil of resistance 1 kΩ and a pull-in current of 10 mA?
2. Calculate the pull-in current of a reed switch designed to operate with a coil of resistance 700 Ω on a 6 V battery.
3. In the circuit of Fig. 22.4 the operation of a relay is delayed intentionally for a certain time by using a large capacitor as shown.
   Explain why it is only a few seconds after S is closed that the normally open contacts of the relay close and the lamp comes on (and stays on so long as S is on).
   Why is R included in the circuit?

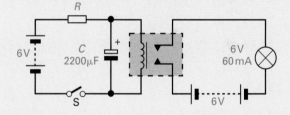

Fig. 22.4

23 ELECTRIC MOTORS

CHECKLIST

After studying this chapter you should be able to:
* state what an electric motor does,
* compare series- and shunt-wound motors,

* explain how the back e.m.f. arises in a motor, and
* state what a stepper motor does and its uses.

Electric motors change electrical energy into mechanical energy.

D.C. ELECTRIC MOTORS

Principle A simple d.c. motor consists of a coil of wire fixed to an axle which can rotate between the poles of a permanent magnet, Fig. 23.1. Each end of the coil is connected to half of a split copper ring, called the *commutator*, which rotates with the coil. Two carbon blocks, the *brushes*, press lightly against the commutator and are connected to a d.c. supply.

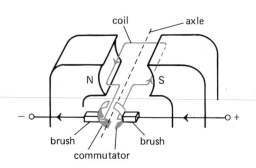

Fig. 23.1

When current passes through the coil, it becomes a 'flat' magnet and rotates so that its N pole is opposite the S pole of the permanent magnet and vice versa. The coil's inertia makes it overshoot the vertical position and the commutator halves change contact from one brush to the other. This reverses the current in the coil, reversing its magnetic polarity and making it carry on for another half-turn in the same direction to face opposite poles again, and so on.

Practical motors Small motors for driving models use permanent magnets. More powerful types have:

(i) an *armature* or *rotor* consisting of several multi-turn coils (each with its own pair of commutator segments) wound in slots in a soft-iron (laminated) core to give smoother running and self-starting (the

two-pole motor in Fig. 23.1 needs a push to start unless it is in the horizontal position, in which the force on it is a maximum); and

(ii) an *electromagnet* instead of a permanent magnet, its windings being called the *field* coils.

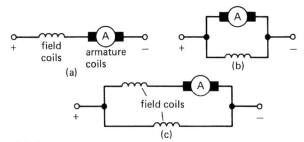

Fig. 23.2

A *series-wound* motor has the armature and field coils in series, Fig. 23.2a, and has a large turning effect (torque) when starting but its speed varies with the load. A *shunt-wound* motor has the armature and field coils in parallel, Fig. 23.2b, and runs at a steady speed when the load varies. A *compound-wound* motor, Fig. 23.2c, has some coils in series and some in parallel and gives the advantages of both types.

Small d.c. motors are only about 10% efficient; larger ones give higher values.

Starting resistor When a motor is running it behaves like a dynamo and produces a *back e.m.f.* which opposes and reduces the current in the armature (Lenz's law, p. 28). On starting, the back e.m.f. is absent and a large, damaging current would flow unless there was a starting resistor in series with the armature to limit it. The resistance is reduced as the motor speeds up.

STEPPER MOTORS

Stepper motors, driven by a series of electrical pulses, rotate by a small exact amount for each pulse. They are used in robots and in magnetic memory disc drives for computers (Chapter 94) where very precise movement is required.

The *rotor* is a permanent cylindrical magnet with a large number of poles round its perimeter. It rotates inside two *stator* (field) coils, Fig. 23.3a, which each have a row of metal 'teeth'. The teeth are magnetized with alternate N and S poles when current is sent through a coil. The rotor poles line up with a pair of opposite poles on the stator teeth, Fig. 23.3b. When the current in one stator coil is reversed, say the upper one, the polarity of each tooth on the upper stator changes. As a result the rotor turns by one tooth as its poles realign themselves with the new arrangement.

A small stepper motor with the integrated circuit (IC) chip needed to drive it is shown in Fig. 23.4. It rotates through angular 'steps' of 7.5°, i.e. it makes 48 'steps' per revolution.

A.C. ELECTRIC MOTORS

The *induction* motor is the commonest a.c. type; its action depends on the production of eddy currents by a moving magnetic field. The large 'squirrel-cage' type is much used in industry; its rotor (armature) is an array of copper rods embedded in the surface of an iron cylinder (like a cage). The small 'shaded-pole' type, in which a copper plate covers part of the magnetic field, is used in record players.

The advantages of induction motors are:

(i) no brushes to wear or cause sparks that give radio interference,
(ii) no starting resistor needed, and
(iii) constant speed controlled by the mains frequency (not voltage).

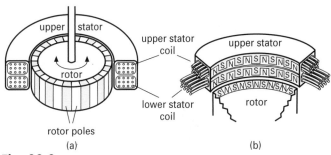
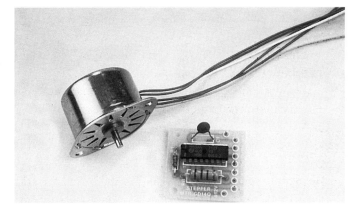

Fig. 23.3

Fig. 23.4

QUESTION

1. a) What is the back e.m.f. of a motor?
b) A d.c. motor has an armature resistance of 0.5 Ω and when working from a 12 V supply the current through it is 2 A. Comment on these figures.

24 PROGRESS QUESTIONS

1. Three types of microphone are in common use: moving coil, piezoelectric and condenser. State which of these (i) requires a battery, (ii) has the highest output voltage, (iii) must be connected to a high input impedance preamplifier? (*N. part qn.*)

2. What is meant by the *sensitivity* of a *microphone*? Describe in detail how you would investigate the variation of sensitivity of a *crystal* microphone with *direction*.

Show by means of a diagram the results you would expect to obtain.

Draw a labelled diagram showing the structure of the crystal microphone. (*C.*)

3. The resistance of a bead thermistor at different temperatures is found to be:

Temperature in °C	20	40	60	80	100
Resistance in kΩ	40	19	8.8	4.2	2.0

a) Plot a graph of the resistance of the thermistor against temperature.

b) Find from your graph the temperature of the thermistor at which it has a resistance of 15 kΩ.

c) Explain why the resistance of the thermistor changes with temperature in this way.

d) Draw a diagram of an electrical circuit suitable for measuring the resistance of the thermistor.

e) Explain what happens to a thermistor when it is connected across the terminals of a battery so that power is converted in it. (*O.L.E.*)

4. In the circuit of Fig. 24.1 the switching on and off of a motor M is controlled by an LDR.

a) Is the motor more likely to be on or off when light falls on the LDR?

b) If the LDR has a resistance of 50 Ω under certain light conditions and the resistance of the relay coil is 200 Ω, calculate the current in the relay coil.

c) Calculate the power dissipated in the relay coil.

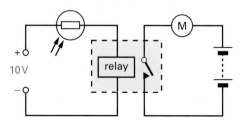

Fig. 24.1

5. Draw a labelled diagram of a cathode ray tube. (Details of associated circuitry are not required.) (*L.*)

6. Explain briefly
a) the piezoelectric effect,
b) how the piezoelectric effect is used in the operation of a crystal microphone.

7. Without capacitor in the circuit of Fig. 24.2, the relay closes and opens normally when the switch S is closed and then opened.

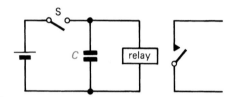

Fig. 24.2

When the switch S is closed with the capacitor in the circuit, the relay closes after a time delay. Explain.

State and explain what happens to the relay when the switch S is now opened. (*O.L.E. part qn.*)

8. Explain how a *back e.m.f.* is generated in the armature of a d.c. electric motor.

Why does this back e.m.f.
a) limit the speed of the motor,
b) necessitate a starting resistance in series with the armature? (*C.*)

SEMICONDUCTOR DIODES

25 SEMICONDUCTORS

CHECKLIST

After studying this chapter you should be able to:
- state that intrinsic semiconduction is due to holes and electrons,
- state how doping produces n- and p-type extrinsic semiconduction, and
- recall that the Hall effect provides evidence for holes and electrons in semiconductors.

Semiconductors are materials whose electrical conductivity lies between that of good conductors, like copper, and good insulators, like polythene. Examples are silicon, germanium, cadmium sulphide and gallium arsenide.

INTRINSIC SEMICONDUCTION

Valence electrons Many of the properties of materials can be explained if we assume that the electrons surrounding the nucleus of an atom are in different energy bands. Those in the highest energy band are most loosely held by the atom. These are known as *valence* electrons because the chemical combining power or valency (as well as many physical properties) of the atom depend on them. The valence electrons form 'bonds' with the valence electrons of neighbouring atoms to produce, in the case of most solids, a regularly repeating three-dimensional pattern of atoms called a crystal lattice.

An atom of a semiconductor such as silicon or germanium has four valence electrons, Fig. 25.1a, and in the lattice each one is shared with a nearby atom to form four covalent bonds, Fig. 25.1b. Every atom has a half-share in eight valence electrons and it so happens that this number of electrons gives a very stable arrangement (most of the inert gases have it). A strong crystal lattice results in which it is difficult for electrons to escape from their atoms. Pure silicon and germanium are therefore very good insulators, being perfect at near absolute zero ($-273\,°C$). Metals, by comparison, are good conductors because the valence electrons escape from their atoms very easily, becoming 'free' electrons.

Electrons and holes At normal temperatures the atoms of a semiconductor are vibrating sufficiently in the lattice for a few bonds to break, setting free some valence electrons. When this happens, a deficit of negative charge, called a vacancy, or *hole*, is left by each free electron in the valence band of the atom from which it came, Fig. 25.2a. The atom, now short of an electron, becomes a positive ion. We think of the hole as a positive charge equal in size to the negative charge on an electron.

The energy needed to produce an electron–hole pair is denoted by E_g; it is greater for an insulator than for a semiconductor because the bonds are stronger in an insulator.

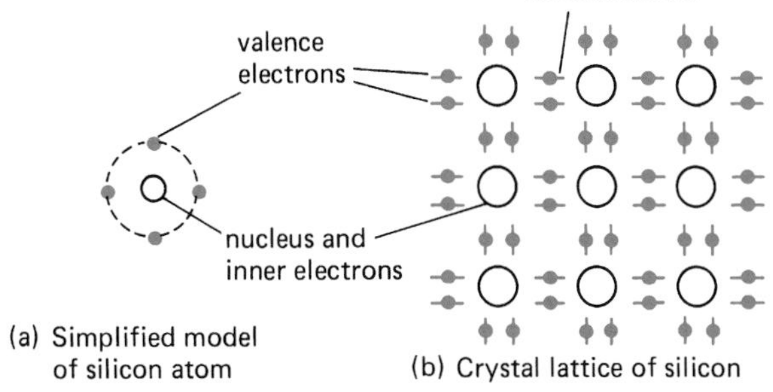

valence electrons

covalent bond

nucleus and inner electrons

(a) Simplified model of silicon atom

(b) Crystal lattice of silicon

Fig. 25.1

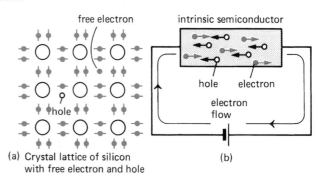

free electron

intrinsic semiconductor

hole   electron

electron flow

hole

(a) Crystal lattice of silicon with free electron and hole

(b)

Fig. 25.2

When a battery is connected across a pure semiconductor, it attracts free electrons to the positive terminal and provides a supply of free electrons at the negative terminal. The free electrons (from broken bonds) drift through the semiconductor by 'hopping' from one hole to another nearer the positive terminal, making it *seem* that positive holes are moving to the negative terminal. The current in a *pure* semiconductor is very small and can be *thought* of as streams of free electrons and holes going in opposite directions as in Fig. 25.2b. It is called *intrinsic* semiconduction because the charge carriers come from inside the material.

If the temperature of a semiconductor rises, more bonds break and intrinsic conduction increases because more free electrons and holes are produced. *The resistance of a semiconductor decreases with temperature rise.*

EXTRINSIC SEMICONDUCTION

The use of semiconductors in diodes, transistors and integrated circuits depends on increasing the conductivity of very pure (i.e. intrinsic) semiconducting materials by adding to them tiny but controlled amounts of certain 'impurities'. The process is called *doping* and the material obtained is known as an *extrinsic* semiconductor because the impurity supplies the charge carriers (which add to the intrinsic ones). The impurity atoms must be about the same size as the semiconductor atoms so that they fit into its crystal lattice without causing distortion. There are two kinds of extrinsic semiconductor.

n-type   This type is made by doping pure semiconductor with, for example, phosphorus. A phosphorus atom has five valence electrons and Fig. 25.3a shows what occurs when one is introduced into the lattice of a silicon crystal. Four of its valence electrons form covalent bonds with four neighbouring silicon atoms but the fifth is spare and, being loosely held, it can take part in conduction. The impurity (phosphorus) atom is called a *donor*.

The 'impure' silicon is an n-type semiconductor since the *majority* charge carriers are negative electrons. (The overall charge in the crystal is zero since every atom present is electrically neutral.) Impurity atoms are added in the amount which produces the required conductivity.

A few positive holes are also present in n-type material (as intrinsic charge carriers formed by broken bonds between silicon atoms); they are *minority* carriers. Fig. 25.3b shows conduction in an n-type semiconductor.

p-type   To make this type, silicon or germanium is doped with, for example, boron. A boron atom has three valence electrons and Fig. 25.4a shows what happens when one is introduced into the lattice of a silicon crystal. Its three valence electrons each share an electron with three of the four silicon atoms surrounding it. One bond is incomplete and the position of the missing electron, i.e. the hole, behaves like a positive charge since it can attract an electron from a nearby silicon atom, so forming another hole. For that reason the impurity (boron) atom is called an *acceptor*.

The 'impure' silicon is a p-type semiconductor since the *majority* charge carriers causing conduction are positive holes. (Note again that the semiconductor as a whole is electrically neutral.) In p-type material a few electrons are present as *minority* carriers. Fig. 25.4b shows conduction in such material.

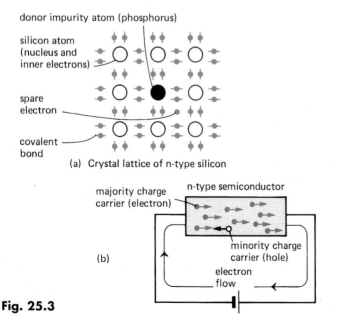

(a) Crystal lattice of n-type silicon

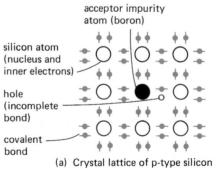

(a) Crystal lattice of p-type silicon

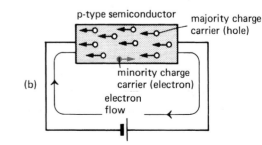

Fig. 25.3

Fig. 25.4

If n- and p-types of semiconductor are heavily doped, they can acquire metallic properties and a temperature rise increases their resistance (p. 9).

HALL EFFECT

This effect gives direct evidence for the existence of positive holes in a p-type semiconductor.

If a slab of current-carrying p-type germanium is in a magnetic field as in Fig. 25.5a, a p.d., called the *Hall p.d.*, arises between sides X and Y, with X being positive. It arises because the positive charge carriers (holes) experience an upwards force at right angles to (and due to) the magnetic field, Fig. 25.5b. As a result the positive holes are pushed towards X, creating an electric field of strength E and p.d. V_H (typically 0.1 V) given by:

$$E = \frac{V_H}{d}$$

where d is the height of the slab.

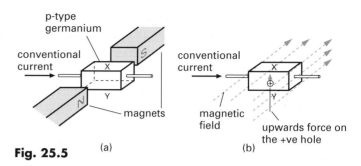

Fig. 25.5 (a)   (b)

If the majority charge carriers were negative, the Hall p.d. would be such that side X would be negative with respect to Y.

QUESTIONS

1. Explain the terms (i) *intrinsic*, (ii) *extrinsic*, semiconductor.
2. What are the majority charge carriers in (i) n-type, (ii) p-type, semiconductors?

26 JUNCTION DIODE

CHECKLIST

After studying this chapter you should be able to:
• state the effect of forward and reverse bias on a p-n junction,
• recognize the symbol for a junction diode,
• draw current–voltage graphs for silicon and germanium diodes,

• recall that a diode conducts in one direction only and state the forward voltages at which silicon and germanium diodes start to conduct, and
• state three uses of junction diodes.

THE p-n JUNCTION

The operation of many semiconductor devices depends on effects which occur at the boundary (junction) between p- and n-type materials formed in the same continuous crystal lattice.

Unbiased junction A p-n junction is represented in Fig. 26.1a. As soon as the junction is produced, free electrons near it in the n-type material are attracted across into the p-type material where they fill holes. At the same time holes pass across the junction from p-type to n-type, capturing electrons there. As a result the n-type

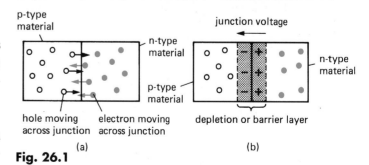

Fig. 26.1   (a)   (b)

material becomes positively charged and the p-type material becomes negatively charged (both previously being neutral).

The exchange of charge soon stops because the negative charge on the p-type material opposes the further flow of electrons and the positive charge on the n-type opposes the further flow of holes. The region on either side of the junction becomes fairly free of majority charge carriers, Fig. 26.1b, and is called the *depletion* or *barrier layer*. It is less than 10^{-6} m wide and is, in effect, an insulator. A p.d. known as the *junction voltage* acts across the depletion layer from n- to p-type, being about 0.6 V for silicon and 0.1 V for germanium.

Reverse-biased junction If a battery is connected across a p-n junction with its positive terminal joined to the n-type side and its negative terminal to the p-type side, it helps the junction voltage. Electrons and holes are repelled farther from the junction and the depletion layer widens, Fig. 26.2a. Only a few minority carriers cross the junction and a tiny current, called the *leakage* or *reverse current*, flows. The resistance of the junction is very high in reverse bias.

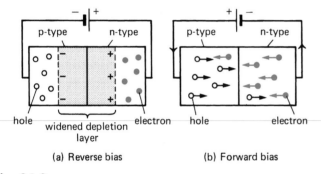

Fig. 26.2

Forward-biased junction If a battery is connected so as to oppose the junction voltage, the depletion layer narrows. When the battery voltage exceeds the junction voltage, appreciable current flows because majority carriers are able to cross the junction. Electrons travel from the n- to the p-side and holes in the opposite direction, Fig. 26.2b. The junction is then forward biased, i.e. the *p-type* side is connected to the *positive* terminal of the battery and the *n-type* side to the *negative* terminal. The resistance of the junction is very low in forward bias.

JUNCTION DIODE

Construction A junction diode consists of a p-n junction with one connection to the p-side, the *anode* A, and another to the n-side, the *cathode* K. Its symbol is shown in Fig. 26.3a. In actual diodes the cathode end is often marked by a band, Fig. 26.3b; it is the end from which conventional current leaves the diode when forward

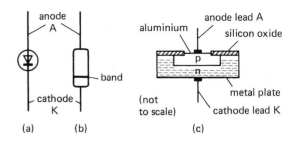

Fig. 26.3

biased. A simplified, much enlarged section of a silicon diode is shown in Fig. 26.3c but in fact the boundaries are neither straight nor well-defined.

Characteristics Typical characteristic curves for silicon and germanium diodes at 25 °C are shown in Fig. 26.4. Conduction does not start until the forward voltage V_F is about 0.6 V for silicon and 0.1 V for germanium; thereafter, a very small change in V_F causes a sudden, large increase in the forward current I_F. When I_F is limited to a value within the power rating of the diode (by a resistor in the circuit), the p.d. across a silicon diode is about 1 V.

The reverse currents I_R are negligible (note the change of scales on the negative axes of the graph) and remain so until the reverse voltage V_R is large enough (from 5 V up to 1000 V depending on the level of doping) to break down the insulation of the depletion layer. I_R then increases suddenly and rapidly and permanent damage to the diode occurs. I_R is smaller and more constant for silicon than for germanium.

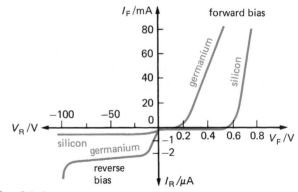

Fig. 26.4

The *average forward current* I_F *(av)* and the *maximum reverse voltage* V_{RRM} (previously called the peak inverse voltage) are usually quoted for a diode. For example, for the 1N4001, I_F *(av)* = 1 A and V_{RRM} = 50 V.

The characteristic for an ideal diode is shown in Fig. 26.5.

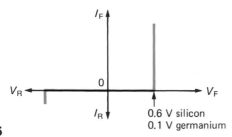

Fig. 26.5

Uses Junction diodes are used in three ways.

(i) As *rectifiers* to change a.c. to d.c. in power supplies (p. 81). Silicon is preferred to germanium because its much lower reverse current makes it a more efficient rectifier, i.e. it provides a more complete conversion of a.c. to d.c. Silicon also has a higher breakdown voltage and can work at higher temperatures.

(ii) *To prevent damage to a circuit by a reversed power supply.* In Fig. 26.6a the battery is correctly connected and forward-biases the diode. In Fig. 26.6b it is incorrectly connected but since it reverse-biases the diode, no current passes and no damage occurs to the circuit.

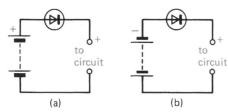

Fig. 26.6

(iii) As a *clamp diode* or *d.c. restorer* to prevent d.c. level shift problems in capacitor coupling circuits like that in Fig. 12.6a (p. 32) where a positive d.c. input, Fig. 12.6b, becomes an a.c. output, Fig. 12.6c, d. In Fig. 26.7 diode D offers a fast discharge path for C and clamps the output to 0 V when the input is zero or negative, thereby stopping a d.c. level shift.

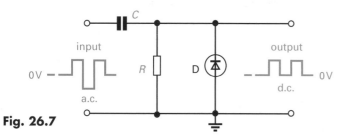

Fig. 26.7

QUESTIONS

1. Which of the 6 V 60 mA lamps are bright, dim or off in the circuits of Fig. 26.8?

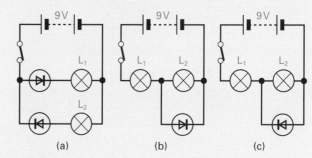

Fig. 26.8

2. The voltage drop across a 1N4001 diode is 1 V when the forward current is 1 A. If it is used on a 6 V supply as in Fig. 26.9, calculate:
a) the value of the safety resistor R,
b) the power dissipated in the diode.

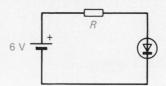

Fig. 26.9

27 OTHER DIODES

CHECKLIST

After studying this chapter you should be able to:
* recognize the symbol for a Zener diode,
* state that a Zener diode has a specified reverse breakdown voltage, called the Zener or reference voltage,
* state that a Zener diode is used in reverse bias,
* draw and interpret a current–voltage graph for a Zener diode,
* calculate the value of the current-limiting series resistor required by a specified Zener diode,

* recognize the symbol for a LED,
* state what a LED does when forward biased,
* state that the forward voltage across a conducting LED is about 2 V and calculate the value of the current-limiting resistor required, and
* describe the action and uses of a point-contact diode, a photodiode, a phototransistor and an opto-isolator.

ZENER DIODE

A Zener diode is a silicon junction diode designed for stabilizing, i.e. keeping steady, the output voltage of a power supply (p. 85). One is shown in Fig. 27.1a with its symbol in b; the cathode end is often marked by a band. It is made so that it can be used in the breakdown region.

(a)   band

anode   cathode

(b)

Fig. 27.1

Characteristic   The characteristic in Fig. 27.2 shows that, when forward biased, a Zener conducts at about 0.6 V, like an ordinary silicon diode. However it is normally used in reverse bias. Then, the reverse current I_R is

negligible until the reverse p.d. V_R reaches a certain value V_Z, called the *Zener* or *reference voltage*, when it increases suddenly and rapidly as the graph indicates. Damage may occur to the diode by overheating unless I_R is limited by a series resistor. If this is connected, the p.d. across the diode remains constant at V_Z over a wide range of values of I_R, i.e. part AB of the characteristic is at right angles to the V_R axis. It is this property of a Zener diode which makes it useful for stabilizing power supplies.

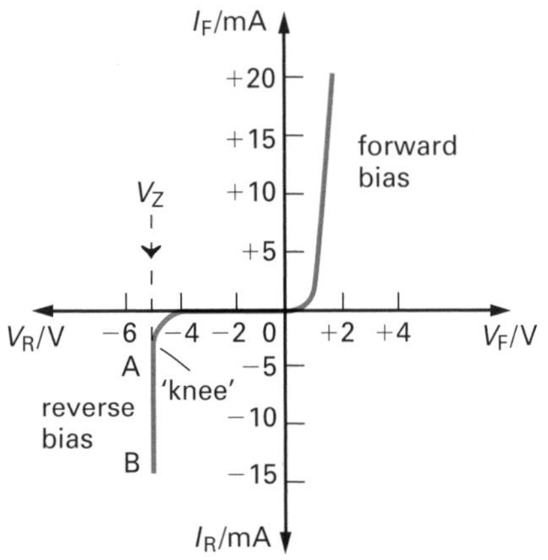

The value of the current-limiting resistor should ensure that the power rating of the diode is not exceeded. For example, a 400 mW (0.4 W) Zener diode for which $V_Z = 10$ V can pass a maximum current I_{max} given by:

$$I_{max} = \frac{\text{power}}{\text{voltage}} = \frac{0.4}{10} = 0.04 \text{ A} = 40 \text{ mA}$$

Zener diodes with specified Zener voltages are made, e.g. 3.0 V, 3.9 V, 5.1 V, 6.2 V, 9.1 V, 10 V, up to 200 V by controlling the doping.

Demonstration   When the input p.d. V_i in Fig. 27.3 is gradually increased we see that:

(i) if $V_i < 6.2$ V (the Zener voltage), $I_R = 0$ and the output p.d. $V_o = V_i$ since there is no current through R and so no p.d. across it, and

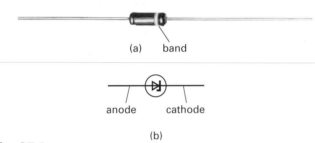

Fig. 27.2

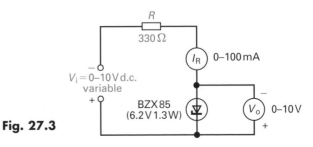

Fig. 27.3

(ii) if $V_i \geq 6.2$ V, the diode conducts, even though it is reverse biased, and $V_o = 6.2$ V, any excess p.d. appearing across R.

The Zener diode thus regulates V_o so long as V_i is not less than V_Z (here 6.2 V), i.e. there must be current through it.

Value of series resistor R In Fig. 27.3 the value of R is given by:

$$R = \frac{V_i - V_o}{I_R}$$

If $V_i = 10$ V, $V_o = V_Z = 6.2$ V and $I_R = 12$ mA = 0.012 A, then:

$$R = \frac{10 \text{ V} - 6.2 \text{ V}}{0.012 \text{ A}} = \frac{3.8 \text{ V}}{0.012 \text{ A}}$$

$$\approx 317 \text{ }\Omega \text{ (preferred value 330 }\Omega\text{)}$$

LIGHT-EMITTING DIODE (LED)

Action A LED, shown in Fig. 27.4 with its symbol, is a junction diode made from the semiconductor gallium arsenide phosphide. When forward biased it conducts and emits red, yellow, orange, blue or green light depending on its composition. No light emission occurs on reverse bias which, if it exceeds 5 V, may damage the LED.

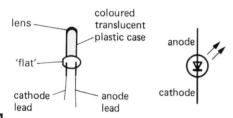

Fig. 27.4

When a p-n junction diode is forward biased, electrons move across the junction from the n-type side to the p-type side where they recombine with holes near the junction. The same occurs with holes crossing from the p-type side. Every recombination results in the release of a certain amount of energy, causing in most semiconductors a temperature rise. In gallium arsenide phosphide some of the energy is emitted as light which escapes from the LED because the junction is very close to the surface of the material.

External resistor Unless a LED is of the 'constant-current' type (incorporating an integrated circuit regulator, p. 86) it must have an external resistor R connected in series to limit the forward current which, typically, may be 10 mA (0.01 A). The voltage drop V_F is greater across a conducting LED than across an ordinary diode and is about 2 V, therefore R can be calculated from:

$$R = \frac{(\text{supply voltage} - 2.0) \text{ V}}{0.01 \text{ A}}$$

For example, on a 5.0 V supply, $R = 3.0/0.01 = 300$ Ω.

Uses LEDs are used as indicator lamps, especially in digital electronic circuits to show whether outputs are 'high' or 'low'. One way of using a LED to test for a 'high' output (9 V in this case) is shown in Fig. 27.5a, and for a 'low' output (0 V) in Fig. 27.5b. In the first case the output acts as the 'source' of the LED current, in the second case the output has to be able to accept or 'sink' the current (but also see p. 135).

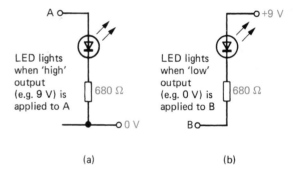

Fig. 27.5

The use of LEDs in seven-segment displays was considered earlier (p. 48).

The advantages of LEDs are small size, reliability, long life and high operating speed.

POINT-CONTACT DIODE

Construction The construction of a germanium point-contact diode is shown in Fig. 27.6. The tip of a gold wire presses on a pellet of n-type germanium. During manufacture a brief current is passed through the diode and produces a tiny p-type region in the pellet around the tip, so forming a p-n junction of very small area.

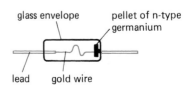

Fig. 27.6

Use Point-contact diodes are used as signal diodes to detect radio signals (a process similar to rectification in which radio frequency a.c. is converted to d.c., p. 198) because of their very low capacitance. When reverse biased, the depletion layer in a junction diode acts as an insulator sandwiched between two conducting 'plates' (the p- and n-regions). It therefore behaves as a capacitor and the larger its junction area and the thinner the depletion layer, the greater is its capacitance (p. 58).

A capacitor 'passes' a.c., and the higher the frequency and the greater the capacitance, the less opposition it offers ($X_C = 1/(2\pi f C)$). At radio frequencies therefore a normal junction diode would not be a very efficient detector (rectifier) because of the comparatively large junction area; its opposition in the reverse direction would not be large enough. A point-contact diode on the other hand is more suited to high frequency signal detection because of its tiny junction area.

Germanium is used for signal diodes because it has a lower 'turn-on' voltage than silicon (about 0.1 V compared with 0.6 V) and so lower signal voltages start it conducting in the forward direction. For the OA91 point-contact diode I_F (av) = 20 mA, $V_F \approx 1$ V and $V_{RRM} = 100$ V.

PHOTODIODE

A photodiode consists of a normal p-n junction in a case with a transparent 'window' through which light can enter. One is shown in Fig. 27.7a, b with its symbol. It is operated in reverse bias and the leakage (minority carrier) current increases in proportion to the amount of light falling on the junction. The effect is due to the light

energy breaking bonds in the crystal lattice of the semiconductor to produce electrons and holes.

Photodiodes are used as 'fast' counters which generate a current pulse every time a beam of light is interrupted.

PHOTOTRANSISTOR

A phototransistor behaves like a photodiode which gives current amplification. It is about 100 times more sensitive than a photodiode. It has a 'window' to allow light to enter; one is shown with its symbol in Fig. 27.8a, b.

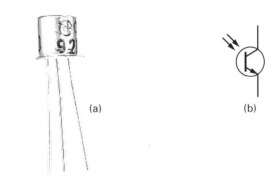

(a) (b)

Fig. 27.8

OPTO-ISOLATOR

An opto-isolator, Figs. 27.9a, b, consists of a LED combined with a phototransistor in the same package. It allows the transfer of signals from one circuit to another that cannot be connected electrically to the first because, for example, it works at a different voltage. Light (or infrared) from the LED falls on the phototransistor which is shielded from outside light.

Slotted opto-isolators like the one shown are used for the detection of liquid levels and as event counters to indicate, for instance, when the end of a tape has passed through a slot.

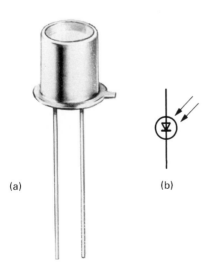

(a) (b)

Fig. 27.7

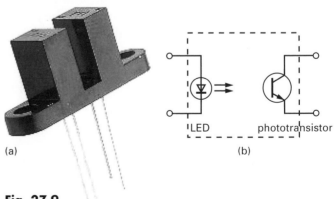

LED   phototransistor

(a) (b)

Fig. 27.9

QUESTIONS

1. In the circuit of Fig. 27.10, a 6 V 1 W Zener diode D in parallel with a 6 V 60 mA lamp L is connected in reverse bias to a battery via a protective resistor R.

Explain how the lamp behaves as the battery voltage is increased gradually from 3 V to 9 V.

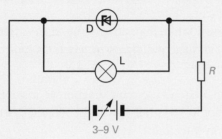

Fig. 27.10

2. Calculate the maximum current a Zener diode can pass without damage if its breakdown voltage is 10 V and its maximum power rating is 5 W.

3. In Fig. 27.11 the LED is bright when a current of 10 mA flows and the forward voltage (V_F) across it is 2 V. Calculate the value (E12 preferred) of R if a 9 V supply is used.

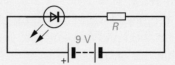

Fig. 27.11

4. State one use for
 a) a Zener diode,
 b) a LED,
 c) a point-contact diode,
 d) a photodiode.

5. a) Name the *two* parts of an opto-isolator.
 b) Explain how it works.
 c) Suggest a use for it.

28 PROGRESS QUESTIONS

1. a) Explain how an electric current is conducted in pure silicon when a p.d. is applied to it.

b) Why does the resistance of pure silicon fall as its temperature rises?

c) If silicon is doped with phosphorus (a pentavalent element) it becomes an n-type semiconductor. Explain what this means.

d) How can silicon be made a p-type semiconductor?

e) What happens to the depletion layer in a p-n junction under (i) reverse bias, (ii) forward bias?

2. The circuit shown in Fig. 28.1 represents a simple remote signalling system. Pressing either switch A or B sounds the buzzer and lights one of the LEDs P and Q.

The diodes have a forward voltage drop of 0.7 V.

The LEDs have a forward voltage drop of 2.0 V and a maximum forward current of 25 mA.

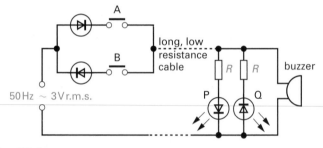

Fig. 28.1

a) (i) State which LED will light when switch B is pressed.

(ii) Explain your answer.

b) Calculate the peak voltage of the a.c. supply.

c) Calculate the minimum preferred value for the resistors R.

d) The buzzer contains a piezoelectric loudspeaker. Briefly explain how a piezoelectric loudspeaker produces sound waves from an electrical signal. (N.)

3. a) Draw the reverse characteristics of a Zener diode, giving typical values on your axes. In what way does its behaviour differ from a normal p-n diode in reverse bias?

b) The Zener diode Z shown in Fig. 28.2 has a breakdown voltage of 6 V. What is the current through it?

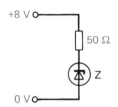

Fig. 28.2

Discuss what happens to the current through the diode, and the p.d. across it when the positive supply is altered to (i) 12 V, (ii) 4 V. (C.)

4. Fig. 28.3 shows a circuit diagram of a stabilized voltage supply together with the characteristic of the diode in the circuit.

(i) What type of diode is represented in the circuit?

(ii) What type of bias is applied to the diode?

(iii) What will be the output voltage?

(iv) Explain why the output voltage is constant for a range of input voltages. (O.L.E.)

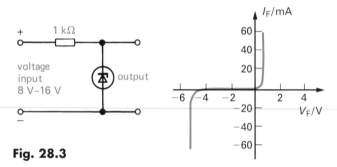

Fig. 28.3

5. a) In which of the circuits in Fig. 28.4 would the LED light when the point Q is raised from 0 V to +9 V?

b) Why is the 680 Ω resistor included in the circuits?

c) Calculate the current through the LED which lights if the forward voltage dropped across the LED is 2 V.

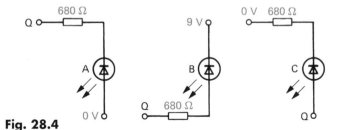

Fig. 28.4

TRANSISTORS AND ICs

29 TRANSISTOR AS A CURRENT AMPLIFIER

CHECKLIST

After studying this chapter you should be able to:
- name the three leads of a junction or bipolar transistor and recognize the symbols for n-p-n and p-n-p types,
- state that a transistor is a current amplifier in which a small base current switches on a much larger collector current when the base–emitter voltage exceeds the turn-on voltage of about 0.6 V,
- state and use the equations $h_{FE} = I_C/I_B$ and $I_E = I_C + I_B$, and
- state how a 'Darlington pair' gives a greater current gain.

ABOUT TRANSISTORS

Transistors are the tiny semiconductor devices which have revolutionized electronics. They have three connections and are made as separate (discrete) components, like those in Fig. 29.1 in their cases, and also as parts of integrated circuits (ICs), where many thousands may be packed on a 'chip' of silicon. Transistors are used as current, voltage and power *amplifiers* and as high-speed *switches*.

Fig. 29.1

There are two basic types, the common *bipolar* or *junction* transistor, to be considered in this chapter and the next two, and the less common *unipolar* or *field effect* transistor (FET), to be discussed in Chapter 32.

JUNCTION (BIPOLAR) TRANSISTOR

A junction transistor consists of two p-n junctions (in effect two diodes back-to-back) in the same semiconductor crystal. A very thin slice of lightly doped p- or n-type material called the *base* B, is sandwiched between two thicker, more heavily doped slices of the opposite type of material, called the *collector* C and the *emitter* E, the latter being smaller. The two possible arrangements, n-p-n and p-n-p, are shown simplified in Figs. 29.2a, b, along with the symbols for each type. The arrows on the symbols indicate the direction in which conventional (positive) current would flow; electron flow is in the opposite direction.

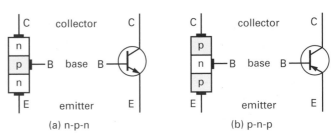

(a) n-p-n   (b) p-n-p

Fig. 29.2

In general, silicon is preferred to germanium for making transistors because it can work at higher temperatures (175 °C compared with 75 °C) and has a lower leakage current (p. 58). Silicon n-p-n types are easier to mass-produce than p-n-p types and so are more common, although it is sometimes useful to have both types in a circuit.

WHAT A JUNCTION TRANSISTOR DOES

Action There are two current paths through a junction transistor. One is the base–emitter path and the other is the collector–emitter (via base) path. The transistor's value arises from the fact that it can link circuits connected to each path so that the current in one controls that in the other.

If a p.d. (e.g. +6 V) is connected across an n-p-n silicon transistor so that the collector becomes positive with respect to the emitter, the base being unconnected, the base–collector junction is reverse biased (since the positive of the supply goes to the n-type collector). Current cannot flow through the transistor.

If the base–emitter junction is now *forward biased* by applying a p.d. V_{BE} of about +0.6 V (+0.1 V for a germanium transistor), Fig. 29.3, electrons flow from the n-type emitter across the junction (as they would in a junction diode) into the p-type base. Their loss is made good by electrons entering the emitter from the external circuit to form the *emitter current* I_E.

In the base, only a small proportion (about 1%) of the electrons from the emitter combine with holes because the base is very thin (less than 10^{-3} mm) and is lightly doped. Most of the electrons pass through the base under the strong attraction of the positive voltage on the collector. They cross the base–collector junction and become the *collector current* I_C in the external circuit.

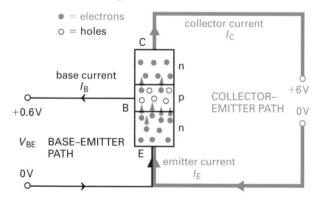

Fig. 29.3

The small amount of electron–hole recombination occurring in the base gives it a momentary negative charge which is immediately compensated by a flow of positive holes from the base power supply. This flow of holes to the base from the external circuit creates a small *base current* I_B and enables the transistor to maintain the much larger collector current.

If we regard I_B as the *input* current to the transistor and I_C as the *output* current from it, then the transistor acts:

(i) *as a switch* in which I_B turns on and controls I_C, i.e. $I_C = 0$ if $I_B = 0$ ($I_B = 0$ until $V_{BE} \approx$ +0.6 V), and increases when I_B does; and

(ii) *as a current amplifier* since I_C is greater than I_B.

Further points Note that for an n-p-n transistor the collector and base must be positive with respect to the emitter; for a p-n-p type they must be negative.

The input (base–emitter) and output (collector–emitter) circuits in Fig. 29.3 have a common connection at the emitter. The transistor is said to be in *common-emitter* mode. Two other less usual modes are *common-collector* and *common-base*.

d.c. current gain Typically I_C is 10 to 1000 times greater than I_B depending on the type of transistor. The *d.c. current gain* h_{FE} is an important property of a transistor and is defined by:

$$h_{FE} = \frac{I_C}{I_B}$$

For example, if $I_C = 5$ mA and $I_B = 0.05$ mA (50 μA), $h_{FE} = 5/0.05 = 100$. Although h_{FE} is approximately constant for one transistor over a limited range of I_C values, it varies between transistors of the same type due to manufacturing tolerances.

Also note that since the current leaving a transistor equals that entering, we have:

$$I_E = I_B + I_C$$

In the above example $I_E = 5.05$ mA.

DEMONSTRATION

Current amplification and control In the circuit of Fig. 29.4 only one power supply is used to provide the positive voltages to the collector and base of the n-p-n transistor and rheostat R controls I_B.

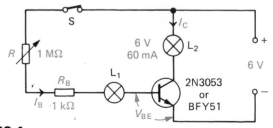

Fig. 29.4

With switch S open $I_B = 0$, therefore $I_C = 0$ and lamps L_1 and L_2 are not lit.

If S is closed and I_B is gradually increased by reducing R, I_C increases and when it is large enough, L_2 lights up but not L_1. In this way a small current I_B (too small to light L_1) produces and controls smoothly the much larger current I_C required to light L_2. The rheostat need only have a low power rating but the transistor must be able to handle I_C (60 mA).

Base resistor R_B The base–emitter junction is, in effect, a forward-biased p-n diode and so the p.d. across it (V_{BE}) cannot rise much above 0.6 V (p. 58). If input p.ds greater than 0.6 V or so were applied *directly* to the base (e.g. by adjusting R to have zero resistance in Fig. 29.4 and omitting R_B), I_B (and I_C) would become excessive and the transistor would be destroyed by over-heating. By limiting I_B, R_B prevents this when the input exceeds 0.6 V and *must always be present*.

DARLINGTON PAIR

Greater current gains can be obtained by connecting two transistors as in Fig. 29.5, called a *Darlington pair*. The emitter current I_E of Tr_1 is nearly the same as its collector current and thus approximately equals h_{FE_1} times its base (input) current I_B. This emitter current forms the base current of Tr_2. The collector (output) current I_C of Tr_2 equals h_{FE_2} times its base current. The overall current gain is:

$$h_{FE} = I_C/I_B \approx h_{FE_1} \times h_{FE_2}$$

This is typically of the order of 10^4.

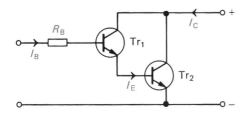

Fig. 29.5

QUESTIONS

1. In the labelled circuit of Fig. 29.6 the lamp L lights when the switch S is closed. Which is
a) the base–emitter current path,
b) the collector–emitter current path?

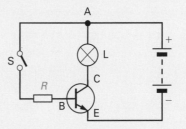

Fig. 29.6

2. In which circuits of Fig. 29.7 will lamp L light? ($B_1 = 3$ V, $B_2 = 6$ V, $R = 1$ kΩ, L = 6 V 60 mA.)

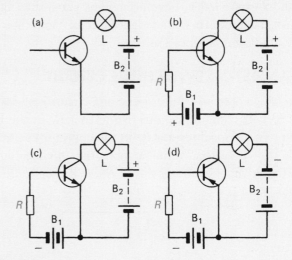

Fig. 29.7

3. For the transistor in Fig. 29.8 what is the value of
a) h_{FE},
b) I_E?

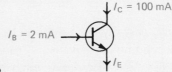

Fig. 29.8

4. A transistor has a d.c. current gain of 200 when the base current is 50 μA. Calculate the collector current in mA.

5. Two transistors have d.c. current gains of 80 and 100. What is their equivalent gain as a Darlington pair?

30 TRANSISTOR AS A SWITCH

CHECKLIST

After studying this chapter you should be able to:
- state the advantages of a transistor as a switch,
- draw a circuit in which a transistor acts as a switch,
- draw and interpret a graph showing how the output voltage V_o varies with the input voltage V_i,
- state that when V_o is 'high', the transistor is 'off' (non-conducting with a high resistance) and when V_o is 'low', the transistor is fully 'on' (conducting with a low resistance and saturated),
- solve numerical problems on transistor switching circuits,
- calculate the value of the current-limiting series base resistor,
- draw and explain the operation of light-operated, temperature-operated and time-operated alarm circuits, and
- explain how a diode connected across a relay in an alarm circuit protects the transistor.

Junction transistors have many advantages over other electrically operated switches such as relays. They are small, cheap, reliable, have no moving parts, their life is almost indefinite (in well-designed circuits) and they can switch millions of times a second.

TRANSISTOR SWITCHING CIRCUIT

The basic common-emitter switching circuit is shown in Fig. 30.1. It contains a protective resistor R_B in the base circuit and a 'load' resistor R_L in the collector circuit. R_L and the transistor form a potential divider across the power supply voltage V_{CC} to the collector (note the double suffix).

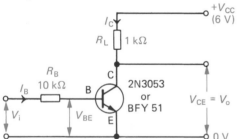

Fig. 30.1

V_i is the input p.d. (usually d.c.) applied to the base–emitter circuit. Depending on its value, V_i causes switching of the output p.d. V_o ($= V_{CE}$ and which is taken off across the collector and emitter) between a 'high' value (e.g. near V_{CC}) and a 'low' value (e.g. near 0 V), as we will now see.

Action If V_i is gradually increased from 0 to 6 V and corresponding values of V_i, V_o, I_C and I_B measured with appropriate meters, voltage and current graphs like those in Figs. 30.2a, b can be plotted.

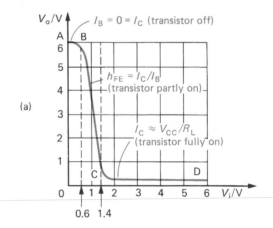

(a)

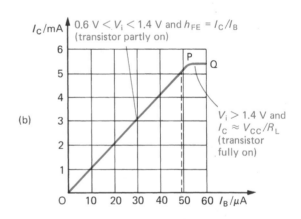

(b)

Fig. 30.2

The graphs show that:

(i) When $V_i < 0.6$ V, $V_o = V_{CC} = 6$ V, i.e. AB on graph (a), and $I_B = I_C = 0$, i.e. point O on graph (b).

(ii) When $V_i > 0.6$ V but <1.4 V, V_o falls rapidly as V_i increases, i.e. BC on graph (a), and I_C increases linearly with I_B, i.e. OP on graph (b).

(iii) When $V_i > 1.4$ V, $V_o \approx 0$ V, i.e. CD on graph (a), and I_C reaches a maximum even though I_B is increased further, i.e. PQ on graph (b).

Explanation The transistor and R_L are in series across V_{CC}; therefore since d.c. voltages can be added directly, we can say for the collector–emitter circuit:

$$V_{CC} = I_C R_L + V_{CE}$$

Rearranging the equation and putting $V_{CE} = V_o$ gives:

$$V_o = V_{CC} - I_C R_L \qquad (1)$$

That is, V_o is always less than V_{CC} by the voltage drop across R_L.

When $V_i < 0.6$ V, $V_o = V_{CC} = 6$ V because $I_B = 0$, therefore from (1), $I_C R_L = 0$ and hence $I_C = 0$. The transistor is *off* and behaves like a very high resistor, i.e. an open switch, Fig. 30.3a, with none of V_{CC} dropped across R_L since there is no current through it.

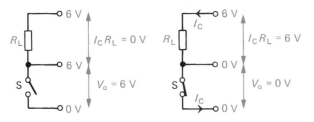

(a) S open = transistor off   (b) S closed = transistor fully on

Fig. 30.3

When $V_i > 0.6$ V but <1.4 V, the transistor is *partly on*, i.e. its resistance is decreasing, I_B and I_C flow and part of V_{CC} is dropped across R_L and part across the transistor, according to (1). In this region I_B and I_C are related by $h_{FE} = I_C/I_B$.

When $V_i > 1.4$ V, $V_o \approx 0$ V, therefore from (1), $V_{CC} \approx I_C R_L \approx 6$ V. Nearly all V_{CC} is now dropped across R_L and very little across the transistor which is behaving as if it had almost zero resistance, i.e. like a closed switch, Fig. 30.3b. The resistance of the collector–emitter circuit is now more or less constant and equal to R_L. Therefore although I_B increases when V_i rises from 1.4 V to 6 V, I_C does not but approaches its maximum value given by V_{CC}/R_L (i.e. 6 V/1 kΩ = 6 mA), as can be seen from (1) by putting $V_o = 0$. In this region the transistor is *fully on* and is said to be *saturated* (or *bottomed*) because increasing I_B does not increase I_C, i.e. $h_{FE} = I_C/I_B$ no longer applies.

Summary Depending on whether V_i is 'low' (0 to 0.6 V) or 'high' (>1.4 V), V_o switches between the two

voltage levels V_{CC} ('high') and 0 V ('low') respectively.

V_i	V_o
'low' (<0.6 V)	'high' (6 V)
'high' (>1.4 V)	'low' (0 V)

FURTHER POINTS

Base–emitter circuit In the circuit of Fig. 30.1, R_B and the base–emitter junction form a potential divider across V_i. Hence:

$$V_i = I_B R_B + V_{BE} \qquad (2)$$

For a silicon transistor $V_{BE} \approx 0.6$ V *whatever the value of* V_i. When $V_i > 0.6$ V, the 'extra' voltage $(V_i - 0.6$ V) is dropped across the protective resistor R_B.

Note that I_B can be calculated from (2) if V_i is known and we assume $V_{BE} = 0.6$ V. It is obtained by rearranging (2), to give:

$$I_B = \frac{V_i - 0.6}{R_B}$$

Power considerations The power P used by a transistor as a switch should be as small as possible. It is given by $P = V_{CE} \times I_C$. When the transistor is off, $I_C = 0$ and so $P = 0$. When the transistor is fully on, $V_{CE} \approx 0$ (typically it is 0.1 V, ideally it should be 0 V), therefore $P \approx 0$.

Power is used only when the transistor is partly on, i.e. along BC in Fig. 30.2a. It is therefore important to ensure (i) it is either fully off or fully on, and (ii) it switches rapidly between these two states.

A two-transistor circuit called a *Schmitt trigger* meets these fast-switching requirements and is considered on p. 159.

WORKED EXAMPLE

a) In the circuit of Fig. 30.4, $h_{FE} = 100$ for the transistor. Calculate the value of R_B for the lamp current I_C to be 60 mA. (Assume the transistor is not saturated and that $V_{BE} = 0.6$ V.)

b) If the lamp has resistance $R_L = 100$ Ω would the transistor be saturated?

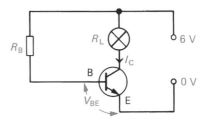

Fig. 30.4

a) If the transistor is not saturated $h_{FE} = I_C/I_B$, therefore:

$$I_B = I_C/h_{FE} = 60/100 = 0.6 \text{ mA}$$

Rearranging equation (2) we get:

$$R_B = (V_i - V_{BE})/I_B$$

where $V_i = V_{CC} = 6$ V (since the input is connected to the collector power supply), $V_{BE} = 0.6$ V and $I_B = 0.6$ mA.

Substitution gives:

$$R_B = (6 - 0.6) \text{ V}/0.6 \text{ mA} = 9 \text{ k}\Omega$$

b) When the transistor is saturated I_C is a maximum, given by:

$$I_C = V_{CC}/R_L = 6 \text{ V}/100 \text{ }\Omega = 0.06 \text{ A} = 60 \text{ mA}$$

The transistor would *just* be saturated.

ALARM CIRCUITS

In many alarm circuits a transducer in a potential divider circuit is used to switch on a transistor which then activates the alarm.

Light-operated   A simple circuit which switches on a lamp L when it gets dark is shown in Fig. 30.5. R and the light-dependent resistor (LDR) form a potential divider across the 6 V supply. The input is the p.d. V_i across the LDR and in bright light is small because the resistance of the LDR is low (e.g. 1 kΩ) compared with that of R (10 kΩ). V_{BE} is less than the 0.6 V or so required to turn on the transistor.

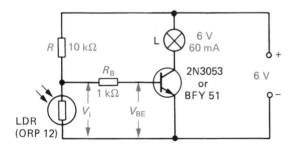

Fig. 30.5

In the dark more of the 6 V supply is dropped across the LDR, due to its greater resistance (e.g. 1 MΩ), and less across R. V_i is then large enough for V_{BE} to reach 0.6 V and switch on the transistor which creates a collector current sufficient to light L. If R is replaced by a variable resistor, the light level at which L comes on can be adjusted.

When R and the LDR are interchanged, L is on in the light and off in the dark and the circuit could act as an intruder alarm.

Temperature-operated   In the high-temperature alarm circuit of Fig. 30.6 an n.t.c. thermistor (p. 46) and resistor R form a potential divider across the 6 V supply. When the temperature of the thermistor rises, its resistance decreases, so increasing V_i and V_{BE}. When $V_{BE} \approx 0.6$ V, the transistor switches on and collector current (too small to ring the bell directly) flows through the relay coil. The relay contacts close, enabling the bell to obtain directly from the 6 V supply, the larger current it needs.

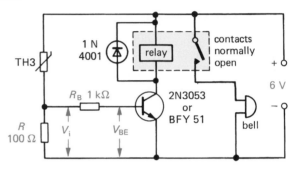

Fig. 30.6

The diode protects the transistor from damage by the large e.m.f. induced in the relay coil (due to its inductance) when the collector current falls to zero (at switch-off). The diode is forward biased to the induced e.m.f. and, because of its low resistance, offers an easy path to it. To the power supply the diode is reverse biased and its high resistance does not short-circuit the relay coil when the transistor is on.

Time-operated   In the circuit of Fig. 30.7 when S$_1$ and S$_2$ are closed, L is on and the transistor is off because $V_{BE} = 0$ (due to S$_2$ short-circuiting C and stopping it charging up). If S$_2$ is opened, C starts to charge through R and, *after a certain time*, $V_{BE} = 0.6$ V causing the transistor

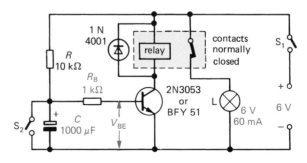

Fig. 30.7

to switch on. This operates the relay whose contacts open and switch off L. The *time delay* between opening S_2 and L going off depends on the time constant CR.

The circuit could be used as a *timer* to control a lamp in a photographic dark room. It is reset by opening S_1 and closing S_2 to let C discharge.

QUESTIONS

1. A transistor circuit is shown in Fig. 30.8.
  a) What is I_B assuming $V_{BE} = 0$?
  b) Calculate I_C if $h_{FE} = 80$.
  c) Is the transistor saturated? Justify your answer.

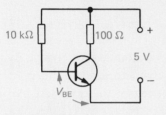

Fig. 30.8   **Fig. 30.9**

2. In the circuit of Fig. 30.9, what is the p.d. across R when (i) S is open, (ii) S is closed?

3. The circuit in Fig. 30.10 is for a silicon transistor, Tr.
  a) When Tr is off what is the p.d. across (i) R_L (ii) Tr?
  b) When Tr is saturated what is the p.d. across (i) R_L, (ii) Tr?
  c) Will R_1/R_2 be large or small when Tr is (i) off, (ii) saturated?
  d) What is the value (or range of values) of V_{BE} when Tr is (i) off, (ii) saturated?
  e) Why is (i) R_B, (ii) R_L, necessary?

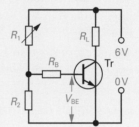

Fig. 30.10

31 MORE ABOUT THE JUNCTION TRANSISTOR

CHECKLIST

After studying this chapter you should be able to:
• define h_{FE}, r_i and r_o, and derive them from the corresponding transistor characteristics, and

• understand transistor parameters.

TRANSISTOR CHARACTERISTICS

Transistor characteristics are graphs, found by experiment, which show the relationships between various currents and voltages and enable us to see how best to use a transistor. A circuit for investigating an n-p-n transistor in common-emitter connection is given in Fig. 31.1. Three characteristics are important.

Transfer characteristic (I_C–I_B) V_{CE} is kept constant (e.g. at the supply voltage) and R_1 varied to give several

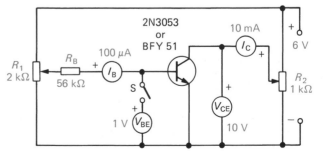

Fig. 31.1

pairs of values of I_B and I_C. If the results are plotted on a graph, a more or less straight line is obtained, like that in Fig. 31.2a, showing that I_C is directly proportional to I_B.

The *a.c. current gain*, h_{fe} (important when the transistor is handling changing currents) is defined by:

$$h_{fe} = \frac{\Delta I_C}{\Delta I_B}$$

where ΔI_C (delta I_C) is the change in I_C produced by a change of ΔI_B in I_B. Since the I_C–I_B graph is almost linear, $h_{fe} \approx h_{FE}$ (the d.c. current gain = I_C/I_B).

Input characteristic (I_B–V_{BE})

Again keeping V_{CE} constant, corresponding values of I_B and V_{BE} are obtained by varying R_1 in the circuit of Fig. 31.1. A typical graph for a silicon transistor is given in Fig. 31.2b. It shows that $I_B = 0$ until $V_{BE} \approx 0.6$ V and thereafter small changes in V_{BE} cause large changes in I_B (but V_{BE} is always near 0.6 V whatever the value of I_B).

The *a.c. input resistance* r_i of the transistor is defined by:

$$r_i = \frac{\Delta V_{BE}}{\Delta I_B}$$

where ΔI_B is the change in I_B due to a change of ΔV_{BE} in V_{BE}. Since the input characteristic is non-linear, r_i varies but is of the order of 1 to 5 kΩ.

Note. The voltmeter for measuring V_{BE} should be an electronic type with a very high resistance (e.g. 10 MΩ). If a moving-coil type is used, allowance must be made for the current it takes due to its 'low' resistance. To do this, the voltmeter is disconnected and R_1 is adjusted to make $I_B = 10$ µA (say). I_C is noted. The voltmeter is connected and R_1 adjusted until I_C has the noted value. The corresponding value of V_{BE} is recorded. The process is repeated for other values of I_B.

Output characteristic (I_C–V_{CE})

In the circuit of Fig. 31.1, I_B is set to a low value (e.g. 10 µA) and I_C is measured as V_{CE} is increased in stages by R_2. This is repeated for different values of I_B, enabling a family of curves like those in Fig. 31.2c to be plotted. They show that I_C hardly changes when V_{CE} does except below the 'knee' of the graphs where V_{CE} is less than a few tenths of a volt.

The *a.c. output resistance* r_o of the transistor is defined by:

$$r_o = \frac{\Delta V_{CE}}{\Delta I_C}$$

where ΔI_C is the change in I_C caused by a change ΔV_{CE} in V_{CE} on the part of the characteristic to the right (i.e. beyond) the 'knee'. In this region the slope ($\Delta I_C/\Delta V_{CE}$) is small, making r_o fairly high, typically 50 kΩ.

TRANSISTOR DATA

Transistors are identified by one of several codes. In the American system they start with 2N followed by a number, e.g. 2N3053. In the Continental system the first letter gives the semiconductor material (A = germanium, B = silicon) and the second letter gives the use (C = audio frequency amplifier, F = radio frequency amplifier). For example, the BC108 is a silicon a.f. amplifier. Some manufacturers have their own code.

While one type of transistor may replace another in many circuits, it is often useful to study the published data when making a choice. The table opposite lists the main ratings, called *parameters*, for five popular n-p-n silicon transistors. All are general-purpose types, suitable for use in amplifying or switching circuits. The BC108 is a low-current, high-gain device, the others are medium-current, medium-gain types.

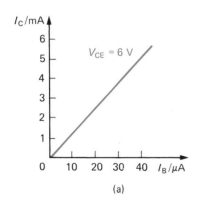

(a)

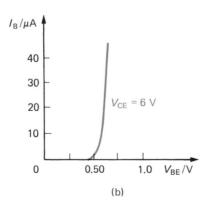

(b)

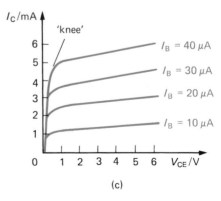

(c)

Fig. 31.2

Type	BC108	ZTX300	2N3705	BFY51	2N3053
I_C max/mA	100	500	500	1000	1000
h_{FE}	110–800	50–300	50–150	40 (min)	50–250
at I_C mA	2	10	50	150	150
P_T/mW	360	300	360	800	800
V_{CEO} max/V	20	25	30	30	40
V_{EBO} max/V	5	5	5	6	5
f_t/MHz	250	150	100	50	100
Outline	TO18	E line	TO92 (A)	TO39	TO39

Current, voltage and power ratings The symbols used have the following meanings.

I_C max   is the maximum collector current

V_{CEO} max   is the maximum collector–emitter p.d. when the base is open-circuited

V_{EBO} max   is the maximum emitter–base p.d. when the collector is open-circuited

P_T max   is the maximum power rating at 25 °C and equals $V_{CE} \times I_C$ approximately

d.c. current gain h_{FE} Owing to manufacturing spreads, h_{FE} is not the same for all transistors of the same type. Usually minimum and maximum values are quoted but sometimes only the former. Also, since h_{FE} decreases at high and low collector currents, the I_C at which it is measured is stated.

In general, when selecting a transistor type we have to ensure that its minimum h_{FE} gives the current gain required by the circuit.

Transition frequency f_T This is the frequency at which $h_{FE} = 1$ and is important in high-frequency circuits.

Outlines   These are given in Fig. 31.3 for the transistors in the table.

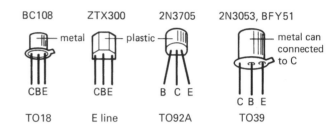

Fig. 31.3

32 FIELD EFFECT TRANSISTORS

CHECKLIST

After studying this chapter you should be able to:
- recall the two main types of field effect transistor (FET), name their three leads and recognize their symbols,
- recall that FETs have a very high input resistance compared with a bipolar transistor,
- explain why precautions have to be taken with MOSFETs, and
- state uses of JUGFETs and MOSFETs.

In a junction transistor, the small input (base) current controls the larger output (collector) current; it is a *current*-controlled amplifier. In a field effect transistor (FET), the input *voltage* controls the output current; the input current is usually negligible (less than $1 \, pA = 1$ pico-ampere $= 10^{-12}$ A). This is very useful when the input is from a device such as a crystal microphone, which cannot supply much current.

A FET consists of a bar or 'channel' of n- or p-type semiconductor having two metal contacts at its ends, called the *drain* D and the *source* S. A third contact, the *gate* G, is connected to a small p- or n-type region between D and S which forms a p-n junction.

There are two main types of FET, (i) the JUGFET (junction-gate FET) and (ii) the MOSFET (metal-oxide-semiconductor FET).

ACTION OF A JUGFET

A simplified diagram of an n-channel JUGFET is shown in Fig. 32.1a. The channel acts as a conductor which is narrower in the middle due to the depletion layer at the p-n junction behaving as an insulator (p. 58). When the drain is made *positive* with respect to the source, electrons flow from source to drain (as the names suggest). Usually the gate is *negative* relative to the source, thus reverse biasing the p-n junction and widening the depletion layer, Fig. 32.1b. This narrows the channel further and reduces the electron flow, i.e. the drain current I_D.

For a given drain–source p.d. V_{DS}, I_D is controlled by the gate–source p.d. V_{GS} (more correctly by the electric field produced by V_{GS}) and decreases as V_{GS} goes more negative.

The symbol for a JUGFET is shown in Fig. 32.1c.

The operation of a junction transistor depends on the flow of both majority and minority carriers (p. 56), hence *bipolar*. In a FET or *unipolar* transistor only majority carriers are involved, these being electrons in an n-channel type. FETs are therefore less upset than bipolar types by a temperature rise since this increases minority carriers.

CHARACTERISTICS OF A JUGFET

Normally only the transfer and output characteristics are plotted because the gate (input) current is negligible. Those for an n-channel JUGFET, such as the general purpose 2N3819, may be found using the circuit of Fig. 32.2.

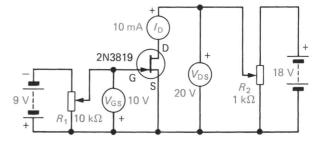

Fig. 32.2

Transfer characteristic $(I_D–V_{GS})$ Since a FET is voltage-operated, a transfer characteristic gives the relation between the input (gate) p.d. V_{GS} and the output (drain) current I_D (for fixed V_{DS}). A typical graph for a JUGFET is given in Fig. 32.3; it is nearly linear.

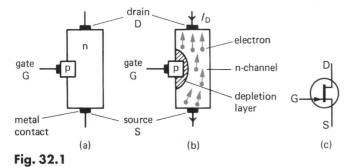

Fig. 32.1

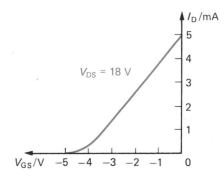

Fig. 32.3

The gain of a FET is measured by its *transconductance*, g_m, defined by:

$$g_m = \frac{\Delta I_D}{\Delta V_{GS}}$$

where ΔI_D is the change in I_D caused by a change ΔV_{GS} in V_{GS}. Its value can be found from a transfer characteristic and is in the range 1 to 10 mA/V.

Output characteristics (I_D–V_{DS}) See Fig. 32.4. They are similar to those of the bipolar transistor, I_D rising sharply initially and then remaining almost constant as V_{DS} increases over a wide range of values. Their slope to the right of the 'knee', however, is less for a JUGFET, indicating a higher *a.c. output resistance* r_o (50 kΩ to 1 MΩ).

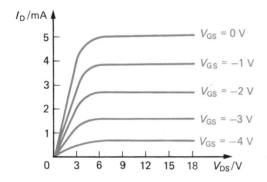

Fig. 32.4

Input resistance The *a.c. input resistance* r_i of a JUGFET is very high, greater than 10^9 Ω. It arises from the fact that the p-n junction is always reverse biased and the only input current flowing in or out of the gate is the tiny one needed to change its potential.

MOSFETs

Action There are various types of MOSFET. Fig. 32.5a and b show the basic structure and symbol of an *n-channel enhancement* type. It has a very thin layer of highly insulating silicon oxide between the metal *gate* G and the n-channel which forms in the p-type substrate (the chip itself) when V_{GS} is positive. Increasing V_{GS} repels positive holes in the substrate away from the insulating layer. This, in effect, increases the n-channel and electron flow between the *drain* D and *source* S, i.e. enhances its conductivity. As a result, the drain current I_D increases.

In a *p-channel enhancement* MOSFET, conduction is due to positive holes and the polarities of the voltages applied are reversed, as is the arrow on its symbol.

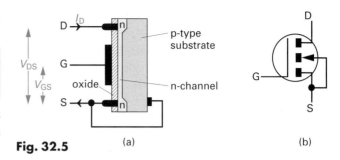

Fig. 32.5   (a)   (b)

Characteristics These can be obtained as for a JUGFET. They show that:

(i) the *transconductance* g_m is about the same as for a JUGFET, i.e. 1 to 10 mA/V;

(ii) the *a.c. output resistance* r_o is in the range 10 to 50 kΩ, i.e. similar to that of a junction transistor but smaller than that of a JUGFET; and

(iii) the *a.c. input resistance* r_i is extremely high, greater than 10^{12} Ω, due to the insulation of the gate from the channel by the oxide layer.

Precautions Because the input resistance is so high and the oxide layer is so thin, MOSFETs can be destroyed if a high voltage, accidentally applied to the gate, ruptures the oxide layer. This could even happen if someone touched the gate of the MOSFET. When someone walks on a nylon carpet, for instance, wearing rubber- or plastic-soled shoes on a dry day, the body can become charged to 10 kV or more.

Hence, the short-circuiting metal clips or conducting foam on the leads of the MOSFET (supplied with it) should not be removed until it has been connected in the circuit. (In a junction transistor charge can leak away because of its much smaller input resistance of 1 to 5 kΩ, so the problem does not arise.)

USES OF FETs

JUGFETs When used as *voltage amplifiers* (Chapter 47) they give lower gains than junction transistors but their much greater input resistances make them better as radio frequency amplifiers and as impedance-matching devices (Chapter 50).

As a *switch* a JUGFET has a higher 'on' resistance than a junction transistor and its switching speed is lower.

MOSFETs Power MOSFETs can act as *transducer drivers* for electric motors and lamps which require large currents. The MOSFET can supply these from a tiny input current taken, for example, from an integrated circuit. Their high input resistance also makes them suitable for use as complementary pairs in the output stages of hi-fi *power amplifiers* (Chapter 80), where they produce less distortion than junction transistors due to their more linear characteristics.

MOSFETs can also act as *switches*. Compared with junction transistors the input current (to the gate) is much smaller, but when switched 'on' their resistance tends to be higher but their switching speed is about ten times greater.

MOSFETs lend themselves to *integrated circuit* construction because of their compactness.

QUESTIONS

1. Compare the electrical characteristics of a JUGFET with those of a junction transistor.
2. In Fig. 32.3 the gate–source *cut-off voltage* is −5 V and the *transconductance* is 1 mA/V. Explain these statements.
3. Why is a FET less affected by temperature changes than a bipolar type?
4. **a)** State *two* uses for MOSFETs.
   b) Why can a MOSFET be damaged by 'handling'?

33 INTEGRATED CIRCUITS (ICs)

CHECKLIST

After studying this chapter you should be able to:
* describe an IC, and

* recall the names of the two main groups of IC and the types of signal they handle.

MICROELECTRONICS

One important aspect of microelectronics is concerned with building electronic systems using integrated circuits (ICs). It has been responsible for many of the exciting developments in electronics in recent years.

An IC is a 'densely populated' miniature electronic circuit. It contains transistors and often diodes, resistors and small capacitors, all made from, and connected together on, a 'chip' of silicon no more than 5 mm square and 0.5 mm thick. A small part of one is shown in Fig. 33.1, much magnified.

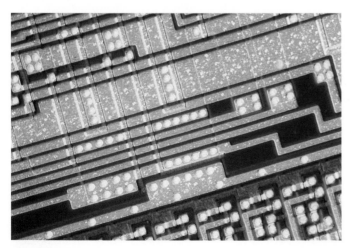

Fig. 33.1

The IC in Fig. 33.2 is in its protective plastic case (its 'package') which has been partly cut away to show the leads radiating from the chip to the pins that enable the IC to communicate with the outside world. This type of package is the popular dual-in-line (d.i.l.) arrangement with the pins (from 6 to 40 in number but usually 14 or 16) 0.1 inch apart, in two lines on either side of the case.

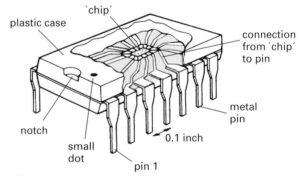

Fig. 33.2

The first ICs were made in the early 1960s and were fairly simple circuits with just a few components per chip; they were small-scale integrated (SSI) circuits. The complexity has increased rapidly through medium-scale integration (MSI) and large-scale integration (LSI) until today very-large-scale integrated (VLSI) circuits may have millions of components.

TYPES OF IC

There are two broad groups.

Digital ICs These contain *switching-type* circuits handling electrical signals which have only one of *two values*, Fig. 33.3a, i.e. their inputs and outputs are either 'high' (e.g. near the supply voltage) or 'low' (e.g. near 0 V). They were the earliest ICs because they were easier to make and large markets were available for them (e.g. as the 'brain' in pocket calculators).

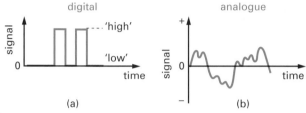

Fig. 33.3

Linear ICs These include *amplifier-type* circuits of many kinds, for both audio and radio frequencies. They handle signals that are often electrical representations, i.e. analogues, of physical quantities such as sound, which change smoothly and continuously over a *range of values*, Fig. 33.3b. One of the most versatile linear ICs and the first of this type to be made (1964), is the operational amplifier (op amp).

MANUFACTURE OF ICs

Silicon which is 99.999 999 9% pure is used and is produced chemically from silicon dioxide, the main constituent of sand. It is first melted and a single, near perfect crystal grown from it in the form of a cylindrical bar (up to 10 cm in diameter and 1 m long). The bar is cut into $\frac{1}{4}$ to $\frac{1}{2}$ mm thick wafers, Fig. 33.4a.

Depending on their size, several hundred identical circuits (the 'chips') may be formed simultaneously on each wafer by the series of operations shown in Fig. 33.4b. This creates 'windows' in an insulating layer of silicon oxide first laid down on the surface of the wafer. Doping then occurs, in one method, by exposing the wafer at high temperature to the vapour of either boron or phosphorus, so that their atoms diffuse through the 'windows' into the silicon. The p- and n-type regions so obtained for the various components are then interconnected to give the required circuit by depositing thin 'strips' of aluminium, Fig. 33.4c.

Integrated diodes and transistors have the same construction as their discrete versions. Integrated resistors are thin layers of p- or n-type silicon of different lengths, cross-sectional areas and degrees of doping. One form of integrated capacitor consists of two conducting areas (of aluminium or doped silicon) separated by a layer of silicon oxide as dielectric.

Every chip is tested, Fig. 33.5a, and faulty ones discarded—up to 70% may fail. The wafers are next cut into separate chips, Fig. 33.5b, and each one is then packaged and connected automatically by gold wires to the pins on the case.

The complete process, which can require up to three months, must be done in a controlled, absolutely clean environment, Fig. 33.6. Although design and development costs are high, volume production makes the whole operation economically viable.

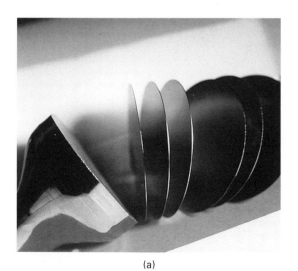

(a)

Fig. 33.4

(a)

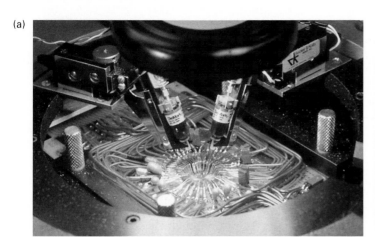

(b)

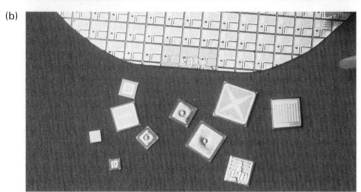

Fig. 33.5

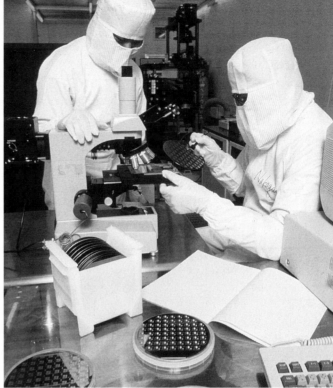

Fig. 33.6

1. State some (i) advantages, (ii) limitations, of building circuits from ICs compared with those built from discrete components.

34 PROGRESS QUESTIONS

1. In the circuit of Fig. 34.1 what is:
 a) the resistance of Tr if it is saturated,
 b) I_C when Tr is saturated,
 c) I_L when Tr is saturated,
 d) I_L when Tr is off (not conducting)?

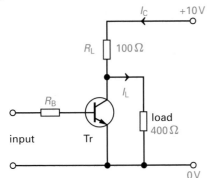

Fig. 34.1

2. Given that the h_{FE} of the transistor in the circuit in Fig. 34.2 is 100, calculate the collector current and the potential difference between the collector and the emitter. (For the purposes of the calculation you should assume that $V_{BE} = 0$.) (L.)

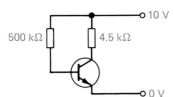

Fig. 34.2

3. An npn transistor has a collector load of 1 kΩ. It is used in a circuit with supply rails of 0 V and 9 V. What is the maximum current that the transistor can conduct?

What is the power dissipation **(a)** in the resistor, **(b)** in the transistor, when the collector current is adjusted to half its maximum value by altering the base current? (O. and C.)

4. a) The input voltage shown is applied to the circuit in Fig. 34.3. Copy the graph and sketch the output voltage immediately below the input. Indicate values of the output voltage on your output graph.
 b) Suggest a value for R_B. (L.)

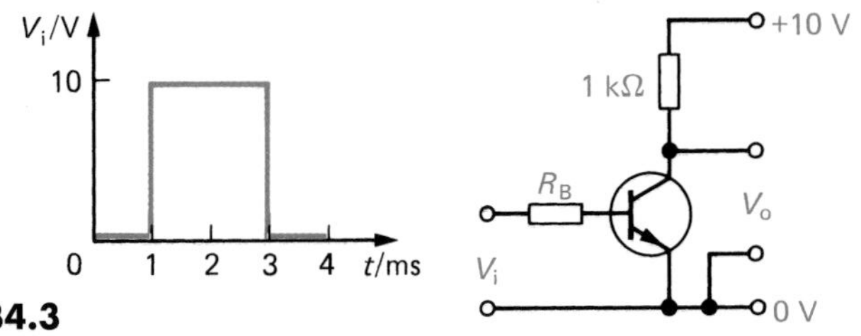

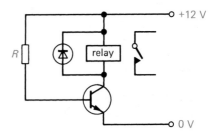

Fig. 34.3

5. Fig. 34.4 shows a relay circuit. The coil of the relay needs a current of 20 mA to close the contacts.
 a) If the current gain h_{FE} of the transistor is 50, calculate the base current needed for the relay to be just switched on.
 b) Neglecting the base–emitter voltage, calculate the value of R for this to happen.
 c) What purpose is served by the diode?
 d) This circuit may be used as the basis of a frost-warning alarm in which a lamp is switched on when the temperature falls below a certain value (which can be altered). Draw a labelled diagram of such a circuit. (O.L.E.)

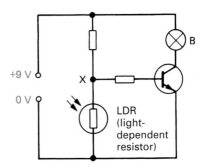

Fig. 34.4

6. For the circuit shown in Fig. 34.5:
 a) state what conditions, light or dark, cause the bulb B to be lit,
 b) explain how, and why, the potential at the point X varies as conditions change from light to dark,
 c) explain how the bulb B becomes lit. (C.)

Fig. 34.5

7. What effect does a change in p.d. between the collector and the emitter of a transistor have on the collector current?

What effect does a change in the p.d. between the base and the emitter of a transistor have on the collector current?

Use a sketch graph to illustrate your answer in each case. (L.)

8. A manufacturer's description of the performance of a bipolar n-p-n transistor includes the following information:

V_{CE} volts (max)	V_{EB} volts (max)	I_C mA (max)	Power mW	h_{fe} at I_C(mA)
45	5	100	300	125–500 at 2

Explain what these entries mean. Describe how you would determine experimentally the value of h_{fe} for a given transistor.

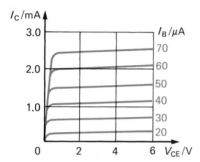

Fig. 34.6

Fig. 34.6 gives an alternative method for quoting some of the data for a transistor. Use these data to estimate a value for h_{fe} for this transistor at (i) $I_C = 0.5$ mA, (ii) $I_C = 2.0$ mA. Explain how your estimate is made. (*O. and C.*)

9. a) Explain why the circuit in Fig. 34.7a switches the transistor *off* a certain time *after* switch S is opened.
b) Explain why the circuit in Fig. 34.7b switches the transistor *on* a certain time *after* switch S is opened.

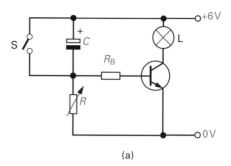

(a)

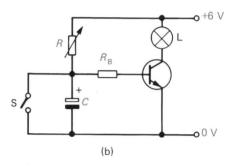

(b)

Fig. 34.7

10. The circuit in Fig. 34.8 is that of a simple burglar alarm. Switches S_1, S_2 and S_3 are normally closed window switches. Assume that the transistors are either saturated or switched off.
a) State the approximate voltage at point P when (i) all the window switches are closed, (ii) when one or more of the window switches are open.
b) What is the function of the diode D?
c) Explain how the circuit behaves when a window switch is opened and then closed again.

N.B. In Fig. 34.8 'n.o.' means 'normally open'. (*N.*)

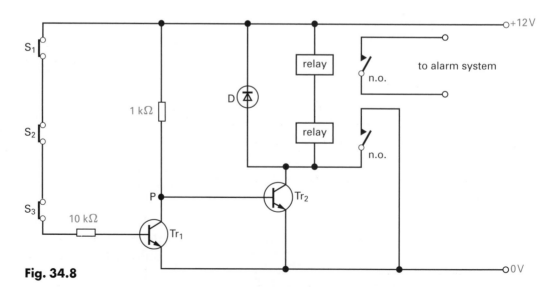

Fig. 34.8

POWER SUPPLIES

35 RECTIFIER CIRCUITS: BATTERIES

CHECKLIST

After studying this chapter you should be able to:
- draw graphs to show the variations of output voltage with time for half- and full-wave rectifiers,
- draw circuits for and explain the action of half-wave, centre-tap full-wave and bridge full-wave rectifiers,
- recall that the voltage drop (of about 1 V) across a forward-biased diode makes the rectified output voltage less than the peak value of the a.c. input voltage, and
- state what a heat sink is used for and what the term 'thermal resistance' means.

POWER SUPPLIES

The voltages required for circuits containing semiconductor devices seldom exceed 30 V and may be as low as 1.5 V. Current demands vary from a few microamperes to many amperes in large systems. Usually the supply must be d.c.

Batteries are suitable for portable equipment but in general power supply units operated from the a.c. mains are employed. In these, a.c. has to be converted to d.c., the process being called *rectification*. Apart from not needing frequent replacement, as batteries may, power supply units are more economical and reliable and can provide more power. They can also supply very steady p.ds when this is required.

In most power supply units a transformer steps down the a.c. mains from 230 V to a much lower voltage. This is then converted to d.c. using one or more junction diodes in a rectifier circuit.

HALF-WAVE RECTIFIER

In the circuit of Fig. 35.1, the load is represented by a resistor R, but in practice it will be a piece of electronic equipment. Positive half-cycles of the alternating input p.d. V_i from the transformer secondary forward-bias diode D which conducts, creating a pulse of current. This produces a p.d. across R having almost the same peak value as V_i, if the small p.d. (about 1 V) across D is ignored.

The negative half-cycles of V_i reverse-bias D, there is little or no current in the circuit and V_o is zero. Fig. 35.2 shows the V_i and V_o waveforms. V_o varies but is unidirectional, i.e. it is d.c.

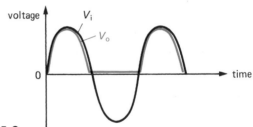

Fig. 35.2

CENTRE-TAP FULL-WAVE RECTIFIER

In full-wave rectification, both halves of every cycle of input p.d. produce current pulses.

In the circuit of Fig. 35.3, two diodes D_1 and D_2 and a transformer with a centre-tapped secondary are used. Suppose that at a certain time during the first half-cycle the input p.d. V_i is 12 V. If we take the centre-tap O as the reference point at 0 V, then when A is at +6 V, B will be at −6 V. D_1 is forward biased and conducts, giving a current pulse via the path AD_1CROA. V_o is again almost equal to V_i. During this half-cycle D_2 is reverse biased by the p.d. across OB (since B is negative with respect to O).

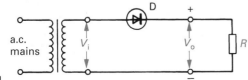

Fig. 35.1

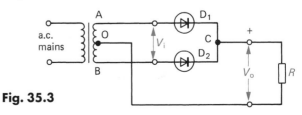

Fig. 35.3

On the other half of the first cycle, B becomes positive relative to O and A negative. D_2 conducts to give current via the path BD_2CROB and D_1 is now reverse biased. In effect, the circuit consists of two half-wave rectifiers working into the same load R on alternate half-cycles of V_i. The current through R is in the same direction during both half-cycles and V_o is a fluctuating direct voltage with a waveform as in Fig. 35.4. It is more continuous than V_o in the half-wave circuit and has a higher average d.c. value.

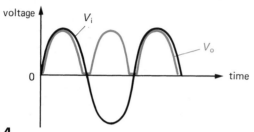

Fig. 35.4

BRIDGE FULL-WAVE RECTIFIER

Only half of the secondary winding is used at any time in the circuit of Fig. 35.3 and the transformer has to produce twice the voltage required. A more popular arrangement using four diodes in a bridge network is shown in Fig. 35.5.

If A is positive with respect to B during the first half-cycle, D_2 and D_4 conduct and current takes the path AD_2RD_4B. On the next half-cycle when B is positive, D_1 and D_3 are forward biased and current follows the path BD_3RD_1A. Once again current through R is unidirectional during both half-cycles of the input V_i. The output V_o is a varying d.c. voltage with a waveform like that in Fig. 35.4.

All four diodes are available in one package, with two a.c. input connections and two output connections.

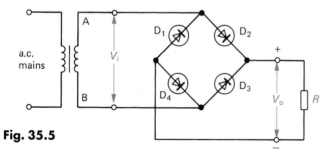

Fig. 35.5

BATTERIES

Batteries fall into two main groups.

Primary cells These are not rechargeable. Examples follow.

(i) *Carbon-zinc* which are cheap and best for low currents or occasional use since their voltage (1.5 V) falls as the current increases. Used in torches and radios. Heavy duty versions are available.

(ii) *Alkaline-manganese* supply a steady voltage (1.5 V) in use and are better for higher currents. They have a long 'shelf' life and last up to four times longer than carbon-zinc cells of the same size. Used in calculators, radios, portable CD players and photographic equipment.

(iii) *Silver oxide* (1.5 V), *Zinc air* (1.4 V) and *Mercuric oxide* (1.3 V) 'button' types have an almost constant voltage until discharged. With a large capacity for their size, they are best for small currents. Used in watches, cameras and hearing aids.

Secondary cells They *can be recharged* repeatedly by sending a current through them in the reverse direction. 'High' continuous currents at constant voltage can be supplied depending on their *capacity* which is measured in milliampere-hours (mA h) for a certain discharge rate. For example, a battery with a capacity of 1500 mA h at the 10 hour rate will sustain a current of 150 mA for 10 hours, smaller currents being supplied for longer and higher currents for a shorter time. Three types are used in portable computers and mobile phones.

(i) *Nickel-Cadmium* (NiCad) which should be fully discharged before recharging (to 1.2 V per cell) from a constant current source to prolong their life.

(ii) *Nickel Metal Hydride* (NiMH) which has no 'memory' effect like some NiCads so can be fully charged each time (to 1.2 V per cell).

(iii) *Lithium ion* is the latest and most efficient type but the most expensive.

HEAT SINKS

High current diode rectifiers are mounted on *heat sinks* with cooling fins (made of aluminium sheet and painted black), as shown in Fig. 35.6, to stop them overheating. Manufacturers state what the *thermal resistance* of the heat sink should be for a particular rectifier. For example, if it is 2 °C W^{-1}, the temperature of the heat sink rises by 2 °C above its surroundings for every watt of power it has to get rid of. To dissipate 10 W, the rise will be 20 °C.

Fig. 35.6

36 SMOOTHING CIRCUITS

CHECKLIST

After studying this chapter you should be able to:
- explain the use of a capacitor in parallel with the load to produce smoothing,
- describe with the aid of a graph what is meant by ripple,
- state the effect on the ripple of changing (i) the value of the smoothing capacitor and (ii) the load current,
- explain why the smoothing capacitor and diodes may be damaged if their voltage rating is not high enough, and
- explain how a capacitor-input filter improves smoothing.

The varying d.c. output voltage from a rectifier circuit can be used to charge a battery but must be 'smoothed' to obtain the steady d.c. required by electronic equipment.

RESERVOIR CAPACITOR

The simplest way to smooth an output is to connect a large capacitor, called a *reservoir capacitor*, across it as in Fig. 36.1 for half-wave rectification. Its value on a 50 Hz supply may range from 100 μF to 10 000 μF, depending on the current and smoothing needed.

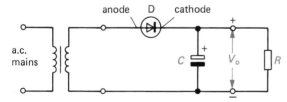

Fig. 36.1

The waveform of the smoothed output voltage V_o for half-wave rectification is shown as a solid red line in Fig. 36.2. The corresponding half-cycles of a.c. during which the diode D conducts are drawn as a black dashed line. The small variation in the smoothed d.c. is called the *ripple voltage*. It has the same frequency as the a.c. supply and causes 'mains hum'.

The smoothing action of C is explained as follows. During part of the half-cycle of a.c. when D is forward biased, there is a current pulse which charges up C to *near* the peak value of the a.c. During the rest of the cycle C

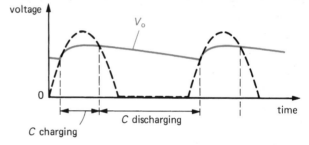

Fig. 36.2

keeps the load R supplied with current by partly discharging through it. While this occurs, the output p.d. V_o falls until the next pulse of rectified current tops up the charge on C. It does this near the peak of the half-cycle. And so for most of each cycle the load current is supplied by C acting as a reservoir of charge.

The ripple voltage in a full-wave rectifier circuit is smaller, giving better smoothing. Waveforms are shown in Fig. 36.3; in this case the ripple frequency is twice that of the a.c. supply.

Fig. 36.3

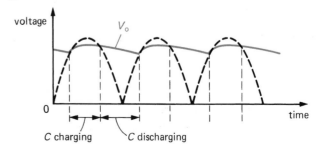

CAPACITOR AND DIODE RATINGS

Capacitor The smoothing action of a reservoir capacitor is due to its large capacitance C making the time constant CR large (p. 31), where R is the resistance of the load, compared with the time for one cycle of the a.c. mains (1/50 s). The larger C is, the better the smoothing, the smaller the fall in V_o but the briefer will be the rectified current pulse to charge up the capacitor again. Consequently, the peak value will have to be greater and damage may occur if the peak current rating of the diode is exceeded. Manufacturers sometimes state the maximum value of reservoir capacitor to be used with a given diode.

The capacitor should have a voltage rating at least equal to the peak value of the transformer's secondary voltage or the dielectric may break down.

Diode Consideration must be given to the maximum reverse voltage V_{RRM} of the diode D (p. 58). *When it is not conducting*, Fig. 36.1 shows that the total voltage across it is the voltage across C plus that across the transformer secondary. For example, if the lower plate of C is at 0 V and the top plate is near the positive peak of the a.c. input, say +12 V, then the cathode of D is at +12 V. Its anode will be at −12 V, i.e. the peak negative value of the voltage across the transformer on a negative half-cycle. The total voltage across D is 24 V in this case. In practice it is wise to use a diode whose V_{RRM} is at least four times the r.m.s. value of the secondary p.d. of the transformer, to allow for voltage 'spikes' picked up by the a.c. mains supply.

CAPACITOR-INPUT FILTER

The smoothing produced by a reservoir capacitor C_1 can be increased and ripple reduced further by adding a *filter circuit*. This consists of an inductor or choke L and another large capacitor C_2, arranged as in Fig. 36.4.

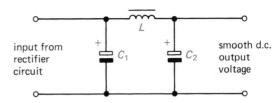

input from rectifier circuit

smooth d.c. output voltage

Fig. 36.4

We can regard the varying d.c. voltage produced across C_1 (V_o in Figs. 36.2 and 36.3) as a steady d.c. voltage (the d.c. component) plus a small ripple voltage (the a.c. component). L offers a much greater impedance than C_2 to the a.c. component and so most of the unwanted ripple voltage is developed across L. For the d.c. component the situation is reversed and most of it appears across C_2. The filter thus acts as a voltage divider, separating d.c. from a.c. (i.e. acting as a filter) and producing a d.c. output voltage across C_2 with less ripple. A resistor may replace the inductor when the current to be supplied is small.

Filter circuits have certain disadvantages which make them less suitable for semiconductor circuits which operate at low voltages and high currents. These are due to (i) the risk of large currents in chokes causing magnetic saturation of the iron cores, (ii) chokes being large and heavy, and (iii) resistors reducing the output voltage.

QUESTIONS

1. What is meant by smoothing?
2. If the load current were zero in a rectifier circuit with a reservoir capacitor, to what p.d. would the capacitor charge up? Illustrate your answer with a graph.
3. In the half-wave rectifier circuit of Fig. 36.1, if the transformer secondary p.d. is 7 V r.m.s., what is the maximum p.d. across (i) C, (ii) D?

37 STABILIZING CIRCUITS

CHECKLIST

After studying this chapter you should be able to:
- explain why the output voltage falls as the output current from an unstabilized power supply increases,
- draw and perform simple calculations on a Zener diode voltage-regulation (stabilization) circuit containing a current-limiting series resistor, and
- draw and explain a stabilized current circuit.

REGULATION

Because of its internal (source) resistance (p. 14), the output voltage from an ordinary power supply (and a battery) decreases as the current it supplies to the load increases, i.e. the 'lost' volts increase and the terminal p.d. decreases. The greater the decrease the worse is the *regulation* of the supply. Good and bad regulation curves are shown in Fig. 37.1.

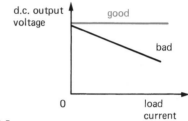

Fig. 37.1

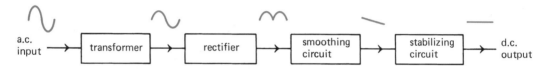

Fig. 37.2

There are many occasions when a d.c. voltage is required which remains constant and is not affected by load current changes. In these cases a stabilizing or regulating circuit is added to the power supply, as shown by the block diagram in Fig. 37.2.

STABILIZED VOLTAGE CIRCUIT

Action When a Zener diode is reverse biased to its breakdown (Zener) voltage, the voltage across it stays almost constant for a wide range of reverse currents, as its characteristic shows (Fig. 27.2, p. 60). Because of this property, a Zener diode can be used to stabilize the output voltage from a smoothing circuit and reduce fluctuations due to load current changes.

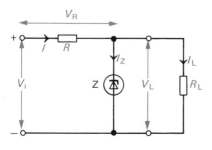

Fig. 37.3

In the circuit of Fig. 37.3, V_i is the smoothed voltage to be stabilized and R_L is the load which changes and draws different currents I_L. The total current I supplied, divides into I_Z through the Zener diode Z and I_L through R_L. Hence:

$$I = I_Z + I_L$$

Since $V_i = V_R + V_L$,

$$I = \frac{V_R}{R} = \frac{V_i - V_L}{R}$$

But $V_L \approx V_Z$, i.e. the output voltage $\approx$ Zener voltage (p. 60),

$$\therefore I = \frac{V_i - V_Z}{R}$$

For a given V_i, I is therefore more or less constant since V_Z and R are constant. This means that if I_L increases, I_Z decreases by the same amount and vice versa. Thus if I_L varies due to R_L varying, V_L is virtually constant (as is V_R).

Further points If $I_L = 0$ (i.e. there is no load drawing current), then $I_Z = I$ and Z has to have the power rating to carry this current safely. On the other hand, when I_L is a maximum, I_Z has its minimum value which must be large enough (at least 5 mA for a small diode) to ensure Z works on the breakdown part of its characteristic (not at or above the 'knee' in Fig. 27.2 where the p.d. across the Zener falls to zero). This is achieved by making V_i several volts greater than the required output voltage.

Stabilization also helps to reduce the effect of variations in V_i due either to any 'ripple' on it after smoothing or to fluctuations of the a.c. mains supply. For example, if V_i rises, the increase in I is such that the rise in p.d. occurs across R, leaving V_L about the same. If V_i varies over a wide range, the value of the safety resistor R must be such that when V_i is a minimum there is enough current available to both Z and R_L.

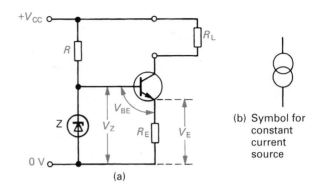

Fig. 37.4

STABILIZED CURRENT CIRCUIT

A constant current supply is sometimes needed, e.g. to recharge Nicad cells. In the circuit of Fig. 37.4a the Zener diode Z and its safety resistor R keep the base of the silicon transistor at voltage V_Z. The p.d. V_E across resistor R_E in the emitter circuit is held constant at $V_Z - V_{BE}$ (where $V_{BE} \approx 0.6$ V). So long as $V_{CC} > V_E$, the current through the load R_L (e.g. a Nicad cell) in the collector circuit, whatever its value, is steady and equal to V_E/R_E. The circuit uses the fact the collector current I_C of a transistor is constant for a wide range of collector–emitter voltages V_{CE}, as its I_C–V_{CE} output characteristic shows, Fig. 31.2c, p. 72.

Fig. 37.4b shows the symbol for a constant current source.

INTEGRATED CIRCUIT VOLTAGE REGULATORS

A range of integrated circuit stabilizers, called *voltage regulators*, are now available in various packages, Fig. 37.5. They use more complex circuits than the simple Zener diode one and protection is also provided against overloading and overheating. Many are designed for one voltage, e.g. 5 V or 15 V.

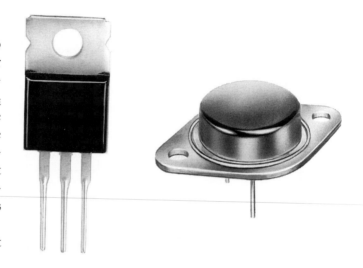

Fig. 37.5

The circuit for a stable 5 V 600 mA supply using the L005 regulator mounted on a 4 °C W^{-1} heat sink, is given in Fig. 37.6. C is the reservoir capacitor. The neon lamp across the primary of the transformer shows when the mains is on and has a very low power consumption. Note that there is a fuse and a switch in the live side of the mains supply.

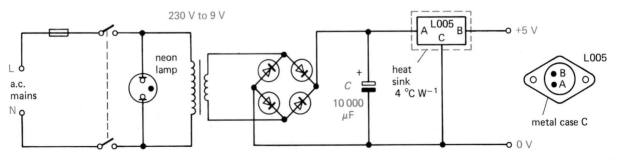

Fig. 37.6

QUESTIONS

1. What is meant by the *regulation* of a power supply?

2. In the circuit of Fig. 37.7, the Zener diode Z has a breakdown voltage of 3 V.

    a) What is the output p.d. V_o when the input p.d. V_i is   (i) 2 V,   (ii) 4 V,   (iii) 6 V?

    b) What is the current through Z when V_i is (i) 2 V,   (ii) 4 V,   (iii) 6 V?

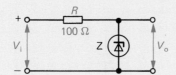

Fig. 37.7

3. In the circuit of Fig. 37.3, the Zener diode Z has a breakdown p.d. of 3 V, the input voltage V_i is 8 V and $R = 50\ \Omega$.

    a) If $R_L = 300\ \Omega$, calculate I, I_L and I_Z.

    b) If $R_L = 100\ \Omega$, calculate I, I_L and I_Z.

    c) If V_L is to be kept steady why must R_L not fall below a certain value? What is this value?

    d) Why is there a maximum value for R?

4. A load is connected across the output of the full-wave bridge rectifier circuit of Fig. 37.6 in which the alternating voltage across the secondary of the transformer is 9.0 V r.m.s.

    a) Calculate the peak value of the secondary voltage.

    b) What is the rectified output voltage to the voltage regulator if the p.d. across a forward-biased silicon diode is taken as 0.85 V?

    c) If a current of 500 mA at 5.0 V is supplied to the load, what is   (i) the voltage across the regulator,   (ii) the power dissipated by it?

38 POWER CONTROL

CHECKLIST

After studying this chapter you should be able to:
- state what a thyristor is and recognize its symbol, and
- describe how a thyristor can control   (i) d.c. power, (ii) a.c. power.

THYRISTOR

A thyristor is a four-layer, three-terminal semiconducting device, Fig. 38.1. It used to be called a silicon-controlled rectifier (SCR) because it is a rectifier which can control the power supplied to a load with little waste of energy.

When forward biased, a thyristor does not conduct until a positive voltage is applied to the *gate*. Conduction continues when the gate voltage is removed and stops only if the supply voltage is switched off or is reversed.

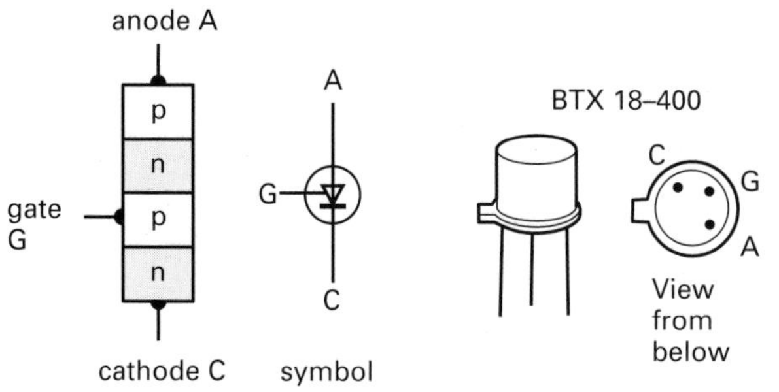

d.c. power control   The circuit of Fig. 38.2 can be used as a simple example. When S_1 is closed, the lamp L stays off. When S_2 is closed as well, gate current flows and the thyristor switches on, i.e. 'fires'. The anode current is large enough to light L, which stays alight if S_2 is opened.

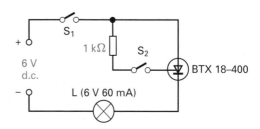

Fig. 38.1

Fig. 38.2

a.c. power control The control of a.c. power can be achieved by allowing current to be supplied to the load during only part of each cycle. A gate pulse is applied automatically at a certain chosen stage during each positive half-cycle of input. This lets the thyristor conduct and the load receives power.

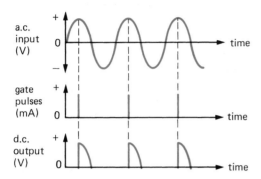

Fig. 38.3

For the case shown in Fig. 38.3, the pulse occurs at the peak of the a.c. input. During negative half-cycles the thyristor is non-conducting and remains so till halfway through the next positive half-cycle. Current flows for only a quarter of a cycle but by changing the timing of the gate pulses this can be decreased or increased. The power supplied to the load is thus varied from zero to that due to half-wave rectified d.c.

TRIAC

The thyristor, being in effect a diode, is a half-wave device and only allows half the power to be available. A *triac* consists of two thyristors connected in parallel but in opposition and controlled by the same gate. It is a two-directional thyristor which is triggered on both halves of

each cycle of a.c. input by either positive or negative gate pulses. The power obtained by a load can therefore be varied between zero and full-wave d.c.

The connections are called *main terminals* 1 and 2 (MT1 and MT2) and *gate* (G), Fig. 38.4. The triggering pulses are applied between G and MT1. The gate current for a triac handling up to 100 A may be no more than about 50 mA. Triacs are used in lamp dimmer circuits and for motor speed control, e.g. in an electric drill.

Fig. 38.4

QUESTIONS

1. Explain the action of a thyristor by referring to Fig. 38.5a, b, which show its transistor equivalent.

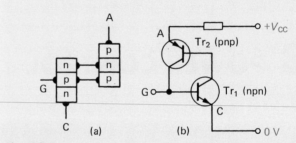

Fig. 38.5

2. What are the advantages of the thyristor in control circuits?

39 PROGRESS QUESTIONS

1. a) Explain the action of the circuit in Fig. 39.1, and draw sketch graphs of the input p.d. and the p.d. across the load. (You should draw the graphs so that they can be compared.) If the peak voltage induced between X and Z is 10 V what is the peak p.d. across (i) R, (ii) D_1?

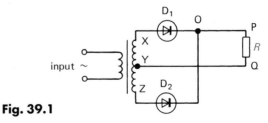

Fig. 39.1

b) Explain, using p.d.–time sketch graphs, the effects of adding to the circuit the smoothing capacitor as shown in Fig. 39.2.

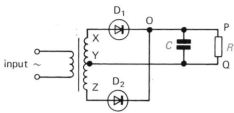

Fig. 39.2

Given that the value of R is 100 Ω and the frequency of the supply is 50 Hz, calculate a suitable value for C. What would be the p.d. between P and Q if R became open circuit? (*L.*)

2. A simple voltage stabilizer using a Zener diode is shown in Fig. 39.3. The Zener voltage is 5.6 V.

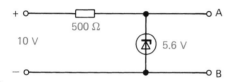

Fig. 39.3

(i) What is the current through the 500 Ω resistor?
(ii) What changes take place in the circuit if a resistive load of 1000 Ω is connected?
(iii) Are there any limitations on the resistive load if the stabilized voltage is to be maintained across AB? (*O. and C.*)

3. a) The Zener diode Z in Fig. 39.4 has a breakdown voltage of 4 V. Calculate the current in Z.

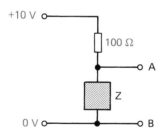

Fig. 39.4

Explain, giving values, the effect on the current in Z of   (i) a rise in supply voltage to 12 V, (ii) connecting a load of 100 Ω across AB (with a supply voltage of 10 V).

If the circuit of Fig. 39.4 is used to provide a stable voltage supply what is the minimum resistance which may be connected across AB?

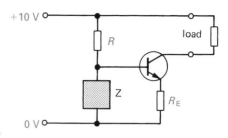

Fig. 39.5

b) The circuit of Fig. 39.5 provides a stabilized *current* source. Explain
(i) why a rise in supply voltage does not lead to a rise in current through the load, and
(ii) how the circuit responds to a change in the resistance of the load.
Discuss briefly what would happen if
(iii) the load were reduced to zero resistance, and
(iv) the load were removed (i.e. increased to infinite resistance). (*L.*)

4. Fig. 39.6 shows a block diagram of a power supply.
a) Draw a circuit diagram to show how a switch, a fuse and an indicator should be connected between the mains and the transformer.
b) The output from the transformer is 15 V r.m.s. Calculate the turns ratio of the primary coil to the secondary coil.
c) (i) Draw a diagram to show how four diodes can be arranged to form a full-wave rectifier. Mark clearly where the transformer is connected and which terminal gives the positive output.
(ii) State, with reasons, a suitable value for the peak inverse voltage rating of the diodes.
d) If the power supply delivers a current of 2 A, calculate the resistance of the load.
e) If the time constant of the smoothing capacitor and load resistance is to be 0.1 s, calculate a value for the smoothing capacitor and state a suitable voltage rating.
f) Sketch a graph of the voltage across the load as a function of time, indicating appropriate scales on the axes.
g) Explain how the capacitor smooths the output from the rectifier. (*N.*)

240 V

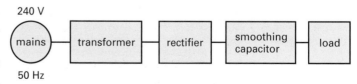

50 Hz

Fig. 39.6

MEASURING INSTRUMENTS

40 MULTIMETERS

CHECKLIST

After studying this chapter you should be able to:
- state what a multimeter measures,
- compare the properties of analogue and digital multimeters, and
- state how to use a multimeter to check for faulty diodes, transistors, capacitors, joints, bulbs, fuses and connecting wires.

A multimeter measures current, voltage and resistance, each on several ranges.

ANALOGUE MULTIMETER

In this type, Fig. 40.1, the deflection of a pointer over a scale represents the value of the quantity being measured. Basically it consists of a *current-measuring* moving-coil meter.

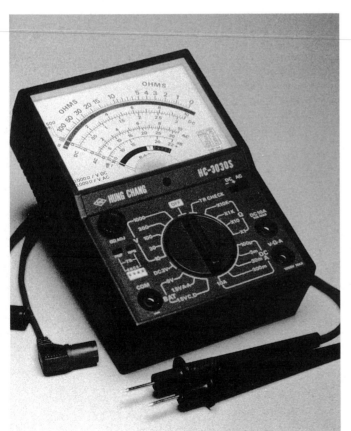

Fig. 40.1

Current and voltage As an ammeter, different low resistance shunts S (p. 11) are switched in for different current ranges, as shown in Fig. 40.2. As a voltmeter, high resistance multipliers M (p. 11) are connected in series with the coil of the meter.

For a.c. measurements a diode rectifier produces pulses of varying d.c. to which the meter responds. Its a.c. scales are calibrated to read r.m.s. values of currents and p.ds that have sine waveforms.

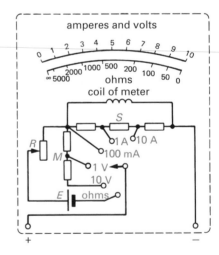

Fig. 40.2

Resistance As an ohmmeter, an internal rheostat R (often marked 'ohms adjust' or 'zero Ω') and a battery E are brought into the circuit.

To take a measurement, the meter leads are first held together, making the external resistance between the terminals zero and the current through the meter a maximum. If necessary, R is adjusted (to allow for battery p.d. changes) until the pointer gives a full-scale deflection to the right, i.e. is on the zero of the ohms scale.

The leads are then connected across the unknown resistance R_X. The current I is now less and the pointer indicates R_X in ohms. If the meter is on open circuit, $I = 0$ and R_X is infinite and the pointer sets on the ∞ mark at the left end of the ohms scale.

An ohmmeter has a non-linear scale because $I = E/(R_M + R_X)$ where E is the e.m.f. of the internal battery (1.5 or 15 V) and R_M is the ohmmeter resistance. The expression shows that as R_X increases, I decreases but not uniformly, i.e. doubling R_X does not halve I.

DIGITAL MULTIMETER

In a digital multimeter the measurement is shown on an LCD or a LED digital decimal display (p. 47), often with four figures, Fig. 40.3. It is made by a high resistance *voltage-measuring* analogue-to-digital (A/D) converter (p. 227).

Fig. 40.3

Voltage and current As a voltmeter, low voltages are measured directly and higher ones via a potential divider which supplies the input to the A/D converter. As an ammeter, a parallel resistor R produces a p.d. V across the input to the converter that is proportional to the input current I, Fig. 40.4.

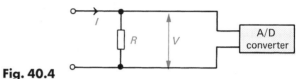

Fig. 40.4

Resistance As an ohmmeter, a stabilized current supply (p. 86) passes a known constant current I through the unknown resistance R_X, Fig. 40.5. The p.d. across R_X is measured by the A/D converter and is proportional to R_X since $V = IR_X$, i.e. the 'scale' is linear.

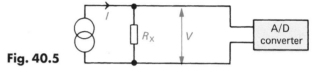

Fig. 40.5

PRACTICAL POINTS

1. Before use, check that the reading is zero; if it isn't, correct it by the 'zero adjustment' screw.
2. For d.c. currents and p.ds, the positive terminal must be connected to the circuit so that it is nearer the supply's positive terminal than is the negative terminal of the meter.
3. Switch to the highest range first.
4. Check you are reading the correct scale.
5. When finished leave the multimeter on the highest d.c. voltage range.

COMPARISON OF MULTIMETERS

Property	**Analogue** (moving-coil)	**Digital** (electronic)
Reading errors	Can occur especially when pointer between marks	Less likely
Input resistance as a voltmeter	Moderate, e.g. 20 V_R kΩ/V where V_R is the f.s.d. voltage. Varies with range	High, e.g. 10 MΩ on d.c.; 0.5 MΩ on a.c. Same on all ranges
Scale/display	Scale continuous	Display changes by 1 digit
Response to input	Continuous	Samples taken regularly
Power used	None except as an ohmmeter	Small if LCD

FAULT-FINDING

Multimeters are useful for locating faults in electronic circuits and components. Sometimes in resistance measure- ments, use is made of the fact that in analogue meters the negative (black) terminal has positive (+) polarity (due to the internal battery) and the positive (red) terminal has negative (−) polarity. With digital meters the polarity is given in the instruction booklet (but often the terminal marked 'mA' is positive).

Diodes   A diode should have a low resistance when the cathode (the end with the band round it) is connected to the terminal of negative polarity and the anode to the terminal of positive polarity. The connections using an analogue multimeter are shown in Fig. 40.6. Reversing the connections gives a high resistance reading with a 'healthy' diode.

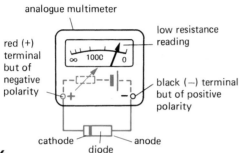

Fig. 40.6

Transistors   In a junction transistor the resistance should be high between the collector and emitter for *both* methods of connection to the multimeter. This enables the base to be identified since it gives a low resistance one way with either the collector or emitter and a high reading the other way.

For an n-p-n transistor the resistance is lower when the multimeter terminal of positive polarity (i.e. the black one marked −) is connected to the base and the other terminal to the emitter or collector than it is with the leads reversed, Fig. 40.7a. For a p-n-p type the opposite is true, Fig. 40.7b. In each case the resistance is low when the p-n junction concerned is forward biased and use is made of the fact that a transistor behaves as two back-to-back diodes.

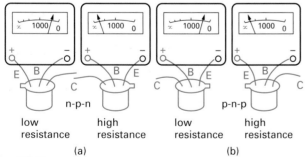

Fig. 40.7

Non-polarized capacitors (p. 24)   If the resistance of the capacitor is less than about 1 MΩ, it is allowing d.c.

to pass (from the battery in the multimeter) i.e. it is 'leaking' and is faulty. With large value capacitors there may be a short initial burst of current as the capacitor charges up.

Polarized capacitors (p. 24)   For the dielectric to form in these, a positive voltage must be applied to the positive lead of the capacitor (marked by a +) from the multimeter terminal of positive polarity. When first connected, the resistance is low but rises as the dielectric forms.

'Dry' joints   These are badly soldered joints (p. 235) which have a high resistance. In the circuit of Fig. 40.8 if there is a 'dry' joint at X, the voltmeter reading to the right of X will be +6 V and to the left 0 V because there is infinite resistance at X.

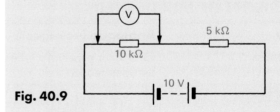

Fig. 40.8

Bulbs, fuses, connecting wire   Their resistance should be low, if not they are faulty.

QUESTIONS

1. Draw *three* basic circuits to show how a moving-coil meter is adapted for use as   (i) an ammeter, (ii) a voltmeter,   (iii) an ohmmeter.
2. What would be the p.d. across the 10 kΩ resistor in the circuit of Fig. 40.9 if the voltmeter were not connected into the circuit?

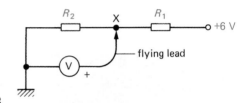

Fig. 40.9

    If the voltmeter has a resistance of 1 kΩ per volt and is set on its 10 V scale, what would it read when connected across the 10 kΩ resistor? What conclusion do you draw from these values?
    (*L. part qn.*)

3. Why does a diode have a low resistance when connected one way to an ohmmeter but a high resistance the other way round?

41 OSCILLOSCOPES

CHECKLIST

After studying this chapter you should be able to:
• state the action of the following: Y-amplifier, Y-shift, X-amplifier, X-shift, time base, trigger circuit, and

• use an oscilloscope to measure direct and alternating voltages, display waveforms and phase relationships, and measure the period and frequency of a repeating waveform.

The cathode ray oscilloscope (CRO) is one of the most important instruments ever to be developed. It is used mainly as a 'graph-plotter' to display the waveform of the voltage applied to its input, i.e. to show how it varies with time. Non-electrical effects can also be studied since almost any measurement can be changed into a voltage by an appropriate transducer. A simple model is shown in Fig. 41.1.

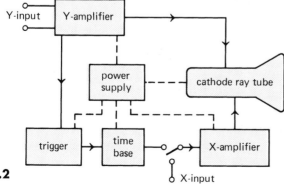

Fig. 41.1

DESCRIPTION

The block diagram in Fig. 41.2 shows that a CRO consists of:

(i) a *cathode ray tube* (CRT), which employs electrostatic deflection and incorporates brilliance and focus controls (p. 49);

(ii) a *power supply*; and

(iii) *circuits to the X- and Y-plates* of the CRT.

Y-input circuit   The input p.d. is connected to the *Y-input* terminals, which may be marked 'high' and 'low'. If the *a.c./d.c. selector* switch is in the 'a.c.' (or 'via C') position, a capacitor blocks any d.c. in the input but allows a.c. to pass; in the 'd.c.' (or 'direct') position, both d.c. and a.c. pass. The input is amplified, via the *Y-amp gain* control (often marked 'volts/cm or div'), by the *Y-amplifier* before it is applied to the Y-plates. It is then large enough to deflect the electron beam vertically.

The *Y-shift* control allows the spot (or waveform) on the screen to be moved 'manually' in the Y-direction by applying a positive or negative voltage to one of the Y-plates.

X-input circuit   The p.d. applied to the X-plates via the *X-amplifier* can either be from an external source connected to the *X-input* terminal or, more commonly, from the *time base circuit* in the CRO.

The *time base* deflects the beam horizontally in the X-direction and makes the spot sweep across the screen from left to right at a steady speed determined by the setting of the time base controls. The coarse control (usually marked 'time or ms/cm or div') gives several preset (fixed) sweep speeds, each variable within limits by the fine control (sometimes marked 'variable'). It must then make the spot 'fly' back very rapidly to its starting position, ready for the next sweep. The time base voltage should therefore have a sawtooth waveform like that in Fig. 41.3.

Since AB is linear, the distance moved by the spot is directly proportional to time and the horizontal deflection becomes a measure of time, i.e. a time axis or base. If the Y-input p.d. represents a quantity which varies with time, its waveform is displayed.

Fig. 41.2

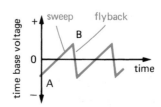

Fig. 41.3

To keep the waveform (trace) steady on the screen, each horizontal sweep must start at the same point on the Y-input waveform. The time base frequency should therefore be that of the input or a sub-multiple of it. Successive traces then coincide and, owing to the persistence of vision and the phosphor on the screen, the trace appears stationary.

One way to obtain synchronization of the time base and input frequencies is to feed part of the Y-amplifier output to a *trigger* circuit. This gives a pulse at a certain point on the input (selected in some CROs by a *trig level* control) which starts the sweep of the time base saw-tooth, i.e. it 'triggers' the time base. In many CROs automatic triggering by the Y-input occurs if they are set on 'auto' (instead of 'ext sync' in the CRO shown) but in some cases, no trace is obtained until there.is an input.

The *X-shift* control allows 'manual' shift in the X-direction.

USES OF THE CRO

Practical points   The *brilliance* (brightness or intensity) control, which is usually the *on/off* switch as well, should be as low as possible when there is just a spot on the screen. Otherwise screen burn occurs and the phosphor is damaged. If possible it is best to defocus the spot or draw it into a line by running the time base.

When preparing the CRO for use, set the *brilliance*, *focus*, *X-* and *Y-shift* controls to their mid positions. The *time base* and *Y-amp gain* controls can then be adjusted to suit the input.

Voltage measurements   A CRO has a large input impedance and can be used as a d.c./a.c. voltmeter if the p.d. to be measured is connected across the Y-input terminals. With d.c., the spot (time base off) or line (time base on) is deflected vertically, Fig. 41.4a, b. With a.c. (time base off) the spot moves up and down producing a vertical line if the motion is fast enough, Fig. 41.4c.

When the *Y-amp gain* control is on, say, 1 V/div, a deflection of 1 division on the screen graticule would be given by a 1 V d.c. Y-input; a line 1 div long would be

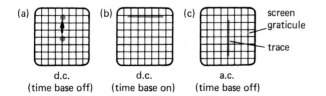

d.c.
(time base off)   d.c.
(time base on)   a.c.
(time base off)

Fig. 41.4

produced by an a.c. input of 1 V peak-to-peak, i.e. peak voltage = 0.5 V and r.m.s. voltage ≈ 0.7 × peak voltage = 0.7 × 0.5 = 0.35 V. On a setting of 0.05 V/div, a Y-input of 0.05 V causes a deflection of 1 div.

While there are more accurate ways of measuring p.ds, the CRO can measure alternating p.ds at frequencies of several megahertz.

Time measurements   The CRO must have a calibrated time base. When set on, for example, 1 ms/div the spot takes 1 millisecond to move 1 division and so travels 10 horizontal divisions in 10 ms. The period and so the frequency of a waveform can be found in this way.

Waveform display   The a.c. voltage whose waveform is required is connected to the Y-input with the time base on. When the time base frequency equals that of the input, one complete wave is displayed; if it is half that of the input, two waves are formed. These and other effects are shown in Fig. 41.5.

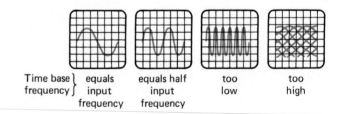

Time base ⎱ equals   equals half   too   too
frequency ⎰ input   input   low   high
                frequency   frequency

Fig. 41.5

Phase relationships   If two sine wave p.ds of the same frequency and amplitude are applied simultaneously to the X- and Y-inputs (time base off), traces like those in Fig. 41.6 are obtained which depend on their phase difference. The method can be used to study the phase relationships between p.ds and currents in a.c. circuits.

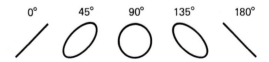

0°   45°   90°   135°   180°

Fig. 41.6

Frequency comparison   When two p.ds of different frequencies f_x and f_y are applied to the X- and Y-inputs (time base off), traces called *Lissajous' figures* are obtained. Some are shown in Fig. 41.7. The frequency ratio is given by:

$$\frac{f_y}{f_x} = \frac{\text{no. of loops touching OX}}{\text{no. of loops touching OY}}$$

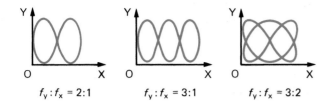

$f_Y : f_X = 2:1$ $f_Y : f_X = 3:1$ $f_Y : f_X = 3:2$

Fig. 41.7

DUAL BEAM AND DUAL TRACE CROs

These are useful for comparing two traces simultaneously. In the dual beam type the CRT has two electron guns, each with its own Y-plates but having the same time base. In the more common dual trace type, electronic switching produces two traces from a single beam. The switching is done by a square wave having a frequency much higher than that of either input. The top of the wave connects one input (Y_1) to the Y-plates and the bottom connects the other (Y_2). This occurs so rapidly that there appears to be two separate traces, one above the other on the screen.

QUESTIONS

1. When the Y-amp gain control on a CRO is set on 2 V/cm, an a.c. input produces a 10 cm long vertical trace. What is (i) the peak voltage, (ii) the r.m.s. voltage, of the input?
2. What is the frequency of an alternating p.d. which is applied to the Y-plates of a CRO and produces five complete waves on a 10 cm length of the screen when the time base setting is 10 ms/cm?

42 SIGNAL GENERATORS

CHECKLIST

After studying this chapter you should be able to:
• state what audio and radio frequency generators do.

A signal generator is an oscillator (p. 112) which produces a signal in the form of a.c. of known but variable frequency. It is used for testing, fault-finding and experimental work. There are two types.

AUDIO FREQUENCY GENERATOR

A typical instrument, Fig. 42.1, covers frequencies from 0.1 Hz to 100 kHz in several ranges. The output may have a sine, square or triangular waveform.

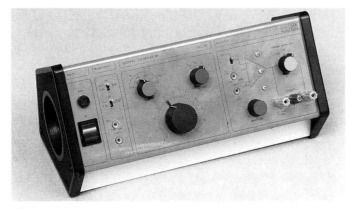

Fig. 42.1

To allow matching to different loads (p. 108), there are usually 'low' and 'high' impedance outputs, e.g. 1 Ω and 100 Ω respectively. The former would be used for direct connection to a 4 Ω loudspeaker and the latter when providing a signal to an oscilloscope (p. 93).

If the maximum output voltage (e.g. 6 V r.m.s.) is not required it can be fed into an attenuator which controls the output by (i) reducing it in steps by a factor of 10 or 100, and (ii) allowing it to be varied continuously from 0 V up to the maximum.

One use for an a.f. signal generator is to test the response of an audio amplifier (p. 186) to different frequencies by supplying it with a square wave, Fig. 42.2a. If all the many (sine wave) frequencies making up the

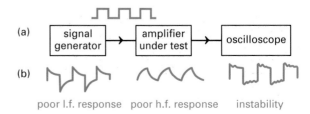

poor l.f. response   poor h.f. response   instability

Fig. 42.2

square wave (p. 17) are amplified equally, the output from the amplifier is seen on the oscilloscope to be a square wave. The waveforms obtained for other responses are given in Fig. 42.2b.

RADIO FREQUENCY GENERATOR

The frequency coverage of this type is from 100 kHz to 450 MHz or so. The output is a pure sine wave which can usually be amplitude or frequency modulated (p. 178) for fault-finding in radio receivers.

An attenuator allows the output, normally about 100 mV, to be reduced when smaller signals are required.

The oscillator producing the r.f. usually contains some type of *LC* circuit (p. 34), coarse adjustment of the frequency being obtained by switching in different coils and fine adjustment by altering a variable capacitor.

A typical instrument is shown in Fig. 42.3.

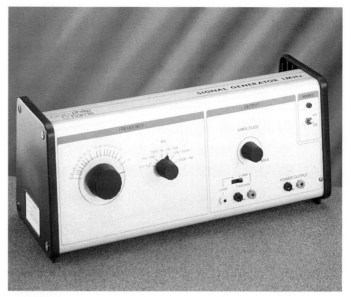

Fig. 42.3

43 PROGRESS QUESTIONS

1. A multimeter measures resistance by incorporating its meter in series with an internal battery of e.m.f. 4.5 V and a variable resistor. When the arrangement is 'zeroed' for a resistance measurement by short-circuiting the leads, 1.0 mA passes through the meter. When a resistor R is connected between the leads, the meter current is 0.5 mA. What is the resistance of R? (*O. and C.*)

2. Explain how you would use an ohmmeter **(a)** to determine which terminal of a transistor is the base, and **(b)** to distinguish between n-p-n and p-n-p types of transistor.

Why is an ohmmeter scale non-linear? (*L.*)

3. An ohmmeter was used to test for a fault in the section of the circuit in Fig. 43.1. Connection of the

meter to A and B indicated a resistance of 10 kΩ whichever way round the meter was connected. There is known to be only one simple fault. Which of the components *could* be faulty and which component or components (if any) can be assumed not to be faulty? Justify your answers. (*A.E.B.*)

4. An oscilloscope has a sensitivity of 5 V/cm and a time base adjusted to give a sweep time of 1 ms/cm. Show, on graticules like that in Fig. 43.2, the trace for **(a)** a 250 Hz sine wave of amplitude 10 V, **(b)** a 500 Hz square wave of amplitude 5 V. (*C.*)

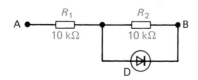

Fig. 43.1

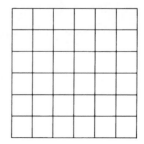

Fig. 43.2

5. Fig. 43.3 shows a waveform observed on the face of a cathode ray oscilloscope. The centimetre markings of the graticule are also shown. The y-deflection sensitivity was set at 0.1 V cm^{-1} and the time base was set at 100 μs cm^{-1}.

Give as much information as you can about the waveform. (O. and C.)

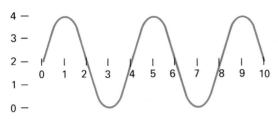

Fig. 43.3

6. The waveform of a p.d. applied internally to the X-plates of an oscilloscope is shown in Fig. 43.4. What is the function of this waveform?

Sketch the appearance of the trace on the screen of the oscilloscope of a sinusoidal alternating p.d. of frequency 1 kHz applied to the Y-input (with the p.d. of Fig. 43.4 on the X-plates).

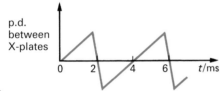

Fig. 43.4

7. a) Explain the terms (i) frequency, (ii) phase difference. State *three* uses of an oscilloscope.

Explain briefly how, in one of these applications, measurements are made.

b) A sinusoidal voltage of frequency 300 Hz is applied to the X-plates of an oscilloscope, and a sinusoidal voltage of frequency f is applied to the Y-plates. What is f when the trace on the screen is that shown in (i) Fig. 43.5a, (ii) 43.5b? (C.)

Fig. 43.5   (a)   (b)

8. a) Explain the action of (i) the time base, (ii) the synchronization control, of a cathode ray oscilloscope.

b) How would you use an oscilloscope to measure (i) a d.c. voltage, (ii) the r.m.s. value of a sinusoidal waveform, and (iii) the period of a square wave?

9. A sine wave p.d. of peak value 6 V is applied to the circuit of Fig. 43.6. An oscilloscope, set for d.c. input, is connected across YE so that when Y is positive relative to E, the spot is deflected upwards.

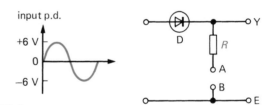

Fig. 43.6

Sketch the waveform you would see when
a) A and B are connected by a wire,
b) A is made 3 V positive relative to B.

Mark voltage values on the waveforms. (Ignore the voltage drop across D when it is conducting.)

ANALOGUE ELECTRONICS: TRANSISTORS

44 TRANSISTOR VOLTAGE AMPLIFIERS (I)

CHECKLIST

After studying this chapter you should be able to:
- state the features of an analogue signal,
- classify amplifiers according to their frequency range,
- define and measure the voltage gain of an amplifier and work it out from a graph of V_o against V_i,
- explain why a transistor amplifier needs bias applied to the base,
- draw and recognize V_o distorted waveforms caused by bias which is too low or too high, or by overloading the input, or by a non-linear characteristic,
- draw the circuit for a common-emitter amplifier and calculate the value of the biasing resistor R_B,
- state that coupling capacitors in the input and output maintain the bias while allowing a.c. signals to pass, and
- define and measure the bandwidth of an amplifier.

INTRODUCTION

Analogue electronics   This is one of the two main branches of electronics; the other is *digital electronics* (p. 130). In analogue electronics, the signals being processed are electrical representations (analogues) of physical quantities which vary continuously over a range of values. The information they carry, e.g. the loudness and pitch of a sound, is in the amplitude and shape of their waveforms, Fig. 44.1. Most audio and radio signals are in this category.

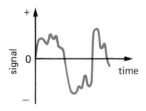

Fig. 44.1

Analogue circuits are *amplifier-type* circuits, designed so that a linear relationship exists between the input and output signals, i.e. input and output are directly proportional. They are also called *linear* circuits.

Amplifiers   In general, the job of an amplifier is to produce an output which is an enlarged copy of the input. Amplification of one or more of the input voltage, current and power occurs. In the last case, the extra power at the output is provided by an external energy source, e.g. a battery. An amplifier differs from a transformer in that it can have a power gain greater than 1.

An *audio frequency* (a.f.) amplifier amplifies a.c. signals in the audio frequency range (20 Hz to 20 kHz), and is the type considered in this and the next two chapters. *Radio frequency* (r.f.) amplifiers (p. 197) operate above 20 kHz, at radio and television signal frequencies.

The symbol for an amplifier is shown in Fig. 44.2; in effect it has three terminals, viz. input, output and common. 'Common' or 'ground' is connected to one side of input, output and the power supply (not shown). It is usually taken as the reference point i.e. 0 V, for all circuit voltages and if connected to earth ($\perp$), it is called 'earth'.

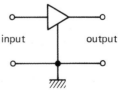

Fig. 44.2

common or ground

VOLTAGE AMPLIFICATION BY A TRANSISTOR

Many transducers (p. 41) produce a.c. voltages which must be amplified before use. The junction transistor, despite being a current-amplifying device, is used mainly as a voltage amplifier.

Action   The basic circuit is the same as the one in which a transistor with a collector load R_L acts as a switch (Fig. 30.1, p. 68) and is given again with its voltage characteristic in Fig. 44.3. The latter shows that if the input voltage V_i changes from 0.8 to 1.2 V, the corresponding variation of the output voltage V_o is from 5 to 1 V. That is, a change of 0.4 V in V_i produces a change of 4 V in V_o.

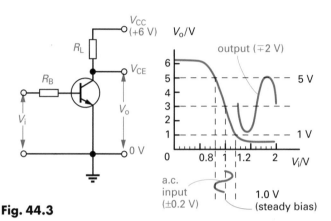

Fig. 44.3

The *voltage gain* A is given by:

$$A = \frac{\text{change in } V_o}{\text{change in } V_i}$$

In this case,

$$A = \frac{4}{0.4} = 10$$

Bias   If small a.c. input voltages are to be amplified, a steady bias voltage must also be applied to the base to forward-bias the base–emitter junction. It should produce a base current which creates a collector current that sets the collector voltage (V_{CE}) at about *half the supply voltage* ($V_{CC} = 6.0$ V) in the absence of an a.c. input voltage (called the 'quiescent' state). This allows V_{CE} to have its maximum 'swing' capability from (in theory) 0 V to V_{CC} when the a.c. input is applied.

In the case we are considering, a base bias of $+1$ V makes $V_{CE} = +3$ V. An a.c. input of ±0.2 V peak superimposed on this bias, would cause V_{CE} to 'swing' down to 1 V and up to 5 V. The output variation would be ∓2 V

and if the characteristic is linear over this range, the output is a ten times amplified undistorted copy of the a.c. input.

The distortion occurring in the output when the bias voltage is too low (or too high) or the a.c. input too large is shown by the graphs in Fig. 44.4a, b.

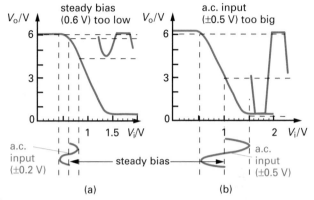

Fig. 44.4

Further points   The transistor and load *together* bring about voltage amplification. It is the small variations of base current which cause the larger variations of collector current to produce varying voltages across the load.

The output is 180° out of phase with the input so that if the input increases, the output decreases, as the sine wave graphs in Fig. 44.3 show.

COMMON-EMITTER AMPLIFIER

In this circuit, Fig. 44.5, the bias for the base is conveniently obtained via a resistor R_B connected to the collector power supply V_{CC}. In the quiescent state a steady d.c. base current I_B flows from V_{CC}, through R_B into the base and back to 0 V via the emitter.

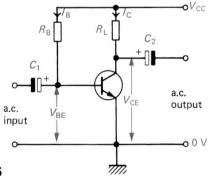

Fig. 44.5

Calculation of R_L and R_B   Suppose (i) the n-p-n transistor is a silicon type which works satisfactorily on a quiescent (no a.c. input) collector current I_C of 3 mA, (ii) $h_{FE} = 100$, and (iii) $V_{CC} = 6$ V.

The collector–emitter circuit equation is (p. 69):

$$V_{CC} = I_C R_L + V_{CE}$$

That is,

$$R_L = (V_{CC} - V_{CE})/I_C \qquad (1)$$

For an undistorted, amplified output V_{CE} must be $\frac{1}{2}V_{CC}$, i.e. 3 V.

Substituting values in (1) we get:

$$R_L = (6-3)V/3 \text{ mA} = 1 \text{ k}\Omega$$

Since $h_{FE} = I_C/I_B$ we have:

$$I_B = I_C/h_{FE} = 3 \text{ mA}/100 = 0.03 \text{ mA}$$

The base–emitter circuit equation is:

$$V_{CC} = I_B R_B + V_{BE}$$

where $V_{BE} = 0.6$ V.

Rearranging gives:

$$R_B = (V_{CC} - V_{BE})/I_B \qquad (2)$$

Substituting values in (2) we get:

$$R_B = (6 - 0.6) \text{ V}/0.03 \text{ mA} = 5.4 \text{ V}/0.03 \text{ mA}$$
$$= 540/3 = 180 \text{ k}\Omega$$

Coupling capacitors When an a.c. input voltage is applied, the collector–emitter voltage becomes a varying direct voltage and may be regarded as an alternating voltage superimposed on a steady direct voltage, i.e. on the quiescent value of V_{CE}, Fig. 44.6. Only the a.c. part is wanted and capacitor C_2 blocks the d.c. part but allows the a.c. part to pass, i.e. it couples the a.c. to the next stage of the circuit. Capacitor C_1 stops any external d.c. voltage upsetting the quiescent base voltage but couples the a.c. input voltage to the transistor (p. 26).

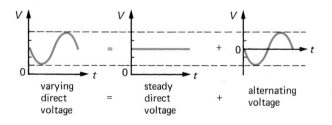

varying direct voltage = steady direct voltage + alternating voltage

Fig. 44.6

C_1 and C_2 have a low reactance ($X_C = 1/(2\pi f C)$) at the audio frequencies involved. For frequencies down to 10 Hz, 10 μF electrolytics are suitable, connected with the polarities shown.

Summing up, a transistor acts as a voltage amplifier if (i) it has a suitable collector load, and (ii) it is biased so that the quiescent value of $V_{CE} \approx \frac{1}{2}V_{CC}$.

Frequency response The *voltage gain* A of a capacitor-coupled amplifier is fairly constant over most of the a.f. range but it falls off at the lower and upper limits. At low frequencies, the reactances of the coupling capacitors increase and less of the low frequency part is passed on. At high frequencies various stray capacitances in the circuit can cause the fall.

A typical *voltage gain–frequency* curve is shown in Fig. 44.7. (To fit in the large frequency range, frequencies are not plotted on the usual linear scale but on a logarithmic scale in which equal divisions represent equal changes in the 'log of the frequency f'.) The *bandwidth* is the range of frequencies within which A does not fall below $1/\sqrt{2}$ (≈ 0.7) of its maximum value A_{max}. The two points P and Q at which this happens are called the '3 dB' points. (The decibel (dB) scale compares signal levels.)

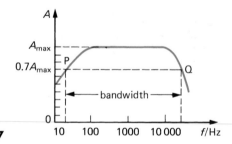

Fig. 44.7

MEASURING THE VOLTAGE GAIN AND BANDWIDTH OF AN AMPLIFIER

Voltage gain The common-emitter amplifier circuit of Fig. 44.5 can be used. An a.c. input of a 10 mV 50 Hz sine wave is taken from a signal generator. The a.c. output voltage from the amplifier is connected to the Y-input terminals of an oscilloscope which is adjusted so that the output waveform is an undistorted sine wave.

The peak-to-peak values V_{in} and V_{out} of the input and output waveforms respectively are measured on the screen graticule. The voltage gain A is then given by:

$$A = \frac{V_{out}}{V_{in}}$$

Bandwidth The above procedure is repeated for different frequencies up to 100 kHz and a graph drawn of A against frequency f using a frequency scale that increases by powers of 10 as in Fig. 44.7, to accommodate the wide range of values.

1. In the circuit of Fig. 44.5, what is the purpose of (i) R_L, (ii) R_B, (iii) C_1, (iv) C_2?

2. In the circuit of Fig. 44.5, if $V_{CC} = 9$ V, $R_L = 2$ kΩ, $I_C = 2$ mA and $h_{FE} = 100$ for the transistor, calculate (i) V_{CE}, (ii) I_B, (iii) R_B, if $V_{BE} = 0.6$ V.

3. Explain the following terms:
  a) analogue electronics,
  b) voltage amplifier,
  c) a.f. amplifier,
  d) 'common' or 'ground',
  e) voltage gain,
  f) quiescent state.

45 TRANSISTOR VOLTAGE AMPLIFIERS (II)

CHECKLIST

After studying this chapter you should be able to:
• draw a load line on the output characteristics of a transistor and select a suitable d.c. operating point,
• explain what 'thermal runaway' is and how it occurs,

• recognize the circuit of a transistor amplifier using collector-to-base bias and explain how it improves stability.

LOAD LINES

When designing a voltage amplifier it is important to ensure that, as well as obtaining the desired voltage gain, there is minimum distortion of the output. The choice of the quiescent or *d.c. operating point* (i.e. the values of I_C and V_{CE}) determines whether these requirements will be met and is made by constructing a *load line*.

Drawing a load line The output characteristics of a transistor (Fig. 31.2c, p. 72) show the relation between V_{CE} and I_C with *no* load in the collector circuit. With a load R_L, the equation connecting them is (p. 69):

$$V_{CC} = I_C R_L + V_{CE}$$

Rearranging, we get:

$$V_{CE} = V_{CC} - I_C R_L \qquad (1)$$

Knowing V_{CC} and R_L this equation enables us to calculate V_{CE} for different values of I_C. If a graph of I_C (on the *y*-axis) is plotted against V_{CE} (on the *x*-axis), the straight line so obtained is the *load line*. It can be drawn knowing just two points, the easiest being the end points A and B where the line cuts the V_{CE}- and I_C-axes.

For A we put $I_C = 0$ in (1) and get $V_{CE} = V_{CC} = 6$ V (say).

For B we put $V_{CE} = 0$ in (1) and get $I_C = V_{CC}/R_L$. If $R_L = 1$ kΩ say, then $I_C = 6$ V/1 k$\Omega = 6$ mA.

In Fig. 45.1, AB is the load line for $V_{CC} = 6$ V and $R_L = 1$ kΩ. It is shown superimposed on the output characteristics of the transistor. We can regard a load line as the output characteristic of the *transistor and load* for particular values of V_{CC} and R_L. Different values of either give a different load line.

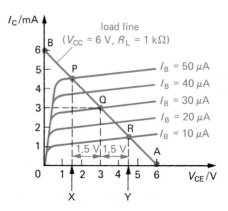

Fig. 45.1

Choosing the d.c. operating point

The best 'position' for the d.c. operating point is near the *middle* of a load line which cuts the output characteristics where they are straight and equally spaced. Otherwise the output has a distorted waveform and its 'swing' capability (from near V_{CC} to near 0 V) is reduced.

If Q is chosen in Fig. 45.1, the quiescent value of V_{CE} is 3 V (i.e. half V_{CC}) and I_C is 3 mA. The value of I_B which gives these values is seen from the output characteristic passing through Q to be 30 µA. R_B can then be calculated as before (p. 100).

Voltage gain A This is given by:

$$A = \frac{\text{output voltage}}{\text{input voltage}} = \frac{\text{change in collector voltage}}{\text{change in base voltage}}$$

It can be obtained from the load line by noting that when the input causes I_B to vary from 10 to 50 µA (from R to P), V_{CE} varies from 4.5 to 1.5 V (from Y to X). If the input characteristic of the transistor (Fig. 31.2b, p. 72) shows that, for example, V_{BE} has to change by 40 mV (0.04 V) to cause a 40 µA change in I_B, then:

$$A = \frac{4.5 - 1.5}{0.04} = \frac{3.0}{0.04} = 75$$

STABILITY

The simple common-emitter voltage amplifier circuit of Fig. 44.5 (p. 99) has two serious defects.

Effect of h_{FE}

The first arises from the fact that it does not work satisfactorily if h_{FE} varies widely (as it does even for transistors of the same type, p. 73) from the value used when designing it.

For example, if h_{FE} is greater, then I_C is greater and causes a larger voltage drop across the collector load R_L. Consequently the quiescent value of V_{CE} is much less than half V_{CC}, thus upsetting the operating point and in this case reducing the swing capability in the negative-going direction. If h_{FE} is smaller, the swing capability in the positive-going direction becomes smaller. In both events the output is distorted.

Thermal runaway

The second problem occurs when the temperature of the transistor (junction) rises, due to the heating effect of I_C or to an increase in the surrounding (ambient) temperature. There is greater vibration of the semiconductor atoms and more free electrons and holes are produced. This increases I_C and causes further

heating and so on until the transistor is destroyed by 'thermal runaway'. Transistors carrying large currents are therefore used with heat sinks (p. 82) to minimize this effect.

In an effort to compensate automatically for increases of I_C (for whatever reason), and so stabilize the d.c. operating point, special bias circuits have been developed.

COLLECTOR-TO-BASE BIAS

This is the simplest circuit but it gives rise to a useful general-purpose voltage amplifier which gives adequate stability. The improvement is achieved by halving the value of R_B in the circuit of Fig. 44.5 and connecting it between the collector and base as in Fig. 45.2, rather than between V_{CC} and base. The quiescent value of I_B then depends on the quiescent value of V_{CE}.

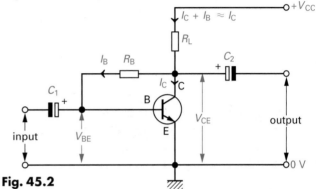

Fig. 45.2

For the circuit in Fig. 45.2 we can write (since I_C is much greater than I_B):

$$V_{CC} = I_C R_L + V_{CE} \tag{2}$$

where

$$V_{CE} = I_B R_B + V_{BE} \tag{3}$$

From (2) you can see that if I_C increases, V_{CE} decreases since V_{CC} is fixed. From (3) it therefore follows that since V_{BE} is constant (0.6 V or so for silicon), I_B must also decrease and in so doing tends to bring back I_C to its original value.

Taking the quiescent conditions as $V_{CE} = \frac{1}{2}V_{CC} = 3$ V, $I_C = 3$ mA and $I_B = 0.03$ mA (as in the 'Calculation of R_L and R_B' on p. 99), the value of R_B in Fig. 45.2 is found by rearranging equation (3) to give:

$$R_B = \frac{V_{CE} - V_{BE}}{I_B} = \frac{(3 - 0.6)\text{ V}}{0.03\text{ mA}}$$

$$= 2.4\text{ V}/0.03\text{ mA} = 80\text{ k}\Omega$$

This is about half the value of 180 kΩ for R_B in the unstabilized circuit of Fig. 44.5.

QUESTIONS

1. The load line for the amplifier in Fig. 45.2 is shown in Fig. 45.3.
a) Use it to find the values of V_{CC} and R_L.
b) If Q is the d.c. operating point, what are the quiescent values of V_{CE}, I_C and I_B?
c) Calculate R_B if the transistor is a silicon type.

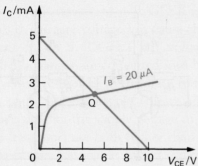

Fig. 45.3

2. The output characteristics of a junction transistor in common-emitter connection are shown in Fig. 45.4. The transistor is used in an amplifier with a 9 V supply and a load resistor of 1.8 kΩ.

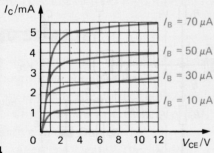

Fig. 45.4

a) What is the value of I_C when $V_{CE} = 0$?
b) Lay a ruler along the load line and choose a suitable d.c. operating point. Read off the quiescent values of I_C, I_B and V_{CE}.
c) What is the quiescent power consumption of the amplifier?
d) If an alternating input voltage varies the base current by ± 20 μA about its quiescent value, what is (i) the variation in the collector–emitter voltage, (ii) the peak output voltage?
e) An input characteristic of the transistor is given in Fig. 45.5. Use it to find the base–emitter voltage variation which causes a change of ± 20 μA in the quiescent base current.

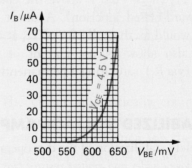

Fig. 45.5

f) Using your answers from **d)** and **e)**, find the voltage gain A of the amplifier.
g) If the amplifier uses the collector-to-base bias circuit of Fig. 45.2, calculate the value of R_B to give the quiescent value of I_B. (Assume $V_{BE} = 0.6$ V.)

48 AMPLIFIERS AND FEEDBACK

CHECKLIST

After studying this chapter you should be able to:
- state the effect on the output of an amplifier of positive and negative feedback,
- perform calculations using the feedback equation,
- recall the advantages of negative feedback (n.f.b.) in an amplifier, and
- recognize and analyse bipolar and FET n.f.b. circuits.

The performance of an amplifier can be changed by feeding part or all of the output back to the input. The feedback is *positive* if it is in phase with the input, i.e. adds to it so that it reinforces changes at the input, as in Fig. 48.1a. As a result the overall gain of the amplifier increases.

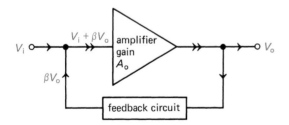

| input | positive feedback | = | greater input | | input | negative feedback | = | smaller input |

(a)   (b)

Fig. 48.1

The feedback is *negative* if it is 180° out of phase (in antiphase) with the input, i.e. subtracts from it, as in Fig. 48.1b. In this case the gain is reduced but, as we shall see presently, other advantages arise.

THE FEEDBACK EQUATION

Suppose the amplifier in Fig. 48.2 has a voltage gain of A_o, called the *open-loop gain*, with no feedback. If a fraction β of the output V_o is fed back so as to add to the *actual* input signal to be amplified V_i (i.e. positive feedback), then the *effective* input to the amplifier becomes $V_i + \beta V_o$. The fraction β is the *feedback factor*.

The gain of the amplifier itself is still A_o and so:

$$V_o = A_o(V_i + \beta V_o)$$

Rearranging,

$$V_o - \beta A_o V_o = A_o V_i$$

That is,

$$V_o(1 - \beta A_o) = A_o V_i$$

$$\therefore \frac{V_o}{V_i} = \frac{A_o}{1 - \beta A_o}$$

The voltage gain A of the whole amplifier with feedback, called the *closed-loop gain*, is given by:

$$A = \frac{V_o}{V_i}$$

Hence:

$$A = \frac{A_o}{1 - \beta A_o} \tag{1}$$

This is the general equation for an amplifier with feedback, and as it is, applies to positive feedback, i.e. β is positive. It shows that (i) $A > A_o$, and (ii) if $\beta A_o = 1$, $A = A_o/0$, i.e. A is infinitely large. The amplifier then gives an output with no input, which is what oscillators (p. 112) do using positive feedback.

For negative feedback, β is negative and the equation becomes:

$$A = \frac{A_o}{1 + \beta A_o} \tag{2}$$

In this case $A < A_o$ and if $\beta A_o \gg 1$, then:

$$A \approx \frac{A_o}{\beta A_o} \approx \frac{1}{\beta} \tag{3}$$

This result has important consequences.

ADVANTAGES OF NEGATIVE FEEDBACK (n.f.b.)

Although n.f.b. reduces the gain of an amplifier and makes more stages of amplification necessary, it has several very desirable results, provided $A_o \gg A$.

(i) *Voltage gain is accurately predictable and constant* because, as equation (3) shows, A does not depend on A_o (which can vary widely from one amplifying device, e.g. a transistor, to another, even of the same type). A depends only on β and this is usually determined by the values of two resistors, often in a potential divider. Resistors of high stability and known accuracy are available (p. 20) and, by using n.f.b., these qualities can also become those of the gain of the whole amplifier.

(ii) *Distortion of the output is reduced*, i.e. the output waveform is a truer copy of the input waveform.

(iii) *Frequency response is better*, i.e. a wider range of frequencies is amplified by the same amount, so increasing the bandwidth (p. 100). (It is roughly true that *gain × bandwidth* is constant.)

(iv) *Stability is increased.*

NEGATIVE FEEDBACK CIRCUITS

Collector-to-base resistor The collector-to-base bias circuit of Fig. 45.2 is shown again in Fig. 48.3. R_B not only provides the correct quiescent d.c. bias to the base–emitter junction (as well as stabilizing the d.c. operating point by what is in effect d.c. negative feedback), but it also supplies negative feedback for the a.c. input as follows.

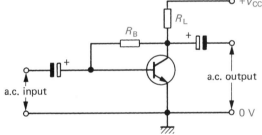

Fig. 48.3

When an alternating input is applied, an alternating current is fed back through R_B to the base, being superimposed on the d.c. base current. It is in anti-phase with the alternating input because the output voltage in a common-emitter amplifier is 180° out of phase with the input (see p. 99), i.e. the feedback is negative.

It can be shown that the feedback factor β is given by:

$$\beta = \frac{R_L}{R_B}$$

If the value of R_B to provide the correct d.c. bias is not the same as that to give the required amount of a.c. negative feedback, the bias is obtained by other means (e.g. by the 'voltage divider and emitter resistor' method) and a capacitor is joined in series with R_B to stop d.c. reaching the base.

Source resistor The circuit is shown in Fig. 48.4 for a FET voltage amplifier. It is the same as Fig. 47.1 but with the source decoupling capacitor C_S omitted.

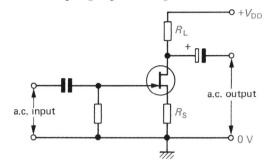

Fig. 48.4

When an alternating input voltage is applied, alternating components are superimposed on the d.c. drain and source currents. The one on the drain current creates the output voltage across R_L; the one on the source current produces an alternating voltage across the source resistor R_S which is fed back to the gate in antiphase with the input voltage (since there is no bypass capacitor to decouple R_S as in Fig. 47.1). The action is briefly as follows.

A positive half-cycle of the input makes the gate less negative, the source current increases, as does the p.d. across R_S. The source therefore becomes more positive with respect to the gate (as explained on p. 105), i.e. the gate tends to go more negative and thereby reduce the effective input to the amplifier.

It can be shown that β is given by:

$$\beta = \frac{R_S}{R_L}$$

With a junction transistor the n.f.b. resistor is in the emitter circuit.

QUESTIONS

1. Explain
   a) positive feedback,
   b) negative feedback.
2. a) State *one* disadvantage and *four* advantages of n.f.b.
   b) Write down the n.f.b. equation, stating the meaning of each symbol.
3. The open-loop gain of an amplifier is 200. What is the closed-loop gain when the negative feedback factor is 1/50?
4. a) Calculate β for Fig. 48.3 if $R_L = 1\,\text{k}\Omega$ and $R_B = 100\,\text{k}\Omega$.
   b) What is β for Fig. 48.4 if $R_L = 10\,\text{k}\Omega$ and $R_S = 1\,\text{k}\Omega$?

49 AMPLIFIERS AND MATCHING

CHECKLIST

After studying this chapter you should be able to:
- explain the terms input impedance (Z_i) and output impedance (Z_o) and state why they are important,
- describe how Z_i and Z_o are measured, and
- perform calculations to match (i) the signal source to the amplifier input and (ii) the amplifier output to the load.

INPUT AND OUTPUT IMPEDANCE

An amplifier has to be 'matched' to the transducer or circuit supplying its input or receiving its output. Usually this means ensuring that either the maximum voltage or the maximum power is transferred to or from the amplifier. In all cases, the input or output impedance (since we are concerned with a.c.) of the amplifier is an important factor.

The *input impedance* Z_i equals V_i/I_i where I_i is the a.c. flowing into the amplifier when voltage V_i is applied to the input. It depends not only on the a.c. input resistance r_i (p. 72) of the transistor but also on the presence of capacitors, resistors, etc. in the amplifier circuit. In effect, the amplifier behaves as if it had an impedance Z_i connected across its input terminals.

The *output impedance* Z_o is the a.c. equivalent of the internal or source resistance of a battery (p. 14). It causes a 'loss' of voltage at the output terminals when the amplifier is supplying current. We can think of an amplifier as an a.c. generator of voltage V which, on open circuit equals the voltage V_o at the output terminals. On closed circuit, when an output current flows, V_o is less than V by the voltage dropped across Z_o, as in the d.c. case.

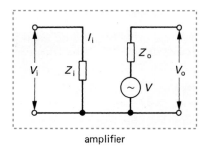

Fig. 49.1

The diagram in Fig. 49.1, called the *equivalent circuit* of an amplifier, is useful when considering matching problems. Z_i and Z_o can be measured (in ohms) by experiment.

MEASURING INPUT AND OUTPUT IMPEDANCES

Input impedance A circuit is shown in Fig. 49.2 which uses a known resistor R connected in *series* with the signal generator supplying an a.c. input signal to the device (e.g. an amplifier) whose input impedance Z_i is required. The voltages V_1 and V_2 are measured by an oscilloscope or high impedance a.c. voltmeter. Then, if I_i is the input current, the voltage drop across R is:

$$V_1 - V_2 = I_iR$$

Hence:
$$I_i = \frac{V_1 - V_2}{R}$$

But,
$$Z_i = \frac{V_2}{I_i}$$

$$\therefore Z_i = \frac{V_2R}{V_1 - V_2}$$

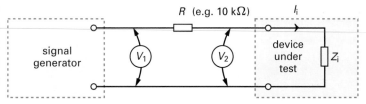

Fig. 49.2

Output impedance In this case the a.c. output voltage from the device (e.g. an amplifier getting its input from a signal generator) is measured first on open circuit (V_1), Fig. 49.3a, and then with a known resistor R in *parallel* with the output, to obtain the reduced output voltage V_2, Fig. 49.3b. The voltage drop (i.e. 'lost' volts) across the output impedance Z_o with R connected is:

$$V_1 - V_2 = I_oZ_o$$

where I_o is the output current. But,

$$I_o = \frac{V_2}{R}$$

$$\therefore Z_o = \frac{(V_1 - V_2)R}{V_2}$$

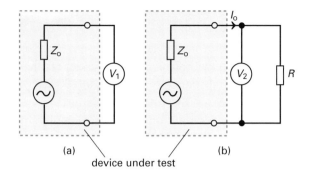

(a)

(b)

device under test

Fig. 49.3

MATCHING TO SIGNAL AND LOAD

Signal The source supplying the a.c. signal to the *input* of an amplifier can also be regarded as an a.c. generator, of voltage V_s, having an output impedance Z_s, as shown by its equivalent circuit in Fig. 49.4. If the input current is I_i, we can write:

$$V_s = I_i(Z_s + Z_i) \quad \text{and} \quad V_i = I_iZ_i$$

It follows that:

$$V_i = V_s\left(\frac{Z_i}{Z_s + Z_i}\right) \tag{1}$$

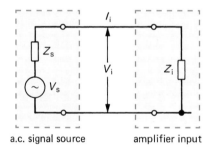

a.c. signal source

amplifier input

Fig. 49.4

The most common requirement is for *maximum voltage transfer* from the signal source to the input of the amplifier. That is, V_i should be as large as possible. Equation (1) shows that if Z_i is large compared with Z_s (say, ten times larger) then $V_i \approx V_s$ and very little of V_s is 'lost' across Z_s.

In practice, things may be very different if, for example, a common-emitter amplifier with an input impedance Z_i of 1 kΩ is supplied with an a.c. input signal from a crystal microphone having an output impedance Z_o of 1 MΩ. Only about 1/1000 of the voltage V_s generated by the microphone would be available at the amplifier input (since $Z_i/(Z_s + Z_i) \approx 1/1000$). Obviously the impedance matching needs to be improved. Ways of doing so will be considered in the next chapter.

Load If the *output* of the amplifier supplies a load of impedance Z_L with current I_o, then from Fig. 49.5 we have:

$$V = I_o(Z_o + Z_L) \quad \text{and} \quad V_o = I_oZ_L$$

Hence:

$$V_o = V\left(\frac{Z_L}{Z_o + Z_L}\right) \tag{2}$$

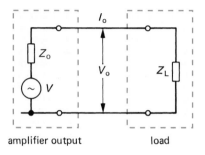

amplifier output

load

Fig. 49.5

For *maximum voltage transfer* equation (2) shows that (as with matching to signal), we need $Z_L \gg Z_o$ if the voltage loss across Z_o is to be small.

For *maximum power transfer* the maximum power theorem (p. 14) states that this occurs when $Z_L = Z_o$. But here too there are problems in practice. For example, a common-emitter amplifier with its output connected directly to a loudspeaker with a typical impedance Z_L of 8 Ω would not deliver maximum power to it because the output impedance Z_o of the amplifier is very much greater. Again steps need to be taken to improve matters.

WORKED EXAMPLE

Calculate **(a)** the voltage across and **(b)** the power developed in the resistive load Z_L in the circuit of Fig. 49.6 if $V_s = 15$ mV, $Z_s = 500$ Ω and $Z_L = 8$ Ω. The amplifier characteristics are $Z_i = 1000$ Ω, $A_o = 100$ and $Z_o = 8$ Ω.

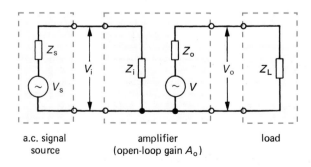

a.c. signal source

amplifier (open-loop gain A_o)

load

Fig. 49.6

(a) From equation (1), the voltage V_i applied to the input is:

$$V_i = 15\left(\frac{1000}{500 + 1000}\right) = \frac{15 \times 2}{3} = 10 \text{ mV}$$

The amplifier output voltage V is given by:

$$V = A_o \times V_i = 100 \times 10 = 1000 \text{ mV} = 1 \text{ V}$$

From equation (2), the voltage V_o across the load Z_L of $8\,\Omega$, is:

$$V_o = 1\left(\frac{8}{8 + 8}\right) = 0.5 \text{ V}$$

(b) Power in load $= \dfrac{V_o^2}{Z_L} = \dfrac{(0.5)^2}{8} \approx 0.03 \text{ W}$

QUESTIONS

1. a) Why does a signal generator designed to supply an alternating voltage *to* a circuit under test have a low output impedance (e.g. less than $100\,\Omega$)?
  b) Why does a CRO intended for studying voltages *from* a circuit under test have a high input impedance (e.g. $1\,\text{M}\Omega$)?

2. In the circuit of Fig. 49.6 calculate:
  a) V_i if $V_s = 100$ mV, $Z_s = 1\,\text{k}\Omega$ and $Z_i = 4\,\text{k}\Omega$,
  b) V if $A_o = 50$,
  c) V_o if $Z_o = Z_L = 4\,\Omega$,
  d) the power developed in Z_L.

50 IMPEDANCE-MATCHING CIRCUITS

CHECKLIST

After studying this chapter you should be able to:
- recognize and explain the action of an emitter-follower circuit,
- compare the emitter-follower with the source-follower, and
- state how a transformer is used to match impedances.

EMITTER-FOLLOWER

An emitter-follower has a *high* input impedance (e.g. $0.5\,\text{M}\Omega$) and a *low* output impedance (e.g. $30\,\Omega$). It can therefore provide 'step-down' impedance matching when connected between a high impedance signal source and a low impedance load. One of its main roles is as the output stage, i.e. the power amplifier, of a multistage amplifier feeding a loudspeaker (p. 188).

The basic circuit is shown in Fig. 50.1; it uses 100% n.f.b.

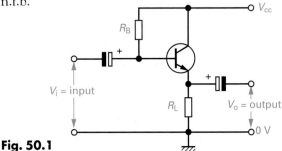

Fig. 50.1

Action The input V_i is applied between base and ground. The emitter resistor R_L, across which the output voltage V_o is developed, is in the input circuit as well as the output circuit. As a result, V_o is in series with V_i but has opposite polarity; *all* V_o thus opposes V_i and acts as the n.f.b. voltage (since R_L is not decoupled) with a feedback factor $\beta = 1$. The input current is reduced (for the same V_i), thereby in effect increasing the input impedance of the circuit. R_B fixes the d.c. operating point.

The circuit gets its name from the fact that the emitter voltage 'follows' the input voltage closely, since any increase in V_i increases V_o (due to the emitter current increasing).

The low output impedance arises from the fact that the input and output are connected by the forward-biased, low resistance base–emitter junction. (In a common-emitter amplifier, the input and output circuits are separated by the high resistance of the reversed-biased collector–base junction, giving it a much higher output impedance.)

Voltage and current gain The *voltage gain* is always just less than one, because V_i is applied across the voltage divider formed by the small resistance r_e of the base–emitter junction in series with the much larger resistance of R_L. V_o on the other hand is the p.d. across R_L alone and so is slightly smaller than V_i.

The *current gain* is high, as in the common-emitter amplifier, since the emitter (output) current is greater than the base (input) current.

Common-collector amplifier

This is the other name given to the circuit. From the point of view of a.c., the collector is common to both input and output circuits, being connected to ground by the negligible resistance of the power supply.

SOURCE-FOLLOWER

This is the FET version of the emitter follower; it is also called the *common-drain amplifier*. Due to the FET it has an even higher input impedance, e.g. 100 MΩ or more; the output impedance is also higher, e.g. 500 Ω. It makes a useful *buffer amplifier*, giving current amplification, to match a very high-impedance input transducer such as a crystal microphone to a conventional amplifier.

THE THREE TRANSISTOR AMPLIFIER CIRCUITS

A transistor can be connected as an amplifier in three ways. The common-emitter (CE) and common-collector (CC) modes have been considered.

The third method is the *common-base* (CB) circuit, which has a *low* input impedance (e.g. 50 Ω) and a *high* output impedance (e.g. 1 MΩ). The voltage gain is high but the current gain is just less than one. Fig. 50.2 shows the basic connections. Because it performs better at high frequencies than the common-emitter amplifier it is used to amplify weak signals from a television aerial.

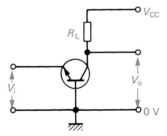

Fig. 50.2

The choice of amplifier for a particular job often depends on the impedance-matching requirements. The main properties of the three one-stage transistor amplifiers are summarized in the table.

Property	CE	CC	CB
Input impedance	medium	high	low
Output impedance	medium	low	high
Voltage gain	high	≈ 1	high
Current gain	high	high	≈ 1
Power gain	high	low	medium
Phase (output w.r.t. input)	180°	0°	0°

TRANSFORMER MATCHING

Transformers can be used either to step-up or step-down impedances.

Suppose a signal, produced by a 'generator' of output impedance Z_1, has to be matched to a load of input impedance Z_2. It can be shown that, by connecting a transformer as in Fig. 50.3, Z_2 can be made to 'look' equal to Z_1, and the maximum power transferred, if the turns ratio n_1/n_2 is chosen so that:

$$\frac{n_1}{n_2} = \sqrt{\frac{Z_1}{Z_2}}$$

Hence if an amplifier with an output impedance Z_1 is to be matched to a loudspeaker of lower impedance Z_2 by this method, a suitable *step-down* transformer is required since $Z_1 > Z_2$. On the other hand, if a transformer is used to match a microphone of low impedance Z_1 (e.g. a ribbon type) to the input of an amplifier of high impedance Z_2, a *step-up* ratio is necessary.

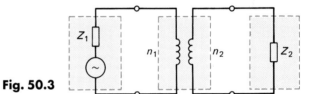

Fig. 50.3

Transformer matching at radio frequencies is much more common than at audio frequencies because a.f. transformers tend to be large, heavy and expensive and also introduce distortion.

QUESTIONS

1. 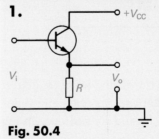 In the circuit of Fig. 50.4, why must the output voltage V_o be less than the input voltage V_i? Why does the circuit present to the input a resistance far greater than R? (L.)

   Fig. 50.4

2. An audio-frequency generator with an output impedance of 2 kΩ is used to test a loudspeaker of impedance 5 Ω. What fraction of the open-circuit signal from the generator appears at the terminals of the loudspeaker if the connection is direct? If connection is made using a transformer as intermediary, what properties should it have to give optimum power transfer to the loudspeaker? (*O. and C. part qn.*)

51 TRANSISTOR OSCILLATOR

CHECKLIST

After studying this chapter you should be able to:
- state what an electronic oscillator does and explain the part played by positive feedback,
- recognize and explain the action of a tuned (LC) oscillator, and
- use the equation $f = 1/(2\pi\sqrt{LC})$ in calculations.

An electronic oscillator generates a.c. having a sine, square or some other shape of waveform, depending on the particular circuit employed. Audio frequency (a.f.) oscillators are used in signal generators (p. 95), as are radio frequency (r.f.) types. The latter are also important in radio and television transmitters and receivers (pp. 197 and 201).

Most oscillators are *amplifiers with a feedback loop* from output to input which ensures that the feedback is (i) in phase with the input, i.e. is *positive*, and (ii) sufficient to make good the inevitable energy losses in the circuit, otherwise the oscillations of the electrons forming the a.c. are damped (p. 35). In effect, oscillators supply their own input and convert d.c. from the power supply into a.c.

TUNED (LC) OSCILLATOR

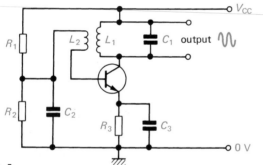

Fig. 51.1

The circuit in Fig. 51.1 is for a tuned sine-wave oscillator. Basically it is that of an amplifier with d.c. bias and stabilization being provided by R_1, R_2 and R_3 and decoupling by C_2 and C_3. The tuned circuit, consisting of inductor L_1 and capacitor C_1, is the collector load (C_1 may be variable) and it determines the frequency f of the oscillations, given in Hz by (p. 35):

$$f = \frac{1}{2\pi\sqrt{L_1 C_1}} \qquad (1)$$

where L_1 is in henrys (H) and C_1 is in farads (F).

In practice, to start oscillations, we do not have to apply a sine-wave input. Switching on the power supply pro-

duces a current pulse which charges C_1 and starts oscillations automatically. Left to themselves these would decay but feedback occurs due to the oscillatory current in L_1 inducing, by transformer action, a voltage of the same frequency in L_2, arranged close to L_1. This voltage is applied to the base of the transistor as the input and is amplified to cause a larger oscillatory current in L_1 and hence a larger voltage in L_2 and so on. The coupling between L_1 and L_2 should be just enough to maintain oscillations.

A single-stage common-emitter amplifier produces a 180° phase shift between its output and input (p. 99). The feedback circuit L_1L_2 must therefore introduce another 180° shift (if it is to be positive) by being connected the 'right way round' to the transistor base.

Tuned oscillators are most suitable for generating radio frequencies. At audio frequencies coils and capacitors to give the large values of L_1 and C_1 required by equation (1) are too bulky. An a.f. sine-wave oscillator will be considered later (p. 126).

When a sine-wave r.f. oscillator with a fixed, very stable frequency is required, use is made of a piezoelectric quartz crystal (p. 44). The crystal behaves like a tuned circuit and a.c. is produced with a frequency which depends on the shape and size of the crystal.

QUESTIONS

1. Calculate the maximum and minimum frequencies of the oscillations which could be generated by a variable capacitor of maximum capacitance 500 pF and minimum capacitance one-tenth of this value connected to a coil of inductance 10 μH. (1 pF = 10^{-12}F)

2. An LC oscillator has a fixed inductance of 50 μH and is required to produce oscillations over the band 1 MHz to 2 MHz. Calculate the maximum and minimum values of the variable capacitor required. (Take $\pi^2 = 10$.)

52 PROGRESS QUESTIONS

1. The circuit for a very simple audio amplifier is shown in Fig. 52.1. State:
a) the type of transistor used,
b) where you would connect a power supply (give polarities),
c) where you would connect a microphone.

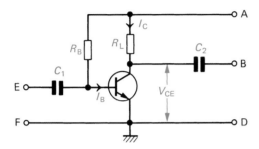

Fig. 52.1

2. Fig. 52.2 shows a load line on the I_C–V_{CE} characteristics of a transistor used in a simple amplifier circuit. Point P represents the steady-state condition of the transistor.

Determine **(a)** the value of the load resistance, **(b)** the power dissipated in the load resistor in the steady state, and **(c)** the power dissipated in the transistor in the steady state.

Is the total power dissipation increased or decreased if the bias is shifted to point Q on Fig. 52.2? Give reasons, or calculations to support your answer. (*O. and C.*)

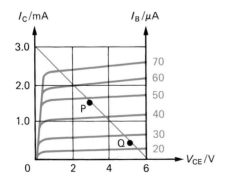

Fig. 52.2

3. The quiescent (no input signal) value of the current through R_s in Fig. 52.3 is 2 mA. What is the voltage with respect to ground (0 V) at
(i) S, (ii) D, (iii) G?

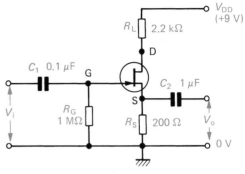

Fig. 52.3

4. Explain what is meant by *negative feedback*.
When negative feedback is used in an amplifier what effect does it have on **(a)** the gain, **(b)** the distortion produced? (C.)

5. a) Draw a circuit with negative feedback. Explain how the feedback is produced and state its effect.
b) Draw a circuit with positive feedback. Explain how the feedback is produced and state its effect. (*L.*)

6. How does the 500 Ω resistor introduce negative feedback into the circuit shown in Fig. 52.4?
What is the gain of the amplifier? (*L.*)

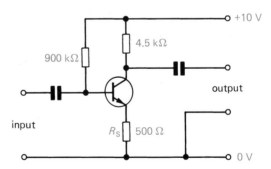

Fig. 52.4

7. a) Transistors are subject to thermal runaway—that is, the collector current causes a slight rise in the temperature which causes greater current to flow and so on. This can lead ultimately to destruction of the transistor. Explain how a low value resistor placed in the emitter lead helps to overcome the problem.
b) In an audio amplifier this resistor is often bypassed with a large value capacitor. Explain the reason for this. (*L.*)

8. An amplifier has the following properties:

open loop voltage gain $= 100$
input impedance $= 1.5 \, k\Omega$
output impedance $= 4 \, \Omega$

Calculate the voltage which would be developed across the load in the circuit shown in Fig. 52.5.

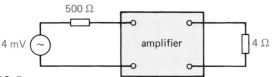

Fig. 52.5

Calculate the power developed in the load and explain the significance of the value of the load shown for this particular amplifier.

By considering the above system, explain the advantage of **(a)** a high input impedance, and **(b)** a low output impedance. (L.)

9. The input resistance of an electronic system is $10 \, k\Omega$. A transducer with a source voltage of 3 V and a source resistance of $2 \, k\Omega$ is connected to the system.
a) Calculate the voltage across the input.
b) Calculate the power transferred to the system.
c) The system is now to be modified. State the new input resistance if the maximum possible power is to be transferred from the transducer to the system. (A.E.B.)

10. a) Copy and correctly complete the characteristics of an emitter follower, using: 50; low; high; 0.95.

Input resistance . . . . . .   Current gain . . . . . .
Output resistance . . . . . .   Voltage gain . . . . . .

b) Explain one use of an emitter follower.

(O.L.E.)

11. If there is no input signal, state the values in Fig. 52.6 of:
a) the voltage relative to ground (0 V) at B,
b) the voltage relative to ground (0 V) at E,
c) the collector current I_C.

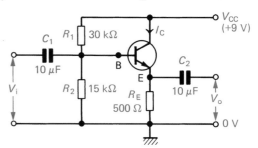

Fig. 52.6

12. The circuit diagram in Fig. 52.7 shows a simple transistor oscillator. Give a full description of how it works, with particular reference to the components labelled A, B, C, D and E.

What extra features are normally added to convert such an oscillator to a *signal generator*? Discuss the uses of the signal generator.

An oscillator has a tuned circuit consisting of a 100 mH inductor and a capacitor. The output frequency is 5 kHz. What order of magnitude of capacitance must be used? (C. *part qn.*)

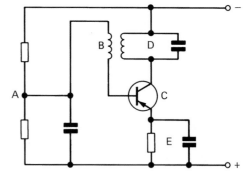

Fig. 52.7

ANALOGUE ELECTRONICS: OP AMPS

53 OPERATIONAL AMPLIFIER

CHECKLIST

After studying this chapter you should be able to:
- state the main properties of an op amp with reference to the open-loop gain A_o, input and output impedances, giving values,
- state that an op amp is a very high gain IC amplifier which has two inputs, called the inverting (−) V_1 and the non-inverting (+) V_2, and one output V_o, and normally operates on a dual power supply,
- recall and use the equation $V_o = A_o(V_2 - V_1)$,
- state that the polarity of a voltage at the (−) input is reversed at the output but not if it is applied at the (+) input,
- draw the transfer characteristic of an op amp,
- state that an op amp has a linear response for a limited range of input voltages, outside which the op amp saturates and the output voltage has a value near the positive or negative of the supply,
- state the advantages of n.f.b. and how it is obtained,
- explain the terms 'input offset voltage' and 'slew rate', and
- recall that because of its high open-loop gain, both inputs of an op amp have almost the same potential, unless it is saturated.

INTRODUCTION

Operational amplifiers (op amps) were originally made from discrete components. They were designed to solve mathematical equations electronically, by performing operations such as addition, etc. in analogue computers (p. 124). Nowadays in IC form they have many uses, one of the most important being as high-gain d.c. and a.c. voltage amplifiers. A typical op amp contains twenty transistors as well as resistors and small capacitors.

Properties The chief properties are:

(i) *a very high open-loop voltage gain* A_o of about 10^5 for d.c. and low frequency a.c., which decreases as the frequency increases;

(ii) *a very high input impedance*, typically 10^6 to 10^{12} Ω, so that the current drawn from the device or circuit supplying it is minute and the input voltage is passed on to the op amp with little loss (p. 108);

(iii) *a very low output impedance*, commonly 100 Ω, which means its output voltage is transferred efficiently to any load greater than a few kilohms.

An *ideal* op amp would have infinite open-loop voltage gain, infinite input impedance and zero output impedance.

Description An op amp has one output and two inputs. The *non-inverting* input is marked + and the *inverting input* is marked −, as shown on the amplifier symbol in Fig. 53.1. Operation is most convenient from a dual balanced d.c. power supply giving equal positive and negative voltages $\pm V_s$, (i.e. $+V_s$, 0 and $-V_s$) in the range ± 5 V to ± 15 V. The centre point of the power supply, i.e. 0 V, is common to the input and output circuits and is taken as their voltage reference level.

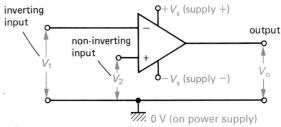

Fig. 53.1

Do not confuse the input signs with those for the power supply polarities, which for clarity, are often left off circuit diagrams.

ACTION AND CHARACTERISTIC

Action If the voltage V_2 applied to the non-inverting $(+)$ input is positive relative to the other input, the output voltage V_o is positive; similarly if V_2 is negative, V_o is negative, i.e. V_2 and V_o are in phase. At the inverting $(-)$ input, a positive voltage V_1 relative to the other input causes a negative output voltage V_o and vice versa, i.e. V_1 and V_o are in antiphase.

Basically an op amp is a *differential* voltage amplifier, i.e. it amplifies the difference between the voltages V_1 and V_2 at its inputs. There are three cases:

  (i) if $V_2 > V_1$, V_o is positive,
 (ii) if $V_2 < V_1$, V_o is negative,
(iii) if $V_2 = V_1$, V_o is zero.

In general, the output V_o is given by:

$$V_o = A_o(V_2 - V_1)$$

where A_o is the open-loop voltage gain.

In most applications, but not all (see p. 121), single-input working is used.

Transfer characteristic A typical voltage characteristic showing how the output V_o in volts (V) varies with the input $(V_2 - V_1)$ in microvolts (μV) is given in Fig. 53.2. It reveals that it is only within the very small input range AOB that the output is directly proportional to the input, i.e. when the op amp behaves more or less *linearly* and there is minimum distortion of the amplifier output. Inputs outside the linear range cause *saturation* and the output is then close to the maximum value it can have, i.e. $+V_s$ or $-V_s$.

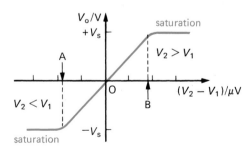

Fig. 53.2

The limited linear behaviour is due to the very high open-loop gain A_o and the higher it is, the greater is the limitation. For example, on a ±9 V supply, since the output voltages can never exceed these values, and have maximum values near either $+9$ V or -9 V, the maximum input voltage swing (for linear amplification) is, if $A_o = 10^5$, ±9 V$/10^5 = \pm90\,\mu$V. A smaller value of A_o would allow a greater input.

Note. The p.d. between the inputs of an op amp is ≈ 0 (unless it is saturated) since the gain is very high. Whatever voltage one input has, the other will also have, more or less.

NEGATIVE FEEDBACK

Op amps almost always use negative feedback (n.f.b.), obtained by feeding back some (or all) of the output to the inverting $(-)$ input. The feedback produces a voltage at the output that opposes the one from which it is taken, thereby reducing the new output of the amplifier, i.e. the one with n.f.b. The resulting closed-loop gain A is then less than the open-loop gain A_o but a wider range of voltages can be applied to the input for amplification. (Feedback applied to the non-inverting $(+)$ input would be positive and would increase the overall output.)

In addition, as stated for discrete component amplifiers (p. 106), n.f.b. gives (provided $A_o \gg A$):

  (i) predictable and constant voltage gain A;
 (ii) reduced distortion of the output;
(iii) better frequency response, i.e. increased bandwidth which depends on the amount of feedback, as shown by the frequency response curve in Fig. 53.3 for a 741 op amp using various values of feedback factor β; for example, when $\beta = 0.1$, the gain is 10 and constant for frequencies up to 10^5 Hz (100 kHz) (note that *gain $\times$ bandwidth* $\approx 10^6$);
 (iv) increased stability.

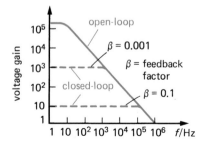

Fig. 53.3

The advantages of using n.f.b. outweigh the accompanying loss of gain which is easily increased by using two or more op amp stages.

DIRECT-COUPLED AMPLIFIERS

In the transistor amplifiers considered before (pp. 98–104), capacitors were used at the input and output of each stage to couple a.c. signals but also to block any d.c. that might upset the operation. However, 'level shifting' circuits have now been developed which omit capacitors and allow direct coupling. Both a.c. and d.c. signals can then be amplified. Op amps are direct-coupled amplifiers.

FURTHER POINTS

Input offset voltage Even when there is no signal on either input of an op amp there may still be a small voltage at the input, called the *differential input offset voltage*. It can arise because of slight differences in internal components. In the 741 op amp it is about 1 mV and in a circuit with a gain of 1000, it would produce 1 V d.c. at the output.

For applications requiring d.c. amplification, having an offset voltage is unacceptable. It is 'removed' as shown in Fig. 53.4 by connecting a 10 kΩ variable resistor to the 'offset null' pins (1 and 5) adjusting it until the output is zero when the input is zero.

For a.c. operation a coupling capacitor at the output removes any d.c. voltage arising from the input offset. Satisfactory amplification then occurs provided the offset has not moved the no-signal output voltage so far from 0 V as to limit the output swing.

Slew rate This is the maximum rate of change of *large amplitude* output voltages that an op amp can allow before it behaves non-linearly. It is measured in volts per microsecond (V/µs) and for the 741 op amp is 0.5 V/µs.

This limitation means that op amps should not be used where signals are changing rapidly as in fast-rising digital pulses (Chapter 33).

The slew rate limit arises because a rapid change of output voltage causes a rapid rate of charge or discharge of capacitors in the op amp. This in turn requires internal transistors to supply large drive currents which they are not designed to do for power-saving reasons.

Op amps with slew rates greater than 10 V/µs are now available.

QUESTIONS

1. Explain the following terms used in connection with op amps:
   a) open-loop gain,
   b) differential input,
   c) saturation output voltage.

2. If A_o for an op amp is 10^5, calculate the maximum input voltage swing that can be applied for linear operation on a ± 15 V supply.

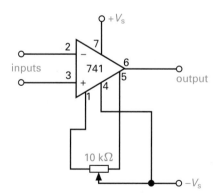

Fig. 53.4

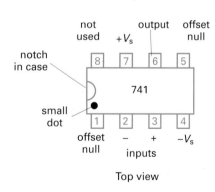

Top view

54 OP AMP VOLTAGE AMPLIFIERS

CHECKLIST

After studying this chapter you should be able to:
- recognize the circuit for an inverting voltage amplifier,
- state and use the relationship for the voltage gain of an inverting amplifier, i.e. $A = -R_f/R_i$,
- recall that a 'virtual earth' point is at or near 0 V though not actually earthed, and arises because the p.ds between the inputs ≈ 0, with one being at 0 V,
- recognize the circuit for a non-inverting voltage amplifier,
- state and use the relationship for the voltage gain of a non-inverting amplifier, i.e. $A = 1 + R_f/R_i$,
- compare the input impedances of inverting and non-inverting amplifiers, and
- recall the properties and use of a voltage-follower.

INVERTING AMPLIFIER

Basic circuit See Fig. 54.1. The input voltage V_i (a.c. or d.c.) to be amplified is applied via resistor R_i to the inverting (−) terminal. The output voltage V_o is therefore in antiphase with the input. The non-inverting (+) terminal is held at 0 V. Negative feedback is provided by R_f, the *feedback resistor*, feeding back a certain fraction of the output voltage to the inverting (−) terminal.

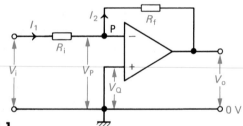

Fig. 54.1

Gain Two assumptions are made to simplify the theoretical derivation of an expression for the gain of the amplifier, i.e. we consider an *ideal op amp*. They are:

(i) each input draws zero current from the signal source (it is less than 0.1 μA for the 741 op amp), i.e. their input impedances are infinite; and

(ii) the inputs are both at the same potential if the op amp is not saturated, i.e. $V_P = V_Q$. (The greater the open-loop gain A_o is, the more true will this be, because as we saw earlier (p. 116), if the supply voltage $V_s = \pm 9$ V and $A_o = 10^5$, then maximum $V_o \approx \pm 9$ V and maximum $V_i \approx \pm 9$ V$/10^5$ = ± 90 μV.)

If A_o is made infinite, we are making the 'infinite gain approximation'.

In the above circuit $V_Q = 0$, therefore $V_P = 0$ and P is called a *virtual earth* (or ground) point, though of course it is not connected to ground. Hence:

$$I_i = (V_i - 0)/R_i \quad \text{and} \quad I_2 = (0 - V_o)/R_f$$

But $I_1 = I_2$ since by assumption (i) the input takes no current, therefore:

$$\frac{V_i}{R_i} = \frac{-V_o}{R_f}$$

The negative sign shows that V_o is negative when V_i is positive and vice versa. The closed-loop gain A is given by:

$$A = \frac{V_o}{V_i} = -\frac{R_f}{R_i} \qquad (1)$$

For example, if $R_f = 100$ kΩ and $R_i = 10$ kΩ, $A = -10$ exactly and an input of 0.1 V will cause an output change of 1.0 V.

Equation (1) shows that the gain of the amplifier depends only on the two resistors (which can be made with precise values) and not on the characteristics of the op amp (which vary from sample to sample).

Input impedance The other versatile feature of this circuit is the way its input impedance can be controlled. Since point P is a virtual earth (i.e. at 0 V), R_i may be considered to be connected between the inverting (−) input terminal and 0 V. The input impedance of the *circuit* is therefore R_i in parallel with the much greater input impedance of the op amp, i.e. effectively R_i, whose value can be changed.

NON-INVERTING AMPLIFIER

Basic circuit See Fig. 54.2. The input voltage V_i (a.c. or d.c.) is applied to the non-inverting (+) terminal of the op amp. This produces an output V_o that is in phase with the input. Negative feedback is obtained by feeding back to the inverting (−) terminal, the fraction of V_o developed across R_i in the voltage divider formed by R_f and R_i across V_o.

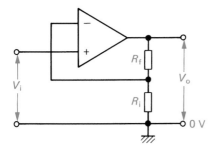

Fig. 54.2

Gain   For the feedback factor β we can write:

$$\beta = \frac{R_i}{R_i + R_f}$$

We saw before (p. 106) that for an amplifier with open-loop gain A_o, the closed-loop voltage gain A is given by:

$$A = \frac{A_o}{1 + \beta A_o}$$

For a typical op amp $A_o = 10^5$, so βA_o is large compared with 1 and we can say $A \approx A_o/\beta A_o = 1/\beta$.
   Hence:

$$A = \frac{V_o}{V_i} = \frac{R_i + R_f}{R_i} = 1 + \frac{R_f}{R_i} \qquad (2)$$

For example, if $R_f = 100\,\text{k}\Omega$ and $R_i = 10\,\text{k}\Omega$, then $A = 110/10 = 11$. As with the inverting amplifier, the gain depends only on the values of R_f and R_i and is independent of the open-loop gain A_o of the op amp.

Input impedance   Since there is no virtual earth at the non-inverting (+) terminal, the input impedance is much higher (typically 50 MΩ) than that of the inverting amplifier. Also it is unaffected if the gain is altered by changing R_f and/or R_i. This circuit gives good matching when the input is supplied by a high impedance source such as a crystal microphone.

VOLTAGE-FOLLOWER

This is a special case of the non-inverting amplifier in which 100% n.f.b. is obtained by connecting the output directly to the inverting (−) terminal, as shown in Fig. 54.3. Thus $R_f = 0$ and R_i is infinite.

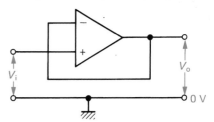

Fig. 54.3

Because all of the output is fed back $\beta = 1$ and since $A = 1/\beta$ when A_o is very large, then $A \approx 1$. The voltage gain is nearly 1 and V_o is the same as V_i to within a few millivolts.
   The circuit is called a *voltage-follower* because, as with its transistor equivalent (the emitter-follower) V_o follows V_i. It has an extremely high input impedance and a low output impedance. Its main use is as a *buffer amplifier*, giving current amplification, to match a high impedance source to a low impedance load. For example, it is used as the input stage of an analogue voltmeter where the highest possible input impedance is required (so as not to disturb the circuit under test) and the output voltage is measured by a relatively low impedance moving-coil meter. In such op amps, FETs replace bipolar transistors in the early stages of the IC.

QUESTIONS

1. In the circuit of Fig. 54.1 the power supply to the op amp is ± 9 V and the input voltage $V_i = +1$ V. What is the value of the output voltage V_o when
   a) $R_f = 20\,\text{k}\Omega$ and $R_i = 10\,\text{k}\Omega$,
   b) $R_f = 200\,\text{k}\Omega$ and $R_i = 10\,\text{k}\Omega$?
2. Repeat question 1 for the circuit of Fig. 54.2.

55 OP AMP SUMMING AMPLIFIER

CHECKLIST

After studying this chapter you should be able to:
- recognize the circuit of a summing amplifier, and
- state and use the equation for a summing amplifier, i.e. $V_o = -(V_1 + V_2 + V_3)R_f/R_i$.

When connected as a multi-input inverting amplifier, an op amp can be used to add a number of voltages (d.c. or a.c.) because of the existence of the virtual earth point. This in turn is a consequence of the high value of the open-loop voltage gain A_o.

Such circuits are employed as 'mixers' in audio applications to combine the outputs of microphones, electric guitars, pick-ups, special effects, etc. They are also used to perform the mathematical process of addition in analogue computing (p. 124).

ACTION

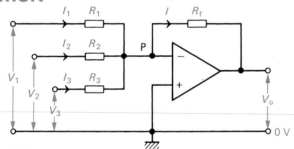

Fig. 55.1

In the circuit of Fig. 55.1 three input voltages V_1, V_2 and V_3 are applied via input resistors R_1, R_2 and R_3 respectively. Assuming that the inverting terminal of the op amp draws no input current, all of it passing through R_f, then:

$$I = I_1 + I_2 + I_3$$

Since P is a virtual earth (i.e. at 0 V), it follows that:

$$\frac{-V_o}{R_f} = \frac{V_1}{R_1} + \frac{V_2}{R_2} + \frac{V_3}{R_3}$$

Therefore:

$$V_o = -\left(\frac{R_f}{R_1} \cdot V_1 + \frac{R_f}{R_2} \cdot V_2 + \frac{R_f}{R_3} \cdot V_3\right)$$

The three input voltages are thus added and amplified if R_f is greater than each of the input resistors, i.e. 'weighted' summation occurs. Alternatively, the input voltages are added and attenuated if R_f is less than every input resistor.

For example, if $R_f/R_1 = 3$, $R_f/R_2 = 1$ and $R_f/R_3 = 2$, and $V_1 = V_2 = V_3 = +1$ V, then addition of 3 V, 1 V and 2 V occurs. That is $V_o = -(3\ \text{V} + 1\ \text{V} + 2\ \text{V}) = -6$ V.

If $R_1 = R_2 = R_3 = R_i$ (say), the input voltages are amplified or attenuated equally, and

$$V_o = -\frac{R_f}{R_i}(V_1 + V_2 + V_3)$$

Further, if $R_i = R_f$, then:

$$V_o = -(V_1 + V_2 + V_3)$$

In this case the output voltage is the sum of the input voltages (but is of opposite polarity).

SUMMING POINT

The virtual earth is also called the *summing point* of the amplifier. It isolates the inputs from one another so that each behaves as if none of the others existed and none feeds any of the other inputs even though all the resistors are connected at the inverting input.

Also the resistors can be selected to produce the best impedance matching with the transducer supplying its input (p. 108), although compromise may be necessary to obtain both the gain required and the correct input impedance.

QUESTION

1. In the circuit of Fig. 55.2 the power supply is ±15 V, $R_f = 30$ kΩ and $R_i = 15$ kΩ. Calculate V_o when
   a) $V_1 = +1$ V and $V_2 = +4$ V,
   b) $V_1 = +1$ V and $V_2 = -4$ V.

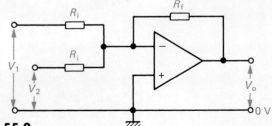

Fig. 55.2

56 OP AMP VOLTAGE COMPARATOR

CHECKLIST

After studying this chapter you should be able to:
- state that as a voltage comparator an op amp is in its saturated state, and
- recognize single-rail and dual-rail supply alarm circuits and explain how they work.

ACTION

If both inputs of an op amp are used simultaneously, then, as we saw earlier (p. 116), the output voltage V_o is given by:

$$V_o = A_o(V_2 - V_1)$$

where V_1 is the inverting ($-$) input, V_2 the non-inverting ($+$) input and A_o the open-loop gain, Fig. 56.1. The voltage difference between the inputs, i.e. ($V_2 - V_1$) is amplified and appears at the output.

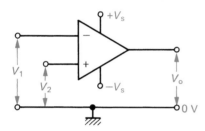

Fig. 56.1

When $V_2 > V_1$, V_o is *positive*, its maximum value being the positive supply voltage $+V_s$, which it has when $(V_2 - V_1) \geqslant V_s/A_o$. The op amp is then saturated. If $V_s = +15$ V and $A_o = 10^5$, saturation occurs when $(V_2 - V_1) \geqslant 15$ V$/10^5$, i.e. when V_2 exceeds V_1 by 150 μV and $V_o \approx 15$ V.

When $V_1 > V_2$, V_o is *negative* and saturation occurs if V_1 exceeds V_2 by V_s/A_o, typically by 150 μV. But in this case $V_o \approx -V_s \approx -15$ V.

A small change in ($V_2 - V_1$) thus causes V_o to switch between near $+V_s$ and near $-V_s$ and enable the op amp to indicate when V_2 is greater or less than V_1, i.e. to act as a *differential amplifier* and compare two voltages. It does this in an electronic digital voltmeter (p. 227).

SOME EXAMPLES

The waveforms in Fig. 56.2 show what happens if V_2 is an alternating voltage (sufficient to saturate the op amp).

In (a), $V_1 = 0$ and $V_o = +V_s$ when $V_2 > V_1$ (positive half-cycle) and $V_o = -V_s$ when $V_2 < V_1$ (negative half-

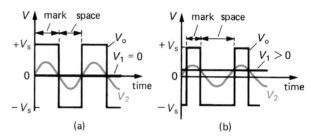

Fig. 56.2

cycle). V_o is a 'square' wave with a mark–space ratio of 1 (p. 16).

In (b), $V_1 > 0$ and switching of V_o occurs when $V_2 \approx V_1$, the mark-to-space ratio now being < 1 and the output a series of 'pulses'.

In effect, the op amp in its saturated condition converts a continuously varying analogue signal (V_2) to a two-state (i.e. 'high'–'low') digital output (V_o).

ALARM CIRCUITS

These were described earlier using a transistor as a switch (Chapter 30). An op amp in its comparator role can also be used and has certain advantages. First, switching is more sensitive (i.e. it occurs for smaller changes in the conditions that set off the alarm) and second, it can be easier to change the reference voltage (i.e. the voltage at which switching occurs).

Two arrangements are considered below.

Single-rail supply In the light-operated alarm circuit of Fig. 56.3, the inputs V_1 and V_2 are supplied by potential dividers (which use just the positive supply rail, hence 'single-rail supply') and the op amp compares them.

In the dark, the resistance of the LDR is much greater than R_1, making V_2 (the voltage across R_1) less than V_1, the reference voltage (set by the ratio of R_2 to R_3). The difference is sufficient to saturate the op amp. Since V_1 (at the inverting input) is positive, V_o will be negative and close to -9 V.

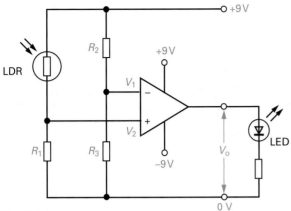

Fig. 56.3

If light falls on the LDR, its resistance decreases and causes V_2 to increase. When V_2 (at the non-inverting input) exceeds V_1, the op amp switches to its other saturated state with V_o being close to $+9$ V. This positive voltage lights the LED, i.e. the alarm.

In effect, in this circuit the op amp changes a continuously varying analogue voltage (V_2) into a two-state digital one (V_o). It is a one-bit *analogue-to-digital* (A/D) converter (Chapter 95).

Dual-rail supply In the temperature-operated alarm circuit of Fig. 56.4, the thermistor TH and the variable resistor R form a potential divider across the ± 15 V supply (i.e. across both positive and negative supply rails, hence 'dual-rail supply').

The op amp compares the voltage V_1 at the potential divider junction X with V_2, which is 0 V.

If the temperature of TH falls, when it is low enough its resistance becomes greater than that of R. More of the 30 V across the potential divider is then dropped across

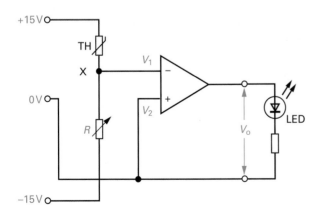

Fig. 56.4

TH and the voltage at X goes negative, i.e. less than 0 V. Therefore $V_2 > V_1$, making V_o switch from near -15 V to near $+15$ V and lighting the LED. The temperature at which this occurs can be varied by altering R.

QUESTIONS

1. In Fig. 56.5, V_1 and V_2 are the p.ds at the $-$ and $+$ inputs respectively of an op amp on a ± 6 V supply. Copy the graphs and draw on the same axes the output p.d. V_o.

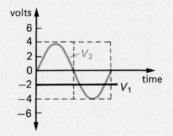

Fig. 56.5

2. An op amp voltage comparator is shown in Fig. 56.6 working on a balanced dual d.c. supply of ± 6 V.

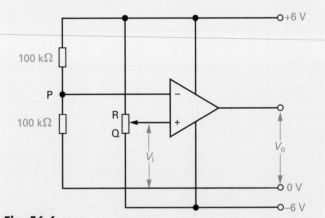

Fig. 56.6

a) What is the voltage at point P?
b) On axes marked in voltages, draw a graph of V_o against V_i as the slider on the voltage divider moves from Q to R.
c) V_o is to be used to drive a relay via a MOSFET buffer amplifier. Draw a suitable circuit to do this.
d) State *two* advantages of MOSFETs over bipolar transistors.

57 OP AMP INTEGRATOR

CHECKLIST

After studying this chapter you should be able to:
- recognize the op amp integrator circuit,
- state that the slope of the ramp output voltage $V_o = V_i t/(CR)$ and use it in calculations, and
- recall how to set the output to 0 V initially.

ACTION

An op amp integrator is at the heart of a 'ramp generator' for producing a voltage with a sawtooth waveform. The circuit is the same as for the op amp inverting amplifier (Fig. 54.1) but feedback occurs via a capacitor C, as in Fig. 57.1, rather than via a resistor. If the input voltage V_i, applied through the input resistor R, is *constant*, and $CR = 1$ s (e.g. $C = 1$ μF, $R = 1$ MΩ), the output voltage V_o after time t (in seconds) is given by:

$$V_o = -V_i t$$

The negative sign is inserted because, when the inverting input is used, V_o is negative if V_i is positive and vice versa.

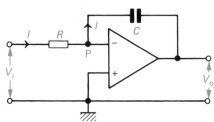

Fig. 57.1

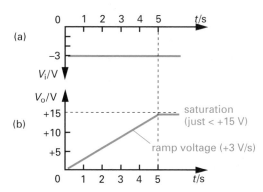

Fig. 57.2

For example, if $V_i = -3$ V, Fig. 57.2a, V_o *rises steadily* by $+3$ V/s and, if the power supply is ± 15 V, V_o reaches about $+15$ V after 5 s, when the op amp saturates,

Fig. 57.2b. V_i is thus 'added up' or integrated over a time t to give V_o a *ramp voltage* waveform whose slope will be shown to be proportional to V_i.

The integrator is set initially to $V_o = 0$ V by shorting out C. To set it to any other initial voltage a battery in series with a switch is connected in parallel with C.

The integrating action of the circuit on V_i is similar to that of a petrol filling station pump which 'operates' on the rate of flow (in litres per second) and the delivery time (in seconds) to produce as its output, the volume of petrol supplied (in litres).

THEORY

Since P is a virtual earth in Fig. 57.1, i.e. at 0 V, the voltage across R is V_i and that across C is V_o. Assuming (as before) that none of the input current I enters the op amp inverting input, then all of I flows 'through' C and charges it up. If V_i is constant, I will be constant, with a value:

$$I = \frac{V_i}{R} \qquad (1)$$

C therefore charges at a *constant rate* and the potential of the output side of C (which is V_o since its input side is at 0 V) changes so that the feedback path absorbs I. If Q is the charge on C at time t and the p.d. across it (i.e. the output voltage) changes from 0 to V_o in that time then, since I is constant, we have:

$$Q = -V_o C = It \quad \text{(pp. 23 and 3)}$$

Substituting for I from (1), we get:

$$-V_o C = \frac{V_i}{R} \cdot t$$

That is,

$$V_o = -\frac{1}{CR} \cdot V_i t$$

If $CR = 1$ s, then:

A more general mathematical treatment shows that if V_i varies then, using calculus notation,

$$V_o = -\frac{1}{CR}\int V_i \, dt$$

where $\int V_i \, dt$ is the 'integral of V_i with respect to time t'.

RESETTING

To reset V_o to zero when it reaches its maximum value (due to the op amp saturating), a transistor is connected across C as in Fig. 57.3a.

When the transistor receives a reset pulse it switches on, short-circuits C and causes V_o to fall rapidly to 0 V.

After each reset pulse, the transistor switches off, allowing V_o to rise again and produce a sawtooth waveform voltage like that in Fig. 57.3b.

A ramp generator is used in a digital voltmeter, as shown in Fig. 95.3, p. 227.

QUESTION

1. In the circuit of Fig. 57.1 (p. 123), if $CR = 1$ s and $V_i = +6$ V,
 a) does V_o rise or fall and at what rate,
 b) if the op amp power supply is ± 18 V, what is the maximum value of V_o and when is it reached?

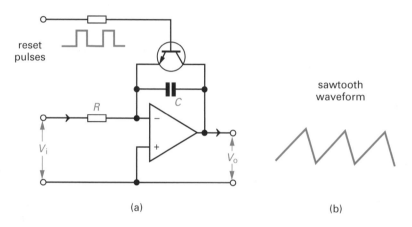

Fig. 57.3
(a)

sawtooth waveform

(b)

58 ANALOGUE COMPUTING

CHECKLIST

After studying this chapter you should be able to:
• state what an analogue computer consists of, and
• state that it solves differential equations.

ELECTRONIC ANALOGUES

Analogue computers are used as electronic models or analogues of mechanical and other systems in cases where conducting experiments on the system itself would be costly or time-consuming or dangerous, or where the system has not yet been built. For example, when designing a bridge or an aircraft wing, information is required before construction starts about how it will react to variable factors like wind speed and temperature.

Such predictions involve solving *differential equations* which are often too complex for ordinary mathematical methods. An equation containing first-order differential

coefficients, e.g. dy/dt (pronounced 'dee-y-by-dee-t') has to be integrated once to obtain a solution, i.e. an expression for y in terms of t. One with second-order differential coefficients, e.g. d²y/dt² (pronounced 'dee-two-y-by-dee-t-squared'), requires two integrations.

An analogue computer, consisting of the appropriate number of op amp integrators, will provide the solution if it is set up with voltages representing differential coefficients of the quantities under investigation. (In recent years analogue computers have been less popular because digital computers can now be programmed to simulate moving physical systems.)

SOLVING A DIFFERENTIAL EQUATION

Consider the very simple second-order differential equation (normally solved mathematically), of

$$\frac{d^2y}{dt^2} = 10$$

We wish to obtain an expression for y in terms of t. Integration can be regarded as the reverse process of differentiation (for both of which there are mathematical rules). For example, the integral of dy/dt is y, that is,

$$\int \frac{dy}{dt} dt = y$$

and the integral of d^2y/dt^2 is dy/dt.

Therefore, if a constant d.c. voltage representing d^2y/dt^2 is supplied as the input to two op amp integrators in series (having $CR = 1$ s for each stage), as in Fig. 58.1, integration (and inversion) occurs twice and the output voltage will represent y in terms of t.

If the output is displayed on an oscilloscope, the graph produced on the screen by the spot shows how y varies with t. In fact, a solution is $y = 5t^2$. Knowing the scales of the voltage and time axes, numerical answers can be obtained.

The equation $d^2y/dt^2 = 10$ represents the motion of an object falling from rest, neglecting air resistance, y being the distance fallen (metres) in time t (seconds). Equations for systems such as car suspensions, spaceship travel and automatic pilots for aircraft and ships are much more complex but they are solved in basically the same way.

1. When and why are analogue computers useful as electronic models of physical systems?

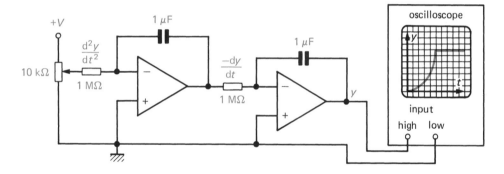

Fig. 58.1

59 MORE OP AMP CIRCUITS

CHECKLIST

After studying this chapter you should be able to:
• state what a relaxation oscillator does, and

• recognize and explain the action of an astable multivibrator, a Wien oscillator, a perfect half-wave rectifier, a sample-and-hold circuit and a constant current generator.

ASTABLE MULTIVIBRATOR

An op amp voltage comparator (p. 121) in its saturated condition can operate as a *relaxation oscillator* if suitable external components are connected and positive feedback used.

A relaxation oscillator produces a non-sine-wave output voltage which, in two important cases, have sawtooth and square waveforms. The former are used as time bases on oscilloscopes and television receivers. The latter are used to produce electronic music since they are rich in

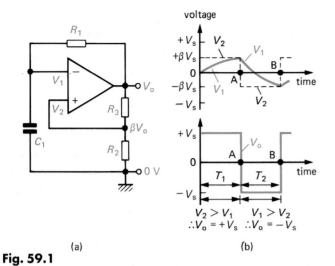

Fig. 59.1

(a) (b)

$V_2 > V_1$ $V_1 > V_2$
$\therefore V_o = +V_s$ $\therefore V_o = -V_s$

harmonics (Chapter 7) and are useful for testing a.f. amplifiers. The action of a relaxation oscillator depends on the charging of a capacitor followed by a period of 'relaxation' when the capacitor discharges through a resistor.

The circuit and various voltage waveforms are shown in Fig. 59.1a, b. Suppose the output voltage V_o is positive at a particular time. A certain fraction β of V_o is fed back as the non-inverting input voltage V_2 which equals βV_o where $\beta = R_2/(R_2 + R_3)$. The output voltage V_o is also fed back via R_1 to the inverting terminal and V_1 rises (exponentially) as C_1 is charged.

After a time which depends on the time constant $C_1 R_1$, V_1 exceeds V_2 and the op amp switches into negative saturation, i.e. $V_o = -V_s$ (point A on graphs). Also, the positive feedback makes V_2 go negative (where $V_2 = \beta V_o = -\beta V_s$).

C_1 now starts to charge up in the opposite direction, making V_1 fall rapidly (during time AB on the graphs) and eventually become more negative than V_2. The op amp therefore switches again to its positive saturated state with $V_o = +V_s$ (point B on graphs). This action continues indefinitely at frequency f given by $f = 1/(T_1 + T_2)$.

The output voltage V_o provides *square* waves and the voltage across C_1 is a source of *triangular* waves. The latter have 'exponential' sides, but they are 'straightened' by extra circuitry which ensures C_1 is charged by a constant current, rather than by the exponential one supplied through R_1.

The term 'astable multivibrator' arises because (i) the circuit has no stable states and switches automatically from a 'high' to a 'low' output (hence 'astable'), and (ii) the output is a square wave which consists of many sine-wave frequencies (p. 17), (hence 'multivibrator').

WIEN OSCILLATOR

Sine waves in the audio frequency range can be generated by an op amp using a Wien network circuit containing resistors and capacitors.

We saw earlier (Fig. 10.11a, p. 26), that if a.c. is applied to a resistor and capacitor in series, the voltages developed across them are 90° out of phase. In the Wien circuit, a network of two resistors R_1, R_2 (usually equal) and two capacitors C_1, C_2 (also usually equal), arranged as in Fig. 59.2, acts as the positive feedback circuit. The network is an a.c. voltage divider and theory shows that the output voltage V_o is *in phase* with the input voltage V_i, i.e. the phase shift is zero, at one frequency f, given in Hz by:

$$f = \frac{1}{2\pi RC}$$

where R is in ohms and C in farads. At all other frequencies, there is a phase shift.

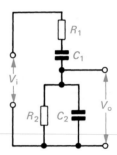

Fig. 59.2

To obtain oscillations the network must therefore be used with a non-inverting amplifier which gives an output to the Wien network that is in phase with its input.

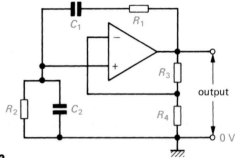

Fig. 59.3

In the circuit of Fig. 59.3, the frequency-selective Wien network, R_1C_1, R_2C_2, applies positive feedback to the non-inverting (+) input. Negative feedback is supplied via R_3 and R_4 to the inverting (−) input. It can be proved that so long as the voltage gain of the amplifier exceeds three, oscillations will be maintained at the desired frequency f.

A variable-frequency output, as in an a.f. signal generator (p. 95), can be obtained if R_1 and R_2 are variable, 'ganged' resistors (i.e. mounted on the same spindle so that they can be altered simultaneously by one control). Also C_1 and C_2 would be pairs of capacitors that are switched in for different frequency ranges.

PERFECT HALF-WAVE RECTIFIER

A way of measuring an a.c. voltage is to first rectify it to d.c. using a silicon diode (Chapter 35). However, inaccuracies occur with low voltages due to the diode not starting to conduct until the forward voltage across it is above 0.5 V or so (see Fig. 26.4, p. 58). Hence a.c. voltages below that value do not produce a rectified output.

The problem can be solved by having a diode D in the feedback loop of an op amp as in Fig. 59.4. This effectively divides the diode's forward voltage by the open-loop gain of the amplifier (e.g. 10^3) and allows D to create d.c. voltages from input voltages of the order of 1 mV.

A perfect full-wave rectifier can be made by including another op amp to use the other half-cycles of the a.c. input.

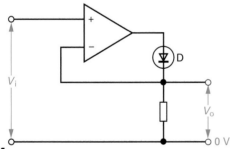

Fig. 59.4

SAMPLE-AND-HOLD CIRCUIT

Increasingly, analogue voltages are having to be changed into digital form. In one type of analogue-to-digital (A/D) converter (used in digital voltmeters and described in Chapter 95), an op amp comparator plays a central role. The analogue voltage to be sampled and converted supplies one input to the op amp and a *sample-and-hold circuit* is used to hold this voltage at a steady level while the conversion is occurring. Otherwise, the A/D converter is trying to convert a continually changing signal.

In the very basic sample-and-hold circuit shown in Fig. 59.5, *closing* S causes C to charge up rapidly until the voltage across it is the analogue input voltage. When S is *opened*, the voltage held by C is the amplitude of the analogue signal at the moment of opening S. This is replicated by the op amp voltage follower (acting as a buffer) at its output and is the voltage to be converted. In practice S would be an electronic switch, often a FET.

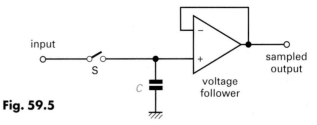

Fig. 59.5

CONSTANT CURRENT GENERATOR

The transistor version was described earlier (Chapter 37). A circuit using an op amp is shown in Fig. 59.6. As before it is based on the fact that the transistor collector current I_C, which charges the battery B, is constant for a wide range of collector voltages. The Zener diode Z and its safety resistor R_Z, along with the feedback loop to the inverting input of the op amp, ensure the p.d. across B remains steady and more or less equal to the Zener voltage V_Z, since the p.d. between the two inputs of the op amp is approximately 0 V. The charging current I_C is given by V_Z/R_E. The transistor is connected as an emitter-follower and acts as a high-gain current amplifier (Chapter 50).

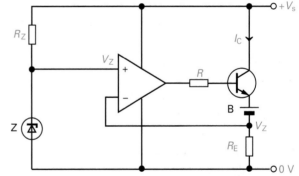

Fig. 59.6

WAVEFORM GENERATORS

Integrated circuit a.f. oscillators, based on the principle outlined above, and called *waveform generators* (e.g. 8038) are now available with outputs having sine, square, triangular, sawtooth and pulse waveforms. The frequency, which is stable over a wide range of temperatures and supply voltages, can be set externally in the range 0.001 Hz to 1 MHz.

60 PROGRESS QUESTIONS

1. For the op amp circuit in Fig. 60.1, write down the equation which connects the input voltage V_i and the output voltage V_o with resistors R_1 and R_2.

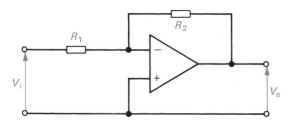

Fig. 60.1

a) If $R_1 = 2.0\,k\Omega$ and $R_2 = 20\,k\Omega$ and the power supply to the op amp is $\pm 9\,V$, calculate V_o when V_i is (i) $+0.3\,V$, (ii) $+1.0\,V$.

b) To show how the op amp circuit in Fig. 60.2 could be used to amplify an input voltage of $+0.2\,V$ to $+3.2\,V$, copy and complete the diagram. Calculate the value of any extra component required.

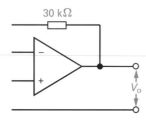

Fig. 60.2

2. The input voltage to the circuit shown in Fig. 60.3 is a sine wave with a peak-to-peak value of 1 V. Using the same set of axes, sketch the voltage waveforms at J, K and L. (You may assume that the open-loop gain of the amplifier is very large.)

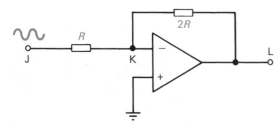

Fig. 60.3

What differences would it make to your graphs if the open-loop gain were somewhat lower, e.g. 1000?

(O. and C.)

3. In the operational amplifier circuit of Fig. 60.4 the power supply voltage is $\pm 15\,V$ and input voltages are applied as shown. Calculate the value of V_o under these conditions. (A.E.B.)

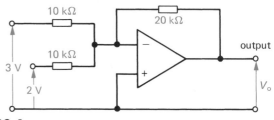

Fig. 60.4

4. What is a differential amplifier?

Explain carefully its main features and how it differs from a simple amplifier.

Given two signal sources A and B, how would you set up a circuit to create (i) $B - A$, (ii) $B + 2A$? (O. and C.)

5. a) Draw a diagram of an inverting amplifier with a gain of 50 designed using an operational amplifier (op amp). (You need not show the power supply connections on your diagram.)

b) Draw a diagram of a non-inverting amplifier with a gain of 50 designed using an op amp. (You need not show the power supply connections on your diagram.)

c) Draw a sketch graph, with labelled axes, of the output voltage against the input voltage for the circuit of Fig. 60.5. (Consider input voltages between $-3\,V$ and $+3\,V$.)

Suggest a use for such a circuit. (L.)

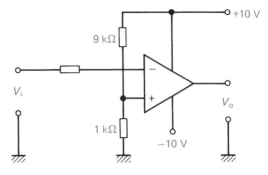

Fig. 60.5

6. This is a question about operational amplifiers and their use.

Explain what is meant by **(a)** feedback, **(b)** virtual earth, **(c)** infinite gain approximation.

Draw a circuit which gives the difference between two input voltages as its output. How is the circuit modified to give an output K times the difference of the inputs, where K is a constant lying between 1 and 100?

Draw another circuit which gives the integral of an input voltage with respect to time. Account for the output voltage which results when a d.c. voltage is applied at the input to this circuit.   (*O. and C.*)

7. A candidate for this examination wished to build a circuit for his project which would operate a relay when the sound level in a room exceeded a particular value. He devised the circuit in Fig. 60.6.

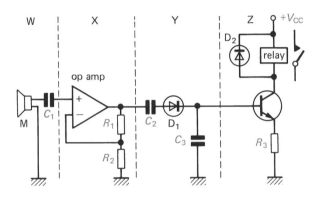

Fig. 60.6

Identify the functions of the sections W, X, Y and Z of the circuit shown.

Explain the purpose of each of the components in section Y.

Explain the purpose of each of the components in section Z.

Explain how a sound above a certain level will cause the relay to operate.

How could the sensitivity of the system be adjusted?   (*L.*)

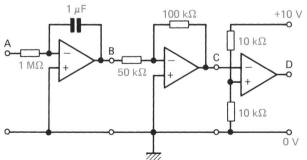

Fig. 60.7

8. The three op amp circuits in Fig. 60.7 are connected as shown. They operate on a ± 10 V power supply.

On the same set of axes draw three graphs to show how the voltages at B, C and D vary with time when $+1$ V is applied to A. Mark clearly the voltage and time scales used.

9. The light-operated switch in Fig. 60.8 uses an op amp as a voltage comparator.
a) How must V_1 and V_2 compare if the op amp output is to be negative in daylight?
b) In darkness what happens to   (i) the LDR, (ii) V_2 compared with V_1,   (iii) the output of the op amp,   (iv) Tr,   (v) the relay?
c) How would you alter the circuit to make the relay be off in the dark and switch on in daylight?

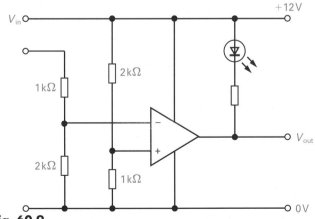

Fig. 60.8

10. a)   Calculate the voltage at the non-inverting input terminal of the operational amplifier in the circuit shown in Fig. 60.9.
b) When the input voltage V_{in} is 0 V, what will be the output voltage V_{out}?
c) Describe what will happen to the LED as V_{in} is increased from 0 to 12 V.   (*N.*)

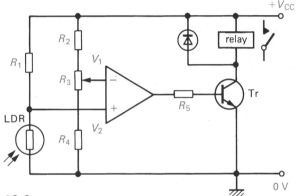

Fig. 60.9

DIGITAL ELECTRONICS: LOGIC CIRCUITS

61 LOGIC GATES

CHECKLIST

After studying this chapter you should be able to:
- state the features of a digital signal,
- recall the truth tables of NOT, NOR, OR, NAND, AND, excl-OR and excl-NOR gates and draw their symbols, and
- recall how other gates are made from NAND gates.

DIGITAL ELECTRONICS

Digital electronics is concerned with two-state, *switching-type* circuits in which signals are in the form of electrical *pulses*, Fig. 61.1. Their outputs and inputs involve only *two* levels of voltage, referred to as 'high' and 'low'. 'High' is near the supply voltage, e.g. +5 V and 'low' is near 0 V and, in the scheme known as 'positive logic', 'high' is referred to as *logic level 1* (or just 1) and 'low' as *logic level 0* (or just 0).

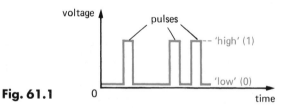

Fig. 61.1

Digital circuits are used in pocket calculators (p. 172), in electronic watches (p. 172), in digital computers (p. 216), in control systems (e.g. for domestic appliances such as washing machines, or for robots or production processes in industry), in television and computer games, and in telecommunications (p. 176).

Today, digital circuits are made as integrated circuits (ICs, p. 76) and are more complex than the discrete component versions from which they developed.

TYPES OF LOGIC GATE

The various kinds of logic gate are switching circuits that 'open' and give a 'high' output depending on the combination of signals at their inputs, of which there is usually

more than one. They are decision-making circuits using what is known as *combinational logic*. The behaviour of each kind is summed up in a *truth table* showing in terms of 1s ('high') and 0s ('low') what the output is (1 or 0) for all possible inputs.

NOT gate or inverter This has one input and one output and the simple transistor switch of Fig. 30.1, p. 68, behaves as one. The circuit is shown again in Fig. 61.2 along with its symbol and the truth table. It produces a 'high' output (e.g. +6 V) if the input is 'low' (e.g. 0 V) and vice versa. That is, the output is 'high' if the input is *not* 'high' and whatever the input, the gate 'inverts' it.

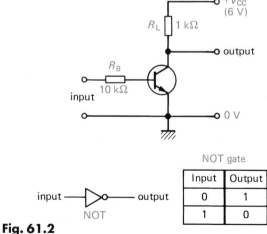

Input	Output
0	1
1	0

Fig. 61.2

The circuit of Fig. 61.2 contains only resistors and a transistor and is said to use 'resistor-transistor logic', or RTL.

NOR gate   This is a NOT gate with two or more inputs. The RTL circuit for two inputs A and B is shown in Fig. 61.3, with symbol and truth table. If either or both of A and B are 'high', the transistor is switched on and saturates and the output F is 'low'. When both inputs are 'low', i.e. neither A *nor* B is 'high', F is 'high'.

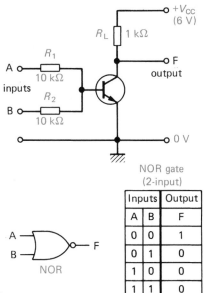

NOR gate (2-input)

Inputs		Output
A	B	F
0	0	1
0	1	0
1	0	0
1	1	0

Fig. 61.3

OR gate   This is a NOR gate followed by a NOT gate; it is shown in symbol form with two inputs in Fig. 61.4. The truth table can be worked out from that of the NOR gate by changing 0s to 1s and 1s to 0s in the output. F is 1 when either A *or* B *or* both is a 1, i.e. if any of the inputs is 'high', the output is 'high'.

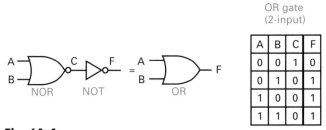

OR gate (2-input)

A	B	C	F
0	0	1	0
0	1	0	1
1	0	0	1
1	1	0	1

Fig. 61.4

NAND gate   A two-input NAND gate is shown in symbol form in Fig. 61.5, along with its truth table.

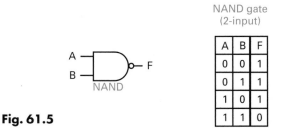

NAND gate (2-input)

A	B	F
0	0	1
0	1	1
1	0	1
1	1	0

Fig. 61.5

When inputs A *and* B are both 'high', the output is *not* 'high'. (The gate gets its name from the AND NOT behaviour.) Any other input combination gives a 'high' output.

AND gate   This is a NAND gate followed by a NOT gate; it is shown with two inputs in symbol form in Fig. 61.6. The truth table may be obtained from that of the NAND gate by changing 0s to 1s and 1s to 0s in the output. The output is 1 only when A *and* B are 1.

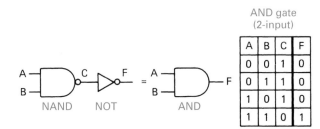

AND gate (2-input)

A	B	C	F
0	0	1	0
0	1	1	0
1	0	1	0
1	1	0	1

Fig. 61.6

Exclusive-OR (excl-OR or X-OR) gate   This is an OR gate with only two inputs which gives a 'high' output when either input is 'high' but not when both are 'high'. Unlike the ordinary OR gate (sometimes called the *inclusive*-OR gate) it excludes the case of both inputs being 'high' for a 'high' output. It is also called the *difference* gate because the output is 'high' when the inputs are different.

The symbol and truth table are shown in Fig. 61.7.

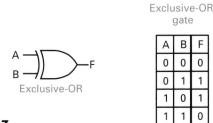

Exclusive-OR gate

A	B	F
0	0	0
0	1	1
1	0	1
1	1	0

Fig. 61.7

Exclusive-NOR (excl-NOR or X-NOR) gate   This is a NOR gate with only two inputs which gives a 'high' output when the inputs are equal, i.e. are both 0's or both 1's, Fig. 61.8. For this reason it is also called a *parity* or *equivalence* gate.

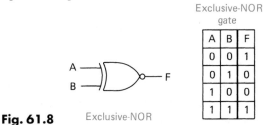

Exclusive-NOR gate

A	B	F
0	0	1
0	1	0
1	0	0
1	1	1

Fig. 61.8

Summary Try to remember the following:

> NOT: output always *opposite* of input
> NOR: output 1 *only* when all inputs 0
> OR: output 1 *unless* all inputs 0
> NAND: output 1 *unless* all inputs 1
> AND: output 1 *only* when all inputs 1
> excl-OR: output 1 when inputs *different*
> excl-NOR: output 1 when inputs *equal*

OTHER GATES FROM NAND GATES

Other logic gates (and circuits) can be made by combining only NAND gates (or only NOR gates), a fact which is often used in logic circuit design, as we will see later (p. 140).

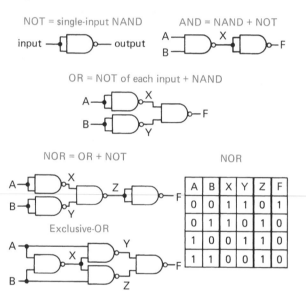

Fig. 61.9

The NAND gate equivalents for different gates are shown in Fig. 61.9. You can check that each one gives the correct output for the various input combinations by constructing a stage-by-stage truth table, as has been done for the NOR gate made from four NAND gates. Note:

> **(i)** NOT = single-input NAND (or NOR), made by joining all the inputs together,
> **(ii)** AND = NAND followed by NOT,
> **(iii)** OR = NOT of each input followed by NAND,
> **(iv)** NOR = OR followed by NOT.

NAND GATES BOARD

There are four CMOS two-input NAND gates on this board, Fig. 61.10, which is useful for constructing different types of logic gate and investigating their properties.

Fig. 61.10

1. The electrical circuits in Fig. 61.11 can be regarded as logic gates. Which type is each one? (Treat A and B as inputs and F as the output.)

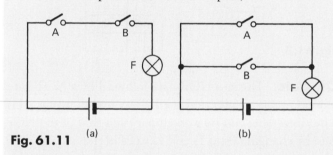

Fig. 61.11 (a)   (b)

2. Write down the truth table for the circuit given in symbols in Fig. 61.12.

Fig. 61.12

3. Voltages with the waveforms in Fig. 61.13 are applied one to each input of a two-input
(i) AND gate, (ii) NOR gate.
Draw the output waveform for each.

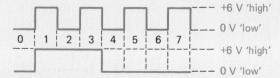

Fig. 61.13

4. Why do single-input NAND or NOR gates act as NOT gates? (Refer to their truth tables if necessary.)

62 LOGIC FAMILIES

CHECKLIST

After studying this chapter you should be able to:
- recall and compare the properties of TTL and CMOS ICs,
- recall the logic levels for TTL and CMOS ICs, and
- describe the interfacing problems arising from mixing TTL and CMOS.

Different types of circuitry or logic can be used to construct logic gates. RTL circuits (p. 130) were the earliest but are now obsolete. The IC logic gates available today fall into two main groups or families and are based on two transistors to avoid problems that arose with the first simple gates.

One family is TTL (standing for transistor-transistor logic) using bipolar transistors; the other is CMOS (standing for complementary metal oxide semiconductor and pronounced 'see-moss') based on FETs.

TTL ICs are made as the '74' series; the figures following '74' indicate the nature of the IC, e.g. the kind of logic gate. CMOS ICs belong to the '4000 B' series; the numbers after the '4' depend on what the IC does. Both types are usually in the form of 14- or 16-pin d.i.l. packages (Fig. 33.2, p. 76), often with several gates on the same chip.

For example, a quad two-input NAND gate contains four identical NAND gates, with two inputs and one output per gate. Every gate therefore has three pins, making 14 pins altogether on the package. This includes two for the positive and negative power supply connnections which are common to all four gates. Pin connections for the 7400 and 4011B quad two-input NAND gates are given in Fig. 62.1.

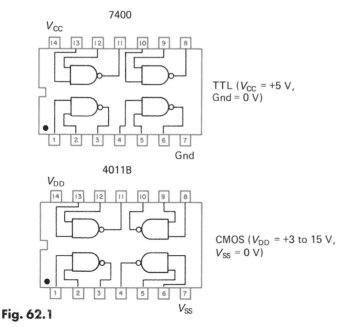

Fig. 62.1

COMPARISON OF TTL AND CMOS ICs

Properties of each family are summarized in the table and explained below.

Property	TTL	CMOS
Power supply	5 V ± 0.25 V d.c.	3 V to 15 V d.c.
Current required	Milliamperes	Microamperes
Input impedance	Low	Very high
Switching speed	Fast	Slow
Fan-out	Ten	Fifty

Power supply TTL requires a stabilized 5 V supply while CMOS will work on any unstabilized d.c. voltage between 3 V and 15 V.

Current requirements In general these are much less (roughly one thousand times) for CMOS than for TTL.

Input impedance The very high input impedance of a CMOS IC (due to the use of FETs) accounts for its low current consumption but it can allow static electric charges to build up on input pins when, for example, they touch insulators, e.g. plastics, clothes, in warm dry conditions. This does not happen with TTL ICs because the low input impedance of bipolar transistors ensures that any charge leaks harmlessly through junctions in the IC.

The static voltages on CMOS ICs can destroy them and so they are supplied in antistatic or conductive carriers. Also, protective circuits are incorporated which operate when the IC is connected to the power supply.

Switching speed This is much faster for TTL than for CMOS, the switching delay time for TTL being about 10 nanoseconds (1 ns = 10^{-9} s) compared with 300 ns for CMOS.

Fan-out For TTL this is ten, which means ten other TTL logic gates can take their inputs from one TTL output and switch reliably, otherwise overloading occurs. For CMOS a fan-out of about fifty is typical for one CMOS IC driving other CMOS ICs (due to their very high input impedance).

Unused inputs Unused TTL inputs assume logic level 1 unless connected to 0 V; it is good practice to connect them to +5 V. Unused CMOS inputs must be connected to supply positive or negative, depending on the circuit and not left to 'float', otherwise erratic behaviour occurs.

LOGIC LEVELS AND INTERFACING

Logic levels Ideally the 'low' and 'high' input and output voltages representing logic levels 0 and 1 respectively should be 0 V and V_{CC} (i.e. +5 V) for TTL. In practice, due to voltage drops across resistors and transistors inside the IC, these values are not achieved.

For example, an input voltage of from 0 to 0.8 V behaves as 'low', while one from 2 V to 5 V is 'high' because it can cause an output to change state. Similarly a 'low' output may be 0 to 0.4 V (since the voltage across a saturated transistor is not 0 V) and 'high' is from 2.4 V to 5 V since any voltage in this range can operate other TTL ICs, Fig. 62.2a.

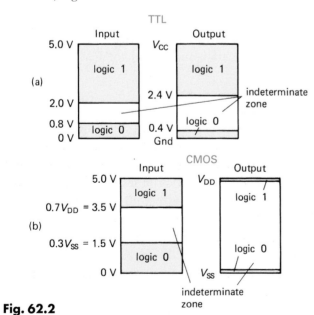

Fig. 62.2

The values for CMOS are given in Fig. 62.2b. The ideal output values are more or less attained in this case and so the output voltage (when unloaded) can swing from zero to near V_{DD}, which is greater than for TTL.

With both families intermediate voltage values (produced for example by overloading outputs) cause faulty switching.

Mixing TTL and CMOS If operated from a common 5 V supply, the logic levels for TTL and CMOS ICs would be different, i.e. the two families are not compatible, even though a TTL IC has the current capability to drive any number of CMOS ICs. For example, the worst-case value of a TTL 'high' output (2.4 V) is lower than the minimum input required by CMOS (3.5 V) to ensure switching. Hence, using TTL and CMOS ICs in the same circuit (to obtain the advantages of each) creates 'interfacing' problems which fortunately can be solved.

NEWER LOGIC FAMILIES

The original TTL and CMOS ranges of ICs have been augmented by newer versions.

The *74LS series* is a *low power* version of the 74 series operating on a 5 V supply. It uses the same identification system as the 74 series. For example, 74LS00 is a quad two-input NAND gate like the 7400.

The *74HC series* is a *high speed* version of the 4000B CMOS series but operates on a 2 to 6 V supply. The same pin connections and numbering system are used as in the equivalent 74LS ICs. For example, the 74HC00 is also a quad two-input NAND gate like the 74LS00.

Eventually the 74HC range, since it combines the advantages of both TTL and CMOS (high switching speed and low current), may replace the other ranges. Its one disadvantage is, being CMOS, it can be damaged by static electricity.

QUESTION

1. a) Possible circuits for a logic level indicator are given in Fig. 62.3(i), (ii), (iii). In which circuit(s) will a logic level 1 (i.e. +5 V) applied to the input light up the LED?

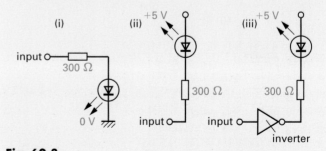

Fig. 62.3

b) The table below, for typical TTL input and output sourcing and sinking currents on a 5 V supply, shows that a TTL gate can sink much more current than it can source (e.g. for an output, 16 mA compared with 400 μA).

TTL	logic 1 (source)	logic 0 (sink)
output	400 μA	16 mA
input	40 μA	1.6 mA

If when checking TTL logic levels we wish to make a logic 1 light a LED (needing typically 10 mA) *and* to avoid overloading a logic 1 output connected to the indicator input (so preventing the logic 1 output voltage from falling into the indeterminate region between 2.4 V and 0.4 V in Fig. 62.2a and producing faulty switching), which circuit should be used for both conditions to be satisfied?

c) From the table shown, work out how many other TTL gate inputs can be driven by one TTL gate output.

d) The output of a CMOS device can sink or source about 10 mA on a 9 V supply. Which circuit(s) in Fig. 62.3(i), (ii), (iii) would satisfy the two requirements in **b)** if a 9 V supply is used and a 680 Ω resistor replaced the 300 Ω one?

63 BINARY ADDERS

CHECKLIST

After studying this chapter you should be able to:
- make code conversions from decimal to binary, hexadecimal, octal and vice versa,
- state what a half-adder does, recognize its circuit and construct a truth table for it,
- state what a full-adder does, recognize its circuit and construct a truth table for it, and
- perform binary subtraction by the 'one's complement method'.

CODES

Normally we count on the scale of ten or decimal system using the ten digits 0 to 9. When the count exceeds 9, we place a 1 in a second column to the left of the units column to represent tens. A third column to the left of the tens column gives hundreds and so on. The values of successive columns starting from the right are 1, 10, 100, etc., or in powers of ten, 10^0, 10^1, 10^2, etc.

Binary code Counting in electronic systems is done by digital circuits on the scale of two or *binary* system. The digits 0 and 1 (called 'bits', from *binary* di*gits*) are used and are represented electrically by 'low' and 'high' voltages respectively. Many more columns are required since the number after 1 in binary is 10, i.e. 2 in decimal but the digit 2 is not used in the binary system.

Successive columns in this case represent, from the right, powers of 2, i.e. 2^0, 2^1, 2^2, 2^3, etc., or in decimal 1, 2, 4, 8, etc. The table below shows how the decimal numbers from 0 to 15 are coded in the binary system; a four-bit code is required. Note that the *least significant bit* (l.s.b.), i.e. the 2^0 bit, is on the extreme right in each number and the *most significant bit* (m.s.b.) on the extreme left.

For large numbers, higher powers of two have to be used. For example:

$$\text{decimal } 29 = 1 \times 16 + 1 \times 8 + 1 \times 4 + 0 \times 2 + 1 \times 1$$
$$= 1 \times 2^4 + 1 \times 2^3 + 1 \times 2^2 + 0 \times 2^1 + 1 \times 2^0$$
$$= 11101 \text{ in binary (a five-bit code)}$$

Decimal		Binary				Hexadecimal
10^1 (10)	10^0 (1)	2^3 (8)	2^2 (4)	2^1 (2)	2^0 (1)	16^0 (1)
0	0	0	0	0	0	0
0	1	0	0	0	1	1
0	2	0	0	1	0	2
0	3	0	0	1	1	3
0	4	0	1	0	0	4
0	5	0	1	0	1	5
0	6	0	1	1	0	6
0	7	0	1	1	1	7
0	8	1	0	0	0	8
0	9	1	0	0	1	9
1	0	1	0	1	0	A
1	1	1	0	1	1	B
1	2	1	1	0	0	C
1	3	1	1	0	1	D
1	4	1	1	1	0	E
1	5	1	1	1	1	F

Binary coded decimal (BCD)   This is a popular, slightly less compact variation of binary but in practice it allows easy conversion back to decimal for display purposes. Each *digit* of a decimal number is coded in binary instead of coding the whole number. For example:

decimal 29 = 0010   1001   in BCD

In binary, decimal 29 requires only five bits. It is obviously essential to know which particular binary code is being used.

Hexadecimal code   One of the obvious disadvantages of binary code is the representation of even quite small numbers by a long string of bits. The *hexadecimal* (*hex*) code is more compact and therefore less liable to error. In it, the range of decimal digits, i.e. 0 to 9, is extended by adding the six letters A to F for the numbers 10 to 15 respectively, as shown in the table above.

For larger numbers, successive columns from the right represent powers of sixteen (hence hexadecimal), i.e. 16^0 (1), 16^1 (16), 16^2 (256), etc. For example:

decimal 16 $= 1 \times 16^1 +\ 0 \times 16^0$ = 10 in hex
decimal 29 $= 1 \times 16^1 + 13 \times 16^0$ = 1D in hex
decimal 407 $= 1 \times 16^2 +\ 9 \times 16^1$
$\qquad\qquad +\ 7 \times 16^0$ = 197 in hex
decimal 940 $= 3 \times 16^2 + 10 \times 16^1$
$\qquad\qquad + 12 \times 16^0$ = 3AC in hex

Octal code   Like the hexadecimal code, the octal code is a shorthand way of expressing binary numbers but to base 8. For example the binary number 10100111, which is the decimal number 167, can be represented by the octal number 247_8 since:

$$\begin{aligned} 167 &= 2 \times 8^2 &+ 4 \times 8^1 &+ 7 \times 8^0 \\ &= 2 \times 64 &+ 4 \times 8 &+ 7 \times 1 \\ &= 128 &+ 32 &+ 7 \\ &= 167 \end{aligned}$$

Hexadecimal and octal codes are useful in microprocessor-based systems where binary 'words' of 8 and 16 bits are used.

HALF-ADDER

Adders are electronic circuits which perform addition in binary and consist of combinations of logic gates.

Adding two bits   A half-adder adds *two bits at a time* and has to deal with four cases (essentially three since two are the same). They are:

$$\begin{array}{cccc} 0 & 0 & 1 & 1 \\ +0 & +1 & +0 & +1 \\ \hline 0 & 1 & 1 & 10 \end{array}$$

In the fourth case, 1 plus 1 equals 2, which in binary is 10, i.e. the right column is 0 and 1 is carried to the next column on the left. The circuit for a half-adder must therefore have two inputs, i.e. one for each bit to be added, and two outputs, i.e. one for the *sum* and one for the *carry*.

Circuit One way of building a half-adder from logic gates uses an exclusive-OR gate and an AND gate. From their truth tables in Fig. 63.1a and b, you can see that the output of the exclusive-OR gate is always the *sum* of the addition of two bits (being 0 for 1 + 1), while the output of the AND gate equals the *carry* of the two-bit addition (being 1 only for 1 + 1). Therefore if both bits are applied to the inputs of both gates at the same time, binary addition occurs.

A	B	Output
0	0	0
0	1	1
1	0	1
1	1	0

(a)

A	B	Output
0	0	0
0	1	0
1	0	0
1	1	1

(b)

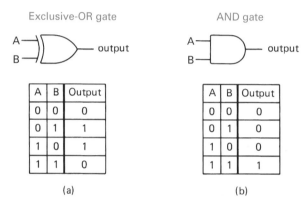

Fig. 63.1

The circuit is shown with its two inputs A and B in Fig. 63.2, along with the half-adder truth table.

A	B	Sum	Carry
0	0	0	0
0	1	1	0
1	0	1	0
1	1	0	1

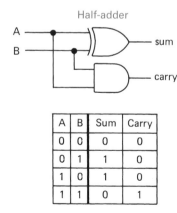

Fig. 63.2

Two NAND gates circuit boards, Fig. 61.10 (p. 132), can be used to build and investigate a half-adder.

FULL-ADDER

Adding three bits This is a necessary operation when two multi-bit numbers are added. For example, to add 3 (11 in binary) to 3 we write:

$$\begin{array}{r} 11 \\ +11 \\ \hline 110 \end{array}$$

The answer 110 (6 in decimal) is obtained as follows. In the least significant (right-hand) column we have:

$$1 + 1 = \text{sum } 0 + \text{carry } 1$$

In the next column three bits have to be added because of the carry from the first column, giving:

$$1 + 1 + 1 = \text{sum } 1 + \text{carry } 1$$

Circuit A full-adder circuit therefore needs three inputs A, B, C (two for the digits and one for the carry) and two outputs (one for the *sum* and the other for the *carry*). It is realized by connecting two half-adders (HA) and an OR gate, as in Fig. 63.3a. We can check that it produces the correct answer by putting A = 1, B = 1 and C = 1, as in Fig. 63.3b.

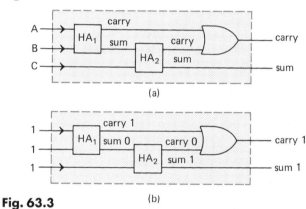

Fig. 63.3

The first half-adder HA_1 has both inputs 1 and so gives a sum of 0 and a carry of 1. HA_2 has inputs 1 and 0 and gives a sum of 1 (i.e. the sum output of the full-adder) and a carry of 0. The inputs to the OR gate are 1 and 0 and, since one of the inputs is 1, the output (i.e. the carry output of the full-adder) is 1. The addition of 1 + 1 + 1 is therefore *sum* 1 and *carry* 1.

The truth table with the other three-bit input combinations is given in Fig. 63.4. You may like to check that the circuit does produce the correct outputs for them.

Full-adder

Inputs			Outputs	
A	B	C	Sum	Carry
0	0	0	0	0
0	0	1	1	0
0	1	0	1	0
1	0	0	1	0
0	1	1	0	1
1	0	1	0	1
1	1	0	0	1
1	1	1	1	1

Fig. 63.4

MULTI-BIT ADDER

Two multi-bit binary *numbers* are added by connecting adders in parallel. For example, to add two four-bit numbers, four adders are needed, as shown in Fig. 63.5 for the addition of 1110 (decimal 14) and 0111 (decimal 7). Follow it through to see that the sum is 10101 (decimal 21). Strictly speaking the full-adder FA_1 need only be a half-adder since it only handles two bits.

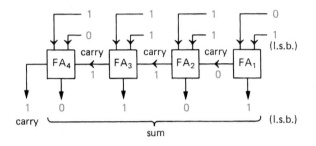

Fig. 63.5

The largest binary numbers that can be added by a four-bit adder are 1111 and 1111, i.e. $15 + 15 = 30$. By connecting more full-adders to the left end of the system, the capacity increases.

In the block diagram for a *four-bit parallel adder* shown in Fig. 63.6, the four-bit number $A_4A_3A_2A_1$ is added to $B_4B_3B_2B_1$, A_1 and B_1 being the least significant bits (l.s.b.). $S_4S_3S_2$ and S_1 represent the sums and C_o the most significant bit (m.s.b.) of the answer, is the carry of the output.

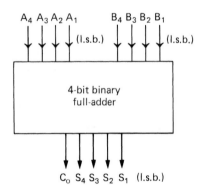

Fig. 63.6

A disadvantage of the adder in Fig. 63.5 is that each stage must wait for the carry from the preceding one before it can decide its own sum and carry, i.e. the carry ripples down from right to left. This is called 'ripple carry addition' and it can be speeded up by including in the adder extra circuits which predict immediately all the carries. In such a 'look-ahead carry adder', all bits are added at the same time giving much faster addition.

BINARY SUBTRACTION

One method, called the *one's complement method*, is best explained by an example. Suppose we have to subtract 0110 (6) from 1010 (10). We proceed as follows.

(i) Form the one's complement of 0110; this is done by changing 1s to 0s and 0s to 1s giving 1001.

(ii) Add the one's complement from (i) to the number from which the subtraction is to be made, i.e. add 1001 and 1010:

$$
\begin{array}{r}
1010 \\
+1001 \\
\hline
10011
\end{array}
$$

(iii) If there is a 1 carry in the most significant position of the total in (ii), remove it and add it to the remaining four bits to get the answer:

$$1\;\boxed{0011} \longrightarrow \begin{array}{r} 0011 \\ +0001 \\ \hline 0100 \quad (4) \end{array}$$

The 1 carry that is added, is called the 'end-around carry' (EAC). When there is an EAC, the answer to the subtraction is positive (as above). If there is no EAC, the answer is negative and is in one's complement form. For example, subtracting 0101 (5) from 0011 (3), we get:

$$
\begin{array}{r}
0011 \\
+1010 \ \text{(one's complement of 0101)} \\
\hline
1101 \rightarrow -0010
\end{array}
$$

The one's complement of 0101 (i.e. 1010) has been added to 0011 to give 1101. There is no EAC and so to get the final answer we take the one's complement of 1101, i.e. 0010 and put a negative sign in front to give -0010 (-2).

Binary subtraction can thus be done simply by addition and is implemented electronically by an adder since the

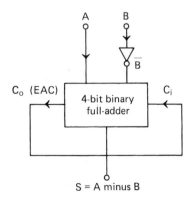

Fig. 63.7

one's complement of a number is easily obtained by an inverter. A simplified circuit of a four-bit 'subtractor' is shown in Fig. 63.7, where the four-bit number B is subtracted from the four-bit number A. There is an EAC which is applied to the carry-in input of the adder and the answer is S.

Multiplication and division can be performed by repeated addition and subtraction respectively.

QUESTIONS

1. Make the following code conversions:

a) decimal numbers 4, 13, 21, 38, 64 to binary,

b) binary numbers 111, 11001, 101010, 110010 to decimal,

c) decimal numbers 9, 17, 28, 370, 645 to BCD,

d) hexadecimal numbers 3, A, D, 1E, 1A5 to decimal,

e) decimal numbers 6, 11, 15, 31, 83, 300 to hexadecimal,

f) decimal numbers 75, 197 to octal,

g) octal numbers 189_8, 231_8 to decimal.

2. Add the following pairs of binary numbers and check your answers by converting the binary numbers to decimal:

a) 10 + 01

b) 11 + 10

c) 101 + 011

d) 1011 + 0111

3. Subtract the following pairs of numbers by the one's complement method and check your answers by converting the binary numbers to decimal:

a) 1001 – 0101

b) 11001 – 00111

c) 0011 – 0110

d) 01100 – 10101

4. a) By drawing up a truth table, identify the logic circuit of Fig. 63.8a.

b) Write down the truth table for a *half-adder*.

c) Identify the logic circuit of Fig. 63.8b by adding another column for G in the truth table you drew up in **a)**.

d) Draw a circuit showing how two half-adders (using the symbol in Fig. 63.8c) and an OR gate can act as a *full-adder*.

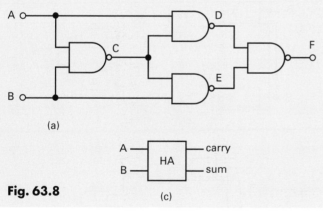

(a)

(c)

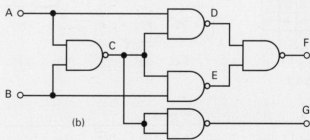

(b)

Fig. 63.8

64 LOGIC CIRCUIT DESIGN (I)

CHECKLIST

After studying this chapter you should be able to:
• design circuits for an excl-OR gate, a half-adder and a full-adder using truth tables and Boolean notation.

Many digital circuits for performing logical operations are built from logic gates. Once it is known what the circuit has to do, a truth table can be drawn up and used as the starting point for the design. The method will be used in this chapter (and the next two) to construct three logic circuits studied previously.

In practice, logic circuits can be made using only NAND (or NOR) gates, as we will see.

EXCLUSIVE-OR GATE CIRCUIT

Lines 2 and 3 of the truth table in Fig. 64.1a show that the output F is 1 only when:

A is 0 *and* B is 1
or
A is 1 *and* B is 0

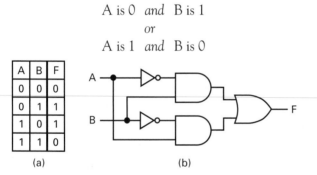

A	B	F
0	0	0
0	1	1
1	0	1
1	1	0

(a)   (b)

Fig. 64.1

The method requires this statement to be rewritten so that both inputs appear as 1s. Therefore if A is 0, inverting A with a NOT gate makes the input (called 'not A' and written $\bar{A}$) a 1. Similarly instead of writing B as 0, we can say $\bar{B}$ (i.e. not B) is 1. The statement, which is the same as before, becomes F is 1 when:

$\bar{A}$ is 1 *and* B is 1
or
A is 1 *and* $\bar{B}$ is 1

Using the notation of the mathematics of logic circuit design, called *Boolean algebra*, we get:

$$F = \bar{A}.B + A.\bar{B} \qquad (1)$$

where a dot (.) represents the AND logic operation and a plus (+) indicates the OR operation.

Using *any* logic gates this means the circuit consists of

two two-input AND gates (each with a NOT gate in one input) feeding a two-input OR gate as in Fig. 64.1b.

To implement it with only NAND gates we use the facts stated earlier (p. 132):

NOT = one-input NAND
AND = NAND followed by NOT
OR = NOT of each input followed by NAND

The circuit becomes that of 64.2a but inverting an input twice gives the original input, i.e. the successive NOT gates 6 and 7, also 8 and 9, cancel each other, so only NAND gates 1, 2, 3, 4 and 5 are needed. The equivalent NAND gate circuit is given in Fig. 64.2b; although different from the exclusive-OR circuit in Fig. 61.9 (p. 132), it performs the same logical operation. Note that $\bar{A}$ and $\bar{B}$ are obtained by inverting A and B with one-input NAND gates (1 and 3), i.e. with NOT gates.

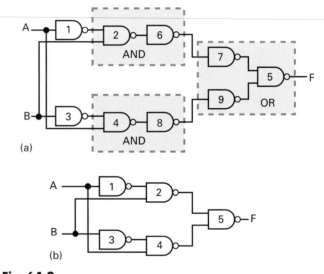

(a)

(b)

Fig. 64.2

HALF-ADDER CIRCUIT

From lines 2 and 3 of the truth table in Fig. 64.3a we see that the SUM output S of the two bits A and B to be added, is 1 only when:

A is 0 *and* B is 1
or
A is 1 *and* B is 0

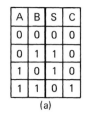

A	B	S	C
0	0	0	0
0	1	1	0
1	0	1	0
1	1	0	1

(a)

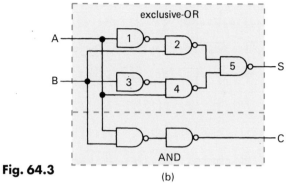

Fig. 64.3

(b)

Following the previous procedure we can therefore write in Boolean notation:

$$S = \bar{A}.B + A.\bar{B}$$

This is the same as equation (1), i.e. S is given by an exclusive-OR gate.

Also, from line 4 of the truth table, the CARRY output C is 1 only when:

A is 1 *and* B is 1

That is,

$$C = A.B$$

This is implemented by a two-input AND gate. The NAND equivalent of the complete half-adder circuit is given in Fig. 64.3b.

FULL-ADDER CIRCUIT

A full-adder for adding three bits can be made from two half-adders and an OR gate (as in Fig. 63.3a, p. 137). It can be designed from first principles using the truth table in Fig. 64.4a.

The Boolean expression for the SUM output is (from lines 2, 3, 4 and 8):

$$SUM = \bar{A}.\bar{B}.C + \bar{A}.B.\bar{C} + A.\bar{B}.\bar{C} + A.B.C$$

The circuit in Fig. 64.4b gives this output. It uses four three-input AND gates (with some inputs inverted) and a four-input OR gate. The simplified equivalent NAND gate circuit is shown on the top half of Fig. 64.5, the cases in which there are two successive NOT gates have been omitted.

A	B	C	Sum	Carry
0	0	0	0	0
0	0	1	1	0
0	1	0	1	0
1	0	0	1	0
0	1	1	0	1
1	1	0	0	1
1	0	1	0	1
1	1	1	1	1

(a)

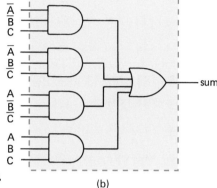

Fig. 64.4

(b)

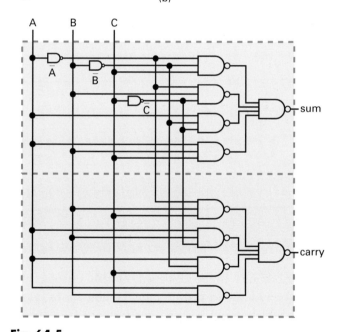

Fig. 64.5

Similarly, the Boolean expression for the CARRY output is obtained from lines 5, 6, 7 and 8 of the truth table and is:

$$CARRY = \bar{A}.B.C. + A.B.\bar{C} + A.\bar{B}.C + A.B.C$$

It can also be implemented using four AND gates and an OR gate. The lower half of Fig. 64.5 gives the simplified NAND gate equivalent: the whole circuit is for a full-adder using only NAND gates.

QUESTIONS

1. Write the truth table for
  a) a two-input NAND gate,
  b) a two-input NOR gate,
  c) the logic circuit in Fig. 64.6.

2. Write a truth table for the circuit in Fig. 64.7, including the states at C, D, E, F and G.

3. Draw a truth table for each of the systems **(a)** and **(b)** shown in Fig. 64.8 and identify the logic function which each possesses.   (*L.*)

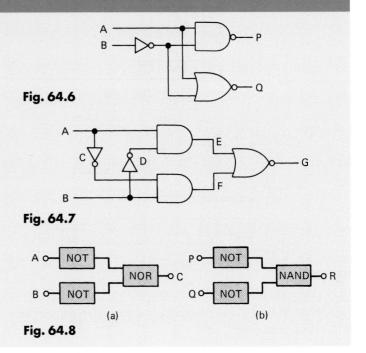

Fig. 64.6

Fig. 64.7

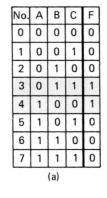
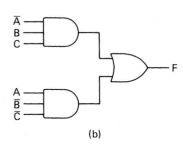

(a)   (b)

Fig. 64.8

65 LOGIC CIRCUIT DESIGN (II)

CHECKLIST

After studying this chapter you should be able to:
• design logic circuits, using truth tables and Boolean notation, which control flashing lamps.

The logic circuits to be designed in this chapter are required to produce, for certain input combinations, logic 1 outputs which could be used, for instance, to control some operation or make lamps flash in a particular order.

CIRCUIT 1

The circuit is to have three inputs A, B, C, fed by a three-bit binary code representing the numbers 0 to 7 (C being the l.s.b.). For example, if A = 1 ('high' input), B = 0 ('low' input) and C = 1, the binary number at the input is 101 (5 in decimal).

Suppose the function of the circuit is to produce a logic 1 at its output F when numbers 3 and 4 occur. The truth

No.	A	B	C	F
0	0	0	0	0
1	0	0	1	0
2	0	1	0	0
3	0	1	1	1
4	1	0	0	1
5	1	0	1	0
6	1	1	0	0
7	1	1	1	0

(a)

(b)

Fig. 65.1

table is given in Fig. 65.1a and lines 3 and 4 show that F is 1 when:

A is 0 *and* B is 1 *and* C is 1
or
A is 1 *and* B is 0 *and* C is 0

This statement is the same if it is rewritten so that all inputs appear, as before (p. 140), as 1s. We then get F is 1 when:

$\overline{A}$ (not A) is 1 *and* B is 1 *and* C is 1
or
A is 1 *and* $\overline{B}$ (not B) is 1 *and* $\overline{C}$ (not C) is 1

In Boolean algebra notation, where a plus (+) indicates the OR logic operation and a dot (.) represents the AND operation, we can write:

$$F = \overline{A}.B.C + A.\overline{B}.\overline{C}$$

In terms of logic gates this means the circuit has two three-input AND gates feeding a two-input OR gate as in Fig. 65.1b. To implement it using only NAND gates we use the facts (see p. 132):

AND = NAND followed by NOT
OR = NOT of each input followed by NAND

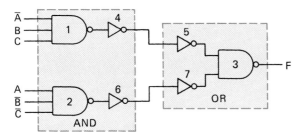

Fig. 65.2

The circuit becomes that of Fig. 65.2 but inverting an input twice gives the original input, i.e. two successive NOT gates cancel each other so only NAND gates 1, 2 and 3 are needed. The equivalent NAND gate circuit is given in Fig. 65.3: note that $\overline{A}$, $\overline{B}$ and $\overline{C}$ are obtained by inverting A, B and C with one-input NAND gates.

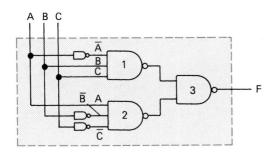

Fig. 65.3

CIRCUIT 2 (TRAFFIC LIGHTS)

The problem is to design a logic circuit with two inputs A, B and three outputs R, Y, G obeying the truth table in Fig. 65.4. If R, Y and G feed red, yellow and green lights (e.g. LEDs), these would flash in the order of British traffic signals when A and B are supplied continuously by a two-bit binary code representing numbers 0 to 3.

State	A	B	R	Y	G
0	0	0	1	0	0
1	0	1	1	1	0
2	1	0	0	0	1
3	1	1	0	1	0

Fig. 65.4

From the truth table we can write the following.

(i) R is 1 when (lines 0 and 1):

A is 0 *and* B is 0 (i.e. $\overline{A}$ is 1 *and* $\overline{B}$ is 1)
or
A is 0 *and* B is 1 (i.e. $\overline{A}$ is 1 *and* B is 1)

(ii) Y is 1 when (lines 1 and 3):

A is 0 *and* B is 1 (i.e. $\overline{A}$ is 1 *and* B is 1)
or
A is 1 *and* B is 1

(iii) G is 1 when (line 2):

A is 1 *and* B is 0 (i.e. A is 1 *and* $\overline{B}$ is 1)

Rewriting **(i)**, **(ii)** and **(iii)** in Boolean notation and factorizing as in ordinary algebra:

$$R = \overline{A}.\overline{B} + \overline{A}.B = \overline{A}(\overline{B} + B)$$
$$Y = \overline{A}.B + A.B = B(\overline{A} + A)$$
$$G = A.\overline{B}$$

Now if the two inputs to an OR gate are an input and its inverse (called its *complement*), one input is a 1, making the output 1 (since for an OR gate, the output is 1 unless all inputs are 0). That is $\overline{A} + A$ is 1 and $\overline{B} + B$ is 1, hence we get:

$$R = \overline{A} \qquad Y = B \qquad G = A.\overline{B}$$

Thus, the logic circuit has to feed R from input A via a NOT gate while Y goes to input B directly and G is supplied by the output of an AND gate having A and the complement of B as its inputs, Fig. 65.5a. Using only NAND gates, the circuit is as in Fig. 65.5b.

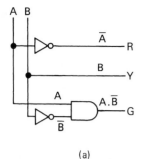
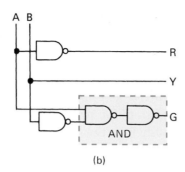

Fig. 65.5

(a)   (b)

QUESTIONS

1. Write the truth table for and, using any gates, design a logic circuit which gives a logic 1 output
a) with a two-bit binary input when the number is 0,
b) with a three-bit binary input when the numbers are 0 and 5.

2. A three-bit binary code is used to represent the numbers 0 to 7 and a logic circuit is required which produces logic 1 outputs when numbers 1 and 3 occur at its inputs.
a) Write the truth table for this logic function.
b) Design a logic circuit to implement this function using NAND gates only.

66 LOGIC CIRCUIT DESIGN (III)

CHECKLIST

After studying this chapter you should be able to:
• design circuits using truth tables and Boolean notation for a binary to decimal decoder, a magnitude comparator and an arithmetic and logic unit.

The design of three logic circuits which are important in computers and other electronic systems will be considered.

BINARY TO DECIMAL DECODER

While digital electronic systems work in binary code, humans prefer the decimal code. A *decoder* (often at the output end of the system) converts a binary input to a decimal output, frequently for display purposes (e.g. on a seven segment display).

Suppose a two-bit binary decoder is required which will produce a logic 1 output at just one of four outputs (depending on the input), the other three remaining at logic 0. The block diagram and truth table are given in Fig. 66.1. We can regard it as part of a binary to decimal decoder with F_0, F_1, F_2 and F_3 representing the decimal outputs of 0, 1, 2 and 3 respectively, while A and B are the binary inputs (A being the l.s.b.) causing the appropriate decimal output to go 'high'.

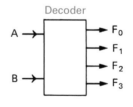

B	A	F_0	F_1	F_2	F_3
0	0	1	0	0	0
0	1	0	1	0	0
1	0	0	0	1	0
1	1	0	0	0	1

Fig. 66.1

From the truth table we can say:

F_0 is 1   when A is 0   *and*   B is 0
F_1 is 1   when A is 1   *and*   B is 0
F_2 is 1   when A is 0   *and*   B is 1
F_3 is 1   when A is 1   *and*   B is 1

Using Boolean notation and making all inputs 1s, as before (p. 140), we get:

$$F_0 = \overline{A}.\overline{B} \quad F_1 = A.\overline{B} \quad F_2 = \overline{A}.B \quad F_3 = A.B$$

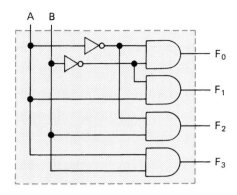

Fig. 66.2

The circuit of the system using AND and NOT gates is shown in Fig. 66.2. The NAND gate only version requires ten gates.

Note. An *encoder* (usually at the input of a system) converts from decimal (e.g. supplied by a numerical keyboard) into binary and works on similar principles using logic gates. Encoders and decoders are called *code converters*.

BINARY TO SEVEN-SEGMENT DECODER

Common cathode display A four-bit binary signal is applied to the four inputs D (m.s.b.), C, B, A (l.s.b.) and is converted into seven outputs a, b, c, d, e, f, g by the decoder (e.g. a CMOS 4511). Each output is connected, via a suitable current-limiting resistor, to the anode of one segment of the seven-segment LED display which it drives when 'high' (i.e. a 1), Fig. 66.3a. The *cathodes* of all seven LEDs in the display are connected to 0 V.

As an example of the decoding action, suppose D = 0, C = 0, B = 1 and A = 1, the binary input is 0011 (3 in decimal) and the five outputs a, b, c, d, g needed to light the five LED segments making a '3', all go 'high'.

Common anode display In this case the decoder must be one whose outputs go 'low' (i.e. a 0) when active (e.g. a TTL 7447), Fig. 66.3b. The *anodes* of all seven segments in the display are connected to +5 V, so when an output on the decoder goes 'low', i.e. falls to 0 V, current passes through the segment to which it is connected and lights it up. In this case the decoder 'sinks' current, with a common cathode display the decoder 'sources' the current.

Decoder and display board See Fig. 66.4. This is useful for experimental work including the display of the hexadecimal code.

Fig. 66.4

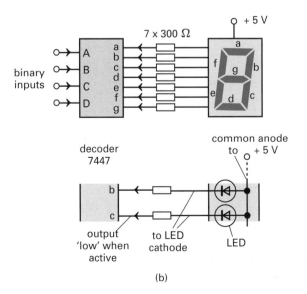

Fig. 66.3

(a)

(b)

MAGNITUDE COMPARATOR

A *magnitude comparator* compares two binary numbers. For example, during a counting operation in say a computer, it may be necessary to know when a certain total has been reached before moving on to the next part of the program. This involves the comparison of two numbers.

Suppose the circuit has to compare two one-bit binary numbers A and B and produce a logic 1 output F when they are equal, i.e. both 0s or both 1s. The block diagram and truth table are shown in Fig. 66.5. Proceeding as before we can write:

$$F = \bar{A}.\bar{B} + A.B$$

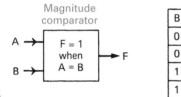

B	A	F
0	0	1
0	1	0
1	0	0
1	1	1

Fig. 66.5

The circuit using AND, OR and NOT gates is given in Fig. 66.6a and using only NAND gates in Fig. 66.6b. It is the same as that for an exclusive-NOR gate (p. 131) and can be built using two of the NAND gate boards shown in Fig. 61.10, p. 132.

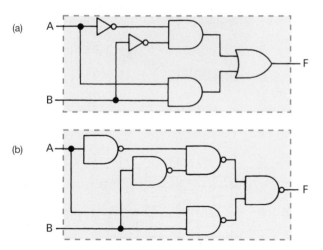

Fig. 66.6

ARITHMETIC AND LOGIC UNIT

An *arithmetic* and *logic unit* (ALU) is at the heart of a digital computer (p. 216). It operates in two alternative ways or modes. In the *arithmetic* mode it performs, in binary, addition, subtraction, multiplication and division, all based on the use of adders (p. 135). In the *logic* mode, logical operations like those done by logic gates are carried out.

Suppose a very simple ALU has two 'ordinary' inputs A and B and a third 'select' input C so that when:

(i) C = 0, it is to be in the arithmetic mode and give an output F = 1 when A = B, i.e. it is a magnitude comparator; and

(ii) C = 1, it is to be in the logic mode and perform the AND operation on A and B, i.e. make F = 1 when A = B = 1.

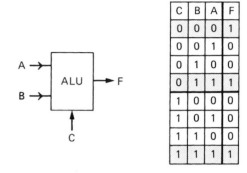

C	B	A	F
0	0	0	1
0	0	1	0
0	1	0	0
0	1	1	1
1	0	0	0
1	0	1	0
1	1	0	0
1	1	1	1

Fig. 66.7

The block diagram and truth table are given in Fig. 66.7. We can write:

$$\begin{aligned} F &= \bar{A}.\bar{B}.\bar{C} + A.B.\bar{C} + A.B.C \\ &= \bar{A}.\bar{B}.\bar{C} + A.B(\bar{C} + C) \\ &= \bar{A}.\bar{B}.\bar{C} + A.B \quad \text{since } \bar{C} + C = 1 \text{ (p. 143)} \end{aligned}$$

The circuit using AND, NOT and OR gates is given in Fig. 66.8a. Note how the need to use a three-input AND gate is avoided by 'ANDing' $\bar{A}$ and $\bar{B}$ first, then 'ANDing' their output $\bar{A}.\bar{B}$ with $\bar{C}$. The simplified equivalent NAND gate circuit is shown in Fig. 66.8b.

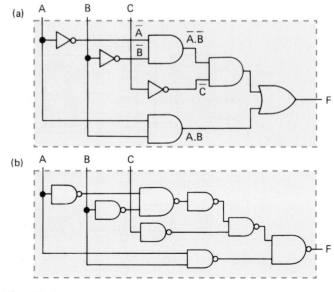

Fig. 66.8

QUESTIONS

1. Design a magnitude comparator with two inputs A and B, which gives a logic 1 at its output F only when A is greater than B.

2. Design a two-way electronic switch (called a *data selector* or a 2-to-1 line *multiplexer*) which connects its output F to input A when the 'command' input C is at logic 0 and to input B when C is at logic 1, Fig. 66.9.

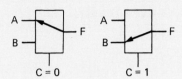

Fig. 66.9

67 MORE BOOLEAN ALGEBRA

CHECKLIST

After studying this chapter you should be able to:
• recall the Boolean expressions for logic gates,
• complete Boolean identities for the OR and AND operations,

• state De Morgan's two theorems and the rule for using them, and
• design, using Boolean algebra, OR, AND and excl-OR gates from universal gates only.

In the previous three chapters logic circuits were designed using truth tables to describe the logical operations required. Some use was also made of Boolean notation. In this chapter, Boolean algebra will be developed more formally as a quicker and neater way of summarizing the action of a digital logic circuit. It also often indicates how simplifications can be made.

Boole, who invented his form of algebra in 1847, showed that logical statements which are either 'true' or 'false' have a binary nature applicable to symbols as well as words. However it was not until nearly 100 years later that the approach was used in digital electronics to analyse and predict the behaviour of telephone switching circuits. They were concerned with on-off signals that can be considered as at logic level 1 if 'on' and at logic 0 if 'off'.

BOOLEAN EXPRESSIONS FOR LOGIC GATES

These are a form of shorthand which describes what a gate does in logical terms. Certain conventions mentioned earlier are used and are restated again.

(i) A plus (+) represents an OR operation and a dot (.) represents an AND operation.

(ii) A bar above the top of a symbol, e.g. $\bar{A}$ means the value of the symbol is inverted or complemented so that $\bar{1} = 0$ and $\bar{0} = 1$.

(iii) A double bar, e.g. $\bar{\bar{B}}$, is a double inversion and cancels out to give the original symbol, i.e. $\bar{\bar{B}} = B$.

The expressions for the various gates can then be written as in Table 67.1, where A and B are inputs that can have logic levels of either 0 or 1, as can the output, usually denoted by Q (or F) in Boolean algebra.

Table 67.1

Gate	Expression	Logic operation
NOT	$Q = \bar{A}$	Q is *not* A, i.e. if A = 1, Q = $\bar{1}$ = 0
OR	$Q = A + B$	Q is the result of OR-ing A and B and is 1 if A *or* B is 1
AND	$Q = A \cdot B$	Q is the result of AND-ing A and B and is 1 if A *and* B are both 1
NOR	$Q = \overline{A + B}$	Q is the complement of the result of OR-ing A and B and is 1 if neither A *nor* B is 1
NAND	$Q = \overline{A \cdot B}$	Q is the complement of the result of AND-ing A and B and is *not* 1 if both A *and* B are 1

Remember that in electronic circuits logic levels 0 and 1 represent 'low' and 'high' voltages and are not the same as mathematical 0 (zero) or 1.

BOOLEAN IDENTITIES

They are statements that arise from the properties of OR and AND operations and frequently allow Boolean equations (and so also logic circuits) to be simplified. Table 67.2 gives the identities that depend on the OR operation.

Table 67.3 sets out the identities that arise from the AND operation.

Some other combinations can be treated as in ordinary algebra. For example:

$$A \cdot B + A \cdot C = A \cdot (B + C)$$
$$A + A \cdot B = A(1 + B) = A$$

since, from identity 2 in Table 67.2, B + 1 = 1.

DE MORGAN'S THEOREMS

These are two useful statements, due to De Morgan, a contemporary of Boole, that enable conversions to be made from one kind of gate to another. They allow circuits to be built from just one type, usually combinations of the commoner and more easily manufactured NAND or NOR gates (which we saw earlier could be done).

The *first theorem* states in symbols that:

$$\overline{A + B} = \bar{A} \cdot \bar{B} \qquad (1)$$

and asserts that the *complement of the output* of OR-ing A and B is equivalent to the output of AND-ing the *complements* of A and B (i.e. $\bar{A}$ and $\bar{B}$).

The *second theorem* in symbols is:

$$\overline{A \cdot B} = \bar{A} + \bar{B} \qquad (2)$$

and states that the *complement of the output* of AND-ing A and B equals the output of OR-ing the *complements* of A and B (i.e. $\bar{A}$ and $\bar{B}$).

Table 67.2 OR operation

No.	Identity	Explanation
1	$A + 0 = A$	For an OR gate if A is 1, the output is 1 since one input is 1; if A is 0, the output is 0 since both inputs are 0, i.e. the output always has the same logic level as A
2	$A + 1 = 1$	If one input is 1 (as here), the output is 1 whether A is 1 or 0
3	$A + \bar{A} = 1$	If A is 1, the output is 1; if A is 0, $\bar{A}$ is 1 making one input 1 and giving a 1 output
4	$A + A = A$	If A is 0, the output is 0 but if A is 1, the output is 1, i.e. output is always A

Table 67.3 AND operation

No.	Identity	Explanation
5	$A \cdot 0 = 0$	For an AND gate if one input is 0 (as here), the output is 0, whatever the other input
6	$A \cdot 1 = A$	If A is 1, both inputs are 1 so the output is 1; if A is 0, the output is 0, i.e. the output is always A
7	$A \cdot \bar{A} = 0$	If A is 1, $\bar{A}$ is 0, so the output is 0; if A is 0, $\bar{A}$ is 1, so the output is again 0
8	$A \cdot A = A$	If A is 1, the output is 1; if A is 0, the output is 0, i.e. the output is always A

Both theorems can be proved by writing down the truth table for each side of the expression, as in Table 67.4.

Table 67.4

A	B	$\bar{A}$	$\bar{B}$	$\overline{A+B}$ NOR	$\bar{A}.\bar{B}$	$\overline{A.B}$ NAND	$\bar{A}+\bar{B}$
0	0	1	1	1	1	1	1
0	1	1	0	0	0	1	1
1	0	0	1	0	0	1	1
1	1	0	0	0	0	0	0

The fifth and sixth columns of Table 67.4 are the same, which proves the first theorem, and the second one follows since the seventh and eighth columns are identical.

The theorems still apply when there are three or more symbols, viz.

$$\overline{A + B + C} = \bar{A}.\bar{B}.\bar{C}$$

and
$$\overline{A.B.C} = \bar{A} + \bar{B} + \bar{C}$$

USING DE MORGAN

Inspection of expressions (1) and (2) for De Morgan's theorems shows that to get from the expression on the left-hand side of the equals sign to that on the right-hand side, you have to first 'break the bar' over the left-hand expression and then 'change the logical operation' connecting its variables.

For example, for equation (1), 'breaking the bar' over $\overline{A + B}$ gives $\bar{A} + \bar{B}$, then 'changing the connecting operation' from OR (+) to AND (.) gives $\bar{A}.\bar{B}$, hence $\overline{A + B} = \bar{A}.\bar{B}$. Equation (2) can be treated in the same way.

UNIVERSAL GATES

Since any combinational logic circuit can be built from either NAND gates only or NOR gates only, these two types are called *universal gates*. Three circuits, which were considered earlier, will now be analysed using De Morgan's theorems.

OR gate from NAND gates The circuit in Fig. 67.1 is for an OR gate made from only NAND gates, (i.e. NOT of each input followed by NAND) as can be checked by drawing up a truth table. However it can be proved more easily using Boolean algebra and De Morgan's theorem.

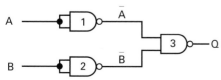

Fig. 67.1

The inputs to NAND gate 3 are $\bar{A}$ and $\bar{B}$ since A and B are inverted by the one-input NAND gates 1 and 2. Hence, for NAND gate 3, we have the Boolean expression:

$$Q = \overline{\bar{A}.\bar{B}}$$

Applying the 'break the bar and change the logic' rule, we get:

$$Q = \bar{\bar{A}} + \bar{\bar{B}}$$
$$= A + B \quad \text{since } \bar{\bar{A}} = A \text{ and } \bar{\bar{B}} = B$$

This is the Boolean expression for an OR gate.

AND gate from NOR gates The circuit is shown in Fig. 67.2 with one-input NOR gates 1 and 2 acting as NOT gates which provide inverted inputs A and B to NOR gate 3. The output Q from NOR gate 3 is given by the Boolean expression:

$$Q = \overline{\bar{A} + \bar{B}}$$

Applying De Morgan's theorem we get:

$$Q = \bar{\bar{A}} . \bar{\bar{B}}$$
$$= A . B \quad \text{since } \bar{\bar{A}} = A \text{ and } \bar{\bar{B}} = B$$

This is the expression for an AND gate.

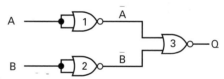

Fig. 67.2

Exclusive-OR gate from NAND gates The excl-OR gate circuit implemented with only NAND gates was derived earlier (Chapter 64) from its truth table; both are shown again in Fig. 67.3. The Boolean expression was deduced from the truth table to be:

$$Q = \bar{A}.B + A.\bar{B}$$

Using the Boolean expression for a NAND gate and De Morgan, this can be proved as follows. The inputs to each NAND gate are shown in Fig. 67.3, from which we can say:

$$Q = \overline{\overline{\bar{A}.B} . \overline{A.\bar{B}}}$$

'Breaking the bar and changing the connecting logic' gives:

$$Q = \overline{\overline{\bar{A}.B}} + \overline{\overline{A.\bar{B}}}$$

But double inversions cancel out, i.e. $\overline{\overline{\bar{A}.B}} = \bar{A}.B$ and $\overline{\overline{A.\bar{B}}} = A.\bar{B}$, hence:

$$Q = \bar{A}.B + A.\bar{B}$$

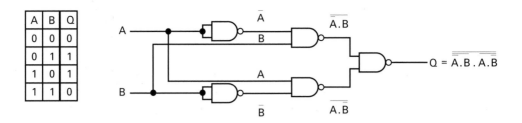

A	B	Q
0	0	0
0	1	1
1	0	1
1	1	0

Fig. 67.3

QUESTIONS

1. Write the Boolean expressions for NOT, OR, AND, NOR and NAND gates.

2. State the result of performing the following logical operations:

   (i) $A + 0$ (ii) $A . 1$ (iii) $A + 1$ (iv) $A . A$
   (v) $A + \bar{A}$ (vi) $\bar{\bar{A}}$ (vii) $0 + 0$ (viii) $0 . 0$
   (ix) $0 + 1$ (x) $1 . 0$ (xi) $1 + 1$ (xii) $1 . 1$

3. a) What are the Boolean expressions for De Morgan's two theorems?

   b) State the rule used to apply them.

4. Prove the following Boolean identities.

   a) $A . (A + B) = A$

   b) $(A + B) . (B + C) = A . C + B$

   c) $\overline{A + B + C} = \bar{A} . \bar{B} . \bar{C}$

   d) $\bar{A} . \bar{B} . C + \bar{A} . B . C + A . C = C$

   e) $\overline{(A + B) . (B + C)} = \bar{B} . (\bar{A} + \bar{C})$

   f) $\overline{\bar{A} . B + B . \bar{C}} = \bar{B} + A . C$

5. a) What are the Boolean expressions for each of the logic circuits in Fig. 67.4(i)–(vi)? For each circuit state the input combination(s) that give an output of $Q = 1$.

   b) Draw up truth tables for circuits (iii) and (iv), and check that they agree with your predictions about the input combinations when $Q = 1$ in each case.

6. a) If $Q = \bar{A} . B + A . \bar{B}$ is the Boolean expression for an exclusive-OR gate, show that for an exclusive-NOR gate $Q = \bar{A} . \bar{B} + A . B$.

   b) Work out the Boolean expression for the circuit in Fig. 67.5. What kind of gate is it?

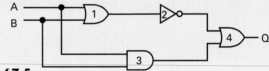

Fig. 67.5

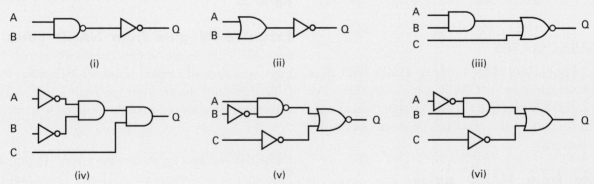

Fig. 67.4

68 PROGRESS QUESTIONS

1. a) Draw the symbols for the following gates: (i) NOT, (ii) AND, (iii) NAND, (iv) NOR, (v) OR, (vi) exclusive-OR and (vii) exclusive-NOR.

b) State what the input conditions must be for each of the gates in **a)** to give a 'high' (logic 1) output.

c) Draw *six* diagrams to show the NAND gate equivalents of the other six gates listed in **a)**.

2. The gas central heating system represented by the block diagram in Fig. 68.1 has digital electronic control. The gas valve turns on the gas supply to the boiler when the output from the logic gate is 'high'. This is so only if *both* of the following conditions hold:

(i) the output from the *thermostat* is 'high', indicating that the room temperature has fallen below that desired and in effect saying 'yes', the room needs more heat, and

(ii) the output from the *pilot flame sensor* is 'high', meaning 'yes', the pilot is lit.

What type of logic gate is required?

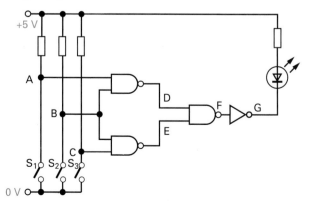

Fig. 68.1

3. In the system of logic gates in Fig. 68.2 the LED lights up when certain switches are closed.

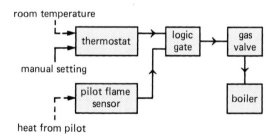

Fig. 68.2

a) Draw up a truth table for the system.

b) What is the logic level at Q when the LED lights?

c) What is the voltage of A when S_1 is (i) open, (ii) closed?

d) State the combinations of switches which, when pressed, light the LED.

4. The circuits in Fig. 68.3a and b show a CMOS 4026 decoder, whose outputs go 'high' when it is active, operating one LED segment of a seven-segment display via a transistor.

One circuit is for a common anode display and the other is for a common cathode display. Which is which? Explain your answers, indicating the part played by the transistor.

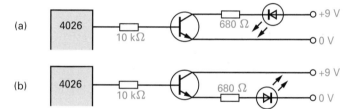

Fig. 68.3

5. Describe the action of an AND gate.

Construct a truth table for a two-input AND gate.

State the output from this AND gate in Boolean form. (C.)

6. Draw a truth table for the logic system shown in Fig. 68.4, including the states at C, D and E. (L.)

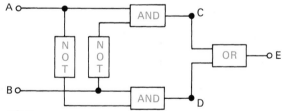

Fig. 68.4

7. Complete a truth table giving the logic states at A, B, C, D and E for the circuit in Fig. 68.5. (O. and C.)

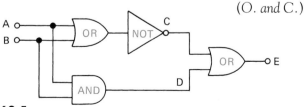

Fig. 68.5

8. Fig. 68.6 represents the apparatus used to measure the rate of flow of a liquid into a graduated flask. The apparatus consists of two reflective opto-switches A and B connected to an exclusive-OR gate. The output from the gate is used to trigger a time clock.

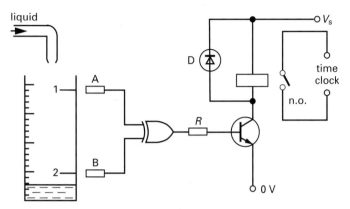

Fig. 68.6

a) (i) Name the *two* parts of an opto-switch.

(ii) In a reflective opto-switch the two parts are placed alongside each other. State how this arrangement enables the level of the liquid to be detected.

(iii) Give *one* advantage of using an opto-switch rather than electrodes immersed in the liquid.

b) Complete the truth table for an XOR gate.

c) Bearing in mind your answers to parts **a)** and **b)** explain how each part of the system enables the time taken for the liquid to rise from level 1 to level 2 to be measured.

d) State the function in the circuit of the resistor R.

e) State the function in the circuit of the diode D. Explain how the diode achieves its function.

N.B. In Fig. 68.6 'n.o.' means 'normally open'. (*N.*)

9. a) Explain in simple terms what is meant by an OR circuit and by a NAND circuit. In each case draw an appropriate circuit diagram and explain how the circuit works.

How are the situations changed if there are three inputs rather than two?

b) Draw a circuit with three inputs A, B and C, which would give a high output if input C were high together with a high voltage at input A or input B, or if the inputs A, B and C were all low.

If the same performance were claimed for a different circuit, how could you test the claim? (*O. and C.*)

10. A game is devised in which rings are thrown to land on hooks A, B, C, D and E mounted on a board as in Fig. 68.7. Each hook is connected to a switch which closes and gives a logic level 1 output when a ring lands on it. The aim of the game is to score exactly 4 with two rings on separate hooks. Hooks A and B each score 1, hook C scores 3 and hooks D and E each score 2. When a player throws two rings so as to score 4 the light is illuminated.

(i) If logic level 1 will turn on the light, write a logic statement of the function which is required to activate the light, considering all possibilities, or list the separate conditions which will activate the light.

(ii) Draw a block diagram of a logic system to fulfil the function given in your answer to (i).

(iii) If you only have NAND gates available, explain how you would construct the other gates you need. (*L. part qn.*)

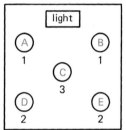

Fig. 68.7

11. The block diagram in Fig. 68.8 is for a system which automatically brings on an electric heater in a greenhouse when it is *cold* at *night*. The switch allows the system to be turned on and off remotely (e.g. from the gardener's home) when it is not required.

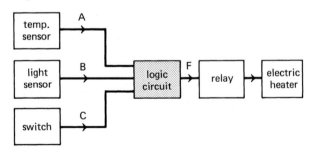

Fig. 68.8

Input A to the logic circuit is 'high' when the temperature sensor is hot; input B is 'high' when light falls on the light sensor; input C is 'high' when the switch is on.

Design the logic circuit which has a 'high' output F when the relay should switch on the heater.

12. The block diagram in Fig. 68.9 is for a system which activates an alarm if the temperature of a hot radiator *falls* during the *day*. The switch allows the operation of the alarm to be tested *at any time*.

Design the logic circuit required. (Inputs A, B and C are 'high' under the conditions given in question 11.)

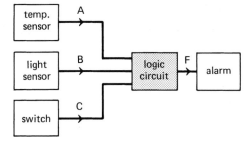

Fig. 68.9

13. The circuit in Fig. 68.10a is known as a 'window comparator'.

a) Calculate the voltage at point A and the voltage at point B.

b) Explain the function of the circuit by considering how V_{out} changes as V_{in} increases from 0 V to 12 V.

c) The thermistor in Fig. 68.10b is connected to the input of the window comparator at the point P.

(i) Calculate the resistance of the thermistor when V_{in} is the same as the voltage at A.

(ii) Calculate the resistance of the thermistor when V_{in} is the same as the voltage at B.

d) The characteristic curve for the thermistor is shown in Fig. 68.10c.

(i) By using the characteristic curve for the thermistor state the range of temperatures for which $V_{out} = 0$ V.

(ii) How could this range of temperature be increased?

e) The complete circuit is used to monitor the temperature of an industrial process. If the temperature goes outside the set range, a mains powered alarm is to sound. Draw a diagram of the interface circuit between the window comparator and the alarm, including a power MOSFET, a relay and any other components you required.

(N.)

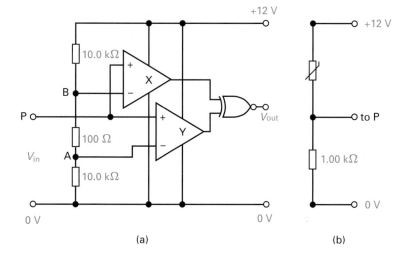

(a)   (b)

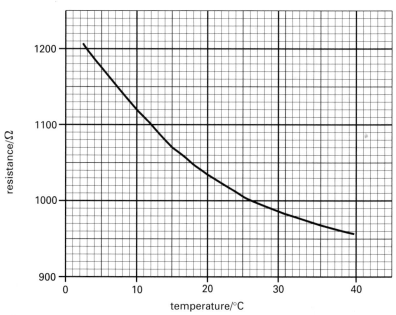

Fig. 68.10   (c)

DIGITAL ELECTRONICS: MULTIVIBRATORS

69 BISTABLES (I)

CHECKLIST

After studying this chapter you should be able to:
- state that a bistable or flip-flop is a switching circuit with two stable states,
- recognize and explain the action of a transistor SR bistable,
- recognize and explain the action of a NAND-gate SR bistable and draw up a truth table showing the outputs for certain input sequences,
- state how a switch can be debounced using an SR bistable, and
- state what a T-type bistable does.

MULTIVIBRATORS

Multivibrators are two-stage switching circuits in which the output of the first stage is fed to the input of the second and vice versa. When one output is 'high', the other is 'low', i.e. their outputs are *complementary*. Switching between the two logic levels is so rapid that the output voltage waveforms are 'square'. The term 'multivibrator' arises from this since a square wave consists of a large number of sine waves with frequencies that are odd multiples of the fundamental.

Multivibrators are of three types. *Bistables* or *flip-flops*, of which there are several varieties, are used in counters (p. 167), shift registers (p. 169) and memories (p. 169) and will be considered first; *astables* and *monostables* are treated later.

TRANSISTOR SR BISTABLE

The basic circuit is shown in Fig. 69.1. The collector of each transistor is coupled to the base of the other by a resistor R_1 or R_2. The output of Tr_1 is thus fed to the input of Tr_2 and vice versa.

When the supply is connected, Tr_1 and Tr_2 both draw base current but, because of slight differences (e.g. in h_{FE}), one, say Tr_1, has a larger collector current and conducts more rapidly than the other. It quickly saturates while Tr_2 is driven to cut-off.

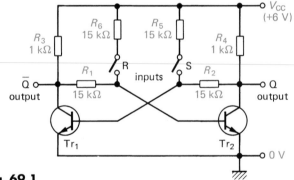

Fig. 69.1

The collector voltage of Tr_1 is therefore 'low' (most of V_{CC} being dropped across R_3) and so no current flows through R_1 into the base of Tr_2. Tr_2 remains off, its collector voltage is thus 'high' (V_{CC}) and causes current to flow via R_2 into the base of Tr_1, reinforcing Tr_1's saturation condition. The *feedback is positive* and the circuit is in a stable state which it can keep indefinitely with output $Q = 1$ (since Tr_2 is cut-off) and its complementary output $\overline{Q}$ (not Q) = 0 (since Tr_1 is saturated).

The state can be changed by applying a positive voltage (>0.6 V for silicon transistors) to the base of the 'off' transistor, i.e. to Tr_2 via R_6 (e.g. by temporarily connecting the *reset* input R to V_{CC}). Tr_2 then draws base current (through R_6), which is large enough to drive it into saturation. Its collector voltage falls from V_{CC} to near zero, so cutting off the base current to Tr_1. Tr_1 switches off, its collector voltage rises from near zero to V_{CC} and is fed via R_1 to the base of Tr_2 to keep it saturated. The circuit is now in its second stable state (hence bistable) but with Tr_1 off ($\overline{Q} = 1$) and Tr_2 saturated ($Q = 0$).

To return to the first state a positive voltage must be applied to the base of the 'off' transistor (Tr_1) at the *set* input S. The output of each transistor can thus be made to 'flip' to V_{CC} or 'flop' to 0 V, according to the truth table in Fig. 69.2 which also gives the SR (set-reset) bistable symbol. Taking the output of the circuit as the collector voltage of Tr_2, each state in effect stores (remembers) one bit of 'information', i.e. a 1 when $Q = 1$ or a 0 when $Q = 0$ and the bistable acts as a *one-bit memory*.

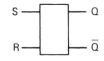

State	S	R	Q	$\overline{Q}$
set	1	0	1	0
reset	0	1	0	1

Fig. 69.2

The state of a bistable can also be changed by briefly connecting to 0 V, either the base of the 'on' transistor or the collector of the 'off' transistor.

NAND-GATE SR BISTABLE

The basic circuit is shown in Fig. 69.3 with its truth table, feedback being from each output to one of the inputs of the other two-input NAND gate.

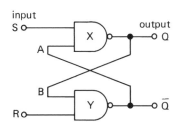

State	S	R	Q	$\overline{Q}$
set	0	1	1	0
	1	1	1	0
reset	1	0	0	1
	1	1	0	1
indeter-minate	0	0	1	1
	1	1	?	?

Fig. 69.3

Set state If $S = 0$ and $R = 1$, NAND gate X has at least one of its inputs at logic 0, therefore its output Q must be at logic 1 (since the output of a NAND is always 1 unless all inputs are 1s). Q is fed back to input B and so both inputs to NAND gate Y are 1s; hence output $\overline{Q} = 0$. This is a stable state, called the *set state*, with $Q = 1$ and $\overline{Q} = 0$ given by $S = 0$ and $R = 1$. (In the transistor SR bistable $S = 1$ and $R = 0$ gives this state.)

If S becomes 1 with R still 1, gate X inputs are now $S = 1$ and $A = 0$ (since $\overline{Q} = 0$), i.e. one input is a 0 and so Q stays at 1. The circuit has thus 'remembered' or 'latched' the state $Q = 1$.

Reset state In this second stable state $Q = 0$ and $\overline{Q} = 1$. It is given by $S = 1$ and $R = 0$, which you can check (see question 2). (In the transistor version the reset state is obtained when $S = 0$ and $R = 1$.)

If R becomes 1 with S still 1, Q remains at 0, showing that the reset state has been latched.

Note. When $S = 1$ and $R = 1$, Q (and $\overline{Q}$) can be either 1 or 0, depending on the state before this input condition existed. The previous output state is retained as shown by the second and fourth rows of the truth table. Hence the logic levels of Q and $\overline{Q}$ depend on the *sequence* which makes both inputs 1.

Indeterminate state When $S = 0$ and $R = 0$, we get $Q = 1$ and $\overline{Q} = 1$. If both inputs are then made 1 *simultaneously*, we cannot predict whether the bistable will return to the 'set' or the 'reset' state. This undesirable situation is avoided by changing the inputs *alternately*.

NAND gates board See Fig. 61.10 (p. 132). Two of the gates on this board can be connected to form an SR bistable and its truth table checked.

DEBOUNCING A SWITCH

If a mechanical switch, for example on a keyboard, is used to change the state of a bistable (or any other logic system), more than one electrical pulse may be produced due to the metal contacts of the switch not staying together at first but bouncing against each other rapidly and creating extra unwanted pulses, Fig. 69.4a. The effect of this 'contact bounce' can be eliminated by using an SR bistable in the 'anti-bounce' circuit of Fig. 69.4b to 'clean up' the switch action.

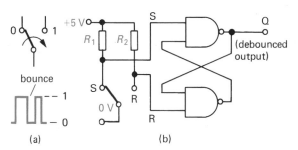

Fig. 69.4

When the switch is in the position shown, S is connected to 0 V and is at logic 0 while R is at logic 1 due to its connection via R_2 to supply positive. Hence $Q = 1$ (from the truth table in Fig. 69.3). If the switch is then operated to make the circuit at the other contact and 'bounces' once,

Switch position		S	R	Q
S○⟍○R	making contact at S	0	1	1
S○⫯○R	moving to make contact at R	1	1	1
S○⟋○R	making contact at R	1	0	0
S○⫯○R	bounces back from contact at R	1	1	0
S○⟋○R	remakes contact at R	1	0	0

Fig. 69.5

the logic levels at S and R for different positions are given in Fig. 69.5. The last three lines show that Q stays at logic 0 despite the bounce.

TRIGGERED (T-TYPE) BISTABLE

A T-type bistable is a modified SR type with extra components that enable successive pulses, applied to an input called the *trigger* T, Fig. 69.6a, to make the bistable switch to and fro, or 'toggle', from one stable state (e.g. $Q = 0$ and $\bar{Q} = 1$) to the other (e.g. $Q = 1$ and $\bar{Q} = 0$).

Two trigger input pulses are required to give one output pulse at either Q or $\bar{Q}$. The frequency of the output

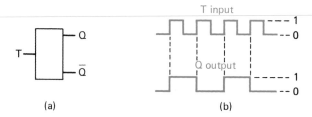

(a)   (b)

Fig. 69.6

pulses is therefore half that of the input pulses as the waveforms in Fig. 69.6b show. A T-type flip-flop *divides the input frequency by two* and forms the basis of binary counters (p. 167). It can be built from transistors or NAND gates but is not available as an IC since other types can be made to toggle, as we will see shortly.

QUESTIONS

1. The circuit in Fig. 69.7 is to be used in a quiz game where there are two competitors. It operates a system whereby a lamp indicates which of them has pressed his switch first. Explain why if switch A is closed first, lamp L_B will not light if the other competitor then closes switch B.   (*L. part qn.*)

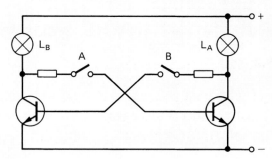

Fig. 69.7

2. In the circuit of Fig. 69.3 (p. 155), if $S = 1$ and $R = 0$ what is the logic level of input   (i) A, (ii) B?

3. a) A T-type flip-flop 'toggles'. Explain this statement.
b) Copy Fig. 69.6b and draw under it the $\bar{Q}$ output.

70 BISTABLES (II)

CHECKLIST

After studying this chapter you should be able to:
- explain the need for 'clocking' and the difference between level and edge triggering,
- draw the block symbol for a D-type bistable and say what each terminal does,
- state when a D-type bistable transfers data from its input to its output and draw a timing diagram,
- state how a D-type bistable acts as a data latch and describe the action of a number of D-type bistables connected as a latch,
- recall how a D-type bistable can act as a T-type,
- recall the properties of a J–K bistable,
- recall the action of a master–slave flip-flop,
- state what a Schmitt trigger does,
- state that a Schmitt trigger is always saturated and that there are two distinct input voltages at which the output changes state, and draw a graph to show its action, and
- state some uses of Schmitt triggers.

The bistables described in this chapter are more versatile than the SR type from which they have been developed. While they can also be made from discrete transistors or NAND gates, it is much more convenient to use them in IC form. Their internal circuits will not be considered. For our purposes it is enough to know what they do and how to connect them, i.e. to treat them as 'black boxes'.

CLOCKED BISTABLES

Clocked logic In large digital systems containing hundreds of interconnected bistables, outputs do not respond immediately to input changes but wait until a *clock* pulse is received (also called a *trigger* or *enabling* pulse). If the same clock pulse is applied to all bistables simultaneously, they change together, i.e. they are synchronized. Otherwise timing sequences can be upset by a 'race condition' in which the output from the system is decided by the speed of operation of individual gates rather than by the logic rules they should obey. In a clocked logic system changes occur in an orderly way, a step at a time, as commanded by the clock pulses.

A clocked SR bistable can be made by adding a two-input AND gate before each of the bistable inputs S_B and R_B, as in Fig. 70.1a. Then, if the clock input CK is at logic 1, S_B is 1 if the upper AND gate input S_A is 1 and R_B is 0 if the lower AND gate input R_A is 0. Data as 1s and 0s thus passes from S_A and R_A to the bistable only when CK is 1. If CK is 0, S_B and R_B cannot change even when S_A and R_A do.

'Clocks' Clock pulses are supplied by some form of pulse generator. It may be a crystal-controlled oscillator with a very steady repetition frequency, e.g. 10 MHz, or

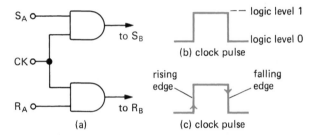

Fig. 70.1

an astable multivibrator (p. 161) or a mechanical switch turning a d.c. supply on and off. The pulses should have fast rise and fall times, i.e. be good 'square' waves, and any switches should be debounced (p. 155).

Level and edge triggering There are two types of clocking or triggering operation. In *level* triggering, bistables change state when the logic *level* of the clock pulse is 1 or 0, Fig. 70.1b. The SR bistables described in chapter 69 are level triggered as is the clocked SR bistable just considered.

In *edge* triggering, a *change* in voltage level causes switching. If it occurs during the rise of the clock pulse from logic level 0 to 1 it is rising or positive-edge triggering and most modern clocked logic is of this type. In falling or negative-edge triggering, switching occurs when the clock pulse falls from 1 to 0, Fig. 70.1c. The T-type bistable (p. 156) is edge triggered.

In general, edge triggering is more satisfactory than level triggering because in the former, output changes occur at an *exact* instant during the clock pulse and any further input changes do not affect the output until the next clock pulse rise (or fall). In level triggering, output changes can occur at any time while CK is 1.

D-TYPE BISTABLE

As a data latch   The symbol for a clocked D-type bistable is given in Fig. 70.2a. CK is the input for *clock* pulses. D is the input to which a bit of *data* (a 0 or a 1) is applied for 'processing' by the bistable, while Q and $\overline{Q}$ are the two outputs, one always being the complement of the other (i.e. if Q = 1, $\overline{Q}$ = 0 and vice versa). S is the *set* input allowing Q to be set to 1 and R is the *reset* input by which Q can be made 0.

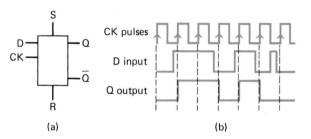

(a)   (b)

Fig. 70.2

Suppose the flip-flop is rising-edge triggered, then the logic level of the D input is transferred to the Q output on the rising edge of a clock pulse, as shown by the *timing diagram* in Fig. 70.2b which you should study carefully. It shows that the Q output 'latches' on to the D input and stores it *at the instant a clock pulse changes its logic level from 0 to 1*. The flip-flop is a *data latch* which acts as a *buffer* between input and output.

Since a D-type bistable has only one data input, the indeterminate state (of two inputs at logic level 0: p. 155) cannot rise.

Quad data latches (i.e. four D-type bistables) are used to obtain a reading on a rapidly changing numerical display that is counting pulses (which would otherwise be seen as a blur) coming at a high rate from a counter. The latches are connected between the counter and the decoder (e.g. binary to decimal, p. 144) feeding the display, as in Fig. 70.3. When the latch is enabled by receiving an appropriate signal, it stores the count (as a four-bit binary number) and holds it for display until the next count enters the latch. In the meantime, the counter can carry on counting.

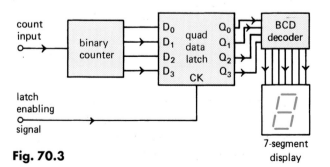

Fig. 70.3

As a T-type bistable   If the $\overline{Q}$ output is connected to the D input as in Fig. 70.4a, successive clock pulses make the flip-flop 'toggle'. If the first clock pulse leaves Q = 1 and $\overline{Q}$ = 0 (because D = 1), then feedback (from $\overline{Q}$ to D) makes D = 0 and during the second clock pulse D is transferred to Q, so now Q = 0 and $\overline{Q}$ = 1. D now becomes 1 and the third clock pulse makes Q = 1 and $\overline{Q}$ = 0 again and so on.

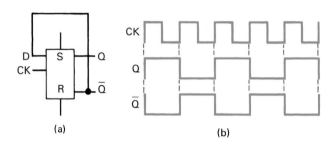

(a)   (b)

Fig. 70.4

Hence, Q = 1 *once* every *two* clock pulses, that is, the output has *half* the frequency of the clock pulses, Fig. 70.4b, and is dividing the clock frequency by two. It is used for this purpose in binary counters (p. 167).

J–K BISTABLE

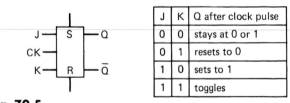

Fig. 70.5

J	K	Q after clock pulse
0	0	stays at 0 or 1
0	1	resets to 0
1	0	sets to 1
1	1	toggles

The clocked J–K bistable is as versatile as a bistable can be and has many uses. Its symbol and truth table are given in Fig. 70.5. There are two inputs, called J and K (for no obvious reason), a clock input CK, two outputs Q and $\overline{Q}$ and set S and reset R inputs. All four input combinations can be used on J and K and there is no disallowed, indeterminate state. When clock pulses are applied to CK it:

(i)   retains its present state if J = K = 0,
(ii) acts as a D-type flip-flop if J and K are different,
(iii) acts as a T-type flip-flop if J = K = 1.

MASTER–SLAVE FLIP-FLOPS

If the duration of a clock pulse is greater than the switching time of a flip-flop, outputs can 'race' round to inputs (because of the feedback connections) causing uncontrolled switching and unreliable operation.

To prevent such undesirable effects, circuits have been developed using two flip-flops, one called the *master* and the other the *slave*. The arrangement for a D-type flip-flop is shown in the block diagram of Fig. 70.6. The action occurs in two stages, controlled by the clock pulse. During the first stage, while the master is enabled and stores the data which is present at its D input, the slave is disabled (by the inverter at its clock input) and thereby isolated from the master. In the second stage, the master is disabled and isolated from the inputs while the slave, now enabled, accepts the data from the master and transfers it to the output. The risk is thus removed of output changes from the slave being fed back to the input of the master before the clock pulse has ended. This solves the 'race' problem.

The circuit changes state reliably after each clock pulse even though the D input changes almost immediately after a clock pulse, i.e. once the clock pulse starts to rise (or fall) the D input is locked out.

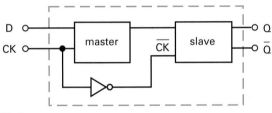

Fig. 70.6

SCHMITT TRIGGER

The Schmitt trigger is a circuit which, like a bistable, uses *positive feedback* and switches its output very rapidly from one state to the other. It is available in IC form as an inverter and as a NAND gate; Fig. 70.7a, b, shows their symbols.

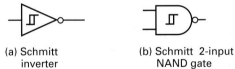

(a) Schmitt inverter

(b) Schmitt 2-input NAND gate

Fig. 70.7

Action   This is shown by the graph in Fig. 70.8. The value of the input voltage V_i which triggers the circuit and makes the output V_o jump sharply from the 'low' to the 'high' state is called the *upper trip point* (UTP), here 2 V. The change from the 'high' to the 'low' state occurs at a lower value of V_i, called the *lower trip point* (LTP), here 1 V. Values of V_i between the UTP and the LTP are in the 'dead' band or hysteresis range and are ignored. Note that the output is always saturated and that there are two distinct input voltages at which the output changes state.

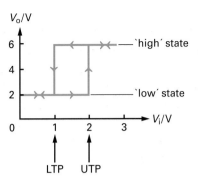

Fig. 70.8

A basic circuit is shown in Fig. 70.9 using two NAND gates, connected as inverters, with a feedback loop. When the input goes positive, it reaches a value where the output of the first inverter begins to go 'low'. As soon as this occurs, the second inverter starts to go 'high' and, due to the feedback loop, makes the input to the first inverter more positive and speeds up the change of state. The opposite action occurs when the input voltage falls. The UTP and LTP are controlled by the ratio R_i/R_f and the supply voltage.

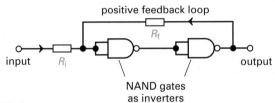

Fig. 70.9

Uses

(i) *Rise-and-fall time improver*. For reliable operation many logic circuits require input pulses with very fast rise and fall times, i.e. good square waves. When switching occurs between 0 and 1 and vice versa, logic gates behave as high-gain amplifiers and are then most likely to become unstable and produce spurious (extra) signals. It is therefore essential for gates to spend only a short time in this critical region.

A Schmitt trigger can convert a slowly changing analogue input from a potential divider like that in Fig. 70.10 (with an LDR, a thermistor or a capacitor between A and B) into one with a fast rise time for application to the NAND gate.

(ii) *Converter of a sine wave into a square wave*. The peaks and troughs of the input to a Schmitt trigger are 'clipped' by it into a good square-wave output, Fig. 70.11 (and see question 3).

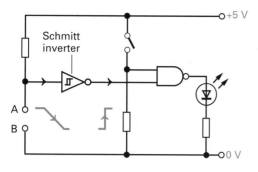

Fig. 70.10

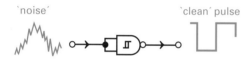

Schmitt NAND gate
as inverter

sine wave   square wave

Fig. 70.11

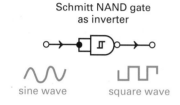

'noise'   'clean' pulse

Fig. 70.12

(iii) *'Noise' remover*. 'Noisy' inputs can be cleaned up. These are inputs which have picked up unwanted voltages or currents which appear as spikes on the waveform, Fig. 70.12.

(iv) *Square-wave generator*, Fig. 70.13. A Schmitt trigger can act as an astable producing a continuous train of pulses with fast rise and fall times suitable for use as a 'clock' in digital circuits. Oscillations occur because C starts to charge up at switch-on and when its voltage reaches the UTP the output of the Schmitt trigger rapidly goes 'low' (since its input is 'high'). C then starts to discharge through R and when its voltage drops to the LTP, the Schmitt output quickly switches to 'high', allowing C to

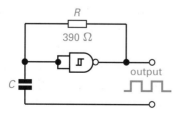

Fig. 70.13

recharge to the UTP and so on. The process is repeated again and again, and 'good' square waves are produced. The output frequency f is given by:

$$f \approx \frac{2000}{C} \, \text{Hz}$$

where C is in µF.

Op amp Schmitt triggers   A Schmitt trigger can be made by applying positive feedback to an op amp, i.e. by a loop from the output to the non-inverting (+) input. An inverting trigger is obtained if the input voltage V_i is applied to the inverting (−) input, Fig. 70.14a. Applying V_i to the non-inverting (+) input gives a non-inverting trigger (i.e. a buffer), Fig. 70.14b.

The switching (trigger) levels depend on the feedback resistor values and the supply voltage.

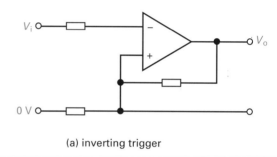

(a) inverting trigger

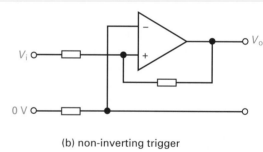

(b) non-inverting trigger

Fig. 70.14

SEQUENTIAL LOGIC

Bistables and Schmitt triggers are 'memory-type' circuits which use *sequential* logic, i.e. their outputs depend not only on *present* inputs (as was the case with combinational logic circuits, p. 130), but on *previous* ones as well. The order or sequence in which inputs are applied is important (p. 155).

The chief requirement of a sequential logic circuit is that it should 'remember' the earlier inputs. This is achieved by feedback connections which ensure that, when the input changes, the effect of the previous input is not lost.

QUESTIONS

1. Copy Fig. 70.15 which refers to a D-type bistable operating on the *rising edges* of clock pulses. Draw under it the waveform for the Q output.

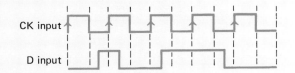

Fig. 70.15

2. Clock pulses of frequency 2 Hz are applied to the CK input of a D-type flip-flop connected to toggle. What is the frequency of the output pulses?

3. a) Draw the characteristic of a Schmitt trigger which has a UTP = 3 V, a LTP = 2 V, a 'high' state = 9 V and a 'low' state = 3 V.

b) Sketch the output voltage waveform from the Schmitt trigger of **a)** if the input is a sine wave of peak value 5 V.

4. The circuit in Fig. 70.16 is suggested for a level-triggered D-type flip-flop. Copy the truth table and by completing it (using the truth table for an SR flip-flop in Fig. 69.3), decide if it is suitable.

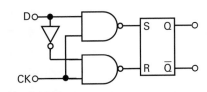

D	CK	S	R	Q
0	1			
0	0			
1	1			
1	0			

Fig. 70.16

71 ASTABLES AND MONOSTABLES

CHECKLIST

After studying this chapter you should be able to:
- state what astables and monostables do,
- recognize the circuits and explain the actions of transistor astables and monostables,
- recognize the circuits and explain the actions of NAND-gate astables and monostables,
- using given formulae, calculate frequency and period for NAND-gate astables and monostables,
- using given formulae, calculate frequency and period for the 555 timer as a monostable and as an astable,
- explain the term 'duty cycle' and state how it can be changed for a 555 timer,
- state the advantages of crystal-controlled oscillators, and
- recognize the circuit and explain the action of an op amp monostable.

INTRODUCTION

Astable or free-running multivibrator This has no stable states. It switches from the 'high' state to the 'low' state and vice versa automatically at a rate determined by the circuit components. Consequently it generates a continuous stream of almost square wave pulses, i.e. it is a square-wave oscillator and belongs to the family of relaxation oscillators (p. 125).

The op amp version was considered earlier (p. 125); here the transistor and NAND-gate versions will be described.

Monostable or 'one-shot' multivibrator This has one stable state and one unstable state. Normally it rests in its stable state but can be switched to the other state by applying an external trigger pulse, where it stays for a certain time before returning to the stable state. It will convert a pulse of unpredictable length (time) from a switch into a 'square' pulse of predictable length and height (voltage) for the input to another circuit.

It can be built from transistors, NAND gates or an op amp.

TRANSISTOR ASTABLE

The circuit, Fig. 71.1, is similar to that of the transistor bistable (Fig. 69.1) but feedback is via capacitors.

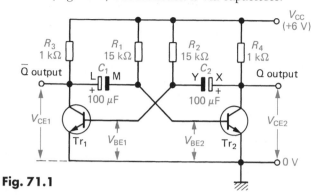

Fig. 71.1

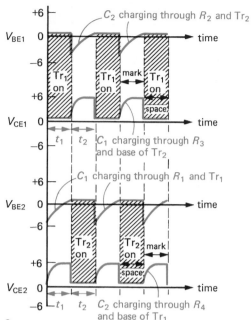

Fig. 71.2

Action When the supply is connected, one transistor quickly saturates and the other cuts off (as with the bistable). Each then switches to its other state, then back to its first state and so on. As a result, the output voltage, which can be taken from the collector of either transistor, is alternately 'high' (+6 V) and 'low' (near 0 V) and is a series of almost square pulses.

To see how these are produced, suppose that Tr_2 was saturated (i.e. on) and has just cut off, while Tr_1 was off and has just saturated. Plate L of C_1 was at +6 V, i.e. the collector voltage of Tr_1 when it was off; plate M was at +0.6 V, i.e. the base voltage of Tr_2 when it was saturated. C_1 was therefore charged with a p.d. between its plates of (6 V − 0.6 V) = +5.4 V.

At the instant when Tr_1 suddenly saturates, the voltage of the collector of Tr_1, and so also of plate L, falls to 0 V (nearly). But since C_1 has not had time to discharge, there is still +5.4 V between its plates and therefore the potential of plate M must fall to −5.4 V (p. 32). This negative potential is applied to the base of Tr_2 and turns it off.

C_1 starts to charge up via R_1 (the p.d. across it is now 11.4 V), aiming to get +6 V, but, when it reaches +0.6 V, it turns on Tr_2. Meanwhile plate X of C_2 has been at +6 V and plate Y at +0.6 V (i.e. the p.d. across it is 5.4 V). Therefore, when Tr_2 turns on, X falls to 0 V and Y to −5.4 V and so turns Tr_1 off. C_2 now starts to charge through R_2 and when Y reaches +0.6 V, Tr_1 turns on again. The circuit thus switches continuously between its two states.

Voltage waveforms The voltage changes at the base and collector of each transistor are shown graphically in Fig. 71.2. We see that V_{CE1} (collector voltage of Tr_1) and V_{CE2} (collector voltage of Tr_2) are complementary (i.e. when one is 'high' the other is 'low'). Also, they are not quite square but have a rounded rising edge. This happens because as each transistor switches off and V_{CE} rises from near 0 V to V_{CC}, the capacitor (C_1 or C_2) has to be charged (via the collector load R_3 or R_4). In doing so it draws current and causes a small temporary voltage drop across the collector load which prevents V_{CE} rising 'vertically'.

Frequency of the square wave The time t_1 for which Tr_1 is on (i.e. saturated with $V_{CE1} \approx 0$) depends on how long C_1 takes to charge up through R_1 from −5.4 V to +0.6 V (i.e. 6.0 V = V_{CC}) and switch on Tr_2. It can be shown that it depends on the time constant C_1R_1 (p. 31) and is given by:

$$t_1 \approx 0.7\,C_1R_1$$

Similarly, the time t_2 for which Tr_2 is on (i.e. the time for which Tr_1 is off) is given by:

$$t_2 \approx 0.7\,C_2R_2$$

The frequency f of the square wave is therefore:

$$f = \frac{1}{t_1 + t_2} \approx \frac{1}{0.7\,(C_1R_1 + C_2R_2)}$$

If $C_1 = C_2$ and $R_1 = R_2$ then $t_1 = t_2$, i.e. the transistors are on and off for equal times and their *mark–space ratio* (p. 16) is 1. We then have, for f in Hz:

$$f \approx \frac{1}{1.4\,C_1R_1} \approx \frac{0.7}{C_1R_1}$$

if C_1 is in farads (F) and R_1 in ohms (Ω). For example, if $C_1 = C_2 = 100 \, \mu F = 100 \times 10^{-6} \, F = 10^{-4} \, F$ and $R_1 = R_2 = 15 \, k\Omega = 1.5 \times 10^4 \, \Omega$, then $f \approx 0.7/(10^{-4} \times 1.5 \times 10^4) = 0.7/1.5 \approx 0.5 \, Hz$.

The mark–space ratio of V_{CE1} and V_{CE2} can be varied by choosing different values for C_1R_1 and C_2R_2. However the base resistors R_1 and R_2 should be low enough to saturate the transistors. (The conditions for satisfactory saturation and cut-off is (as for the bistable) R_1/R_3 and R_2/R_4 both less than h_{FE}. In Fig. 71.2 this means the transistors must have $h_{FE} > 15$.)

TRANSISTOR MONOSTABLE

The circuit is shown in Fig. 71.3; one stage is coupled by a capacitor C_1 and the other by a resistor R_2. When the supply is connected, the circuit settles in its one stable state with Tr_1 off and Tr_2 held on (saturated) by resistor R_1. The output voltage is therefore zero, i.e. $Q = 0$.

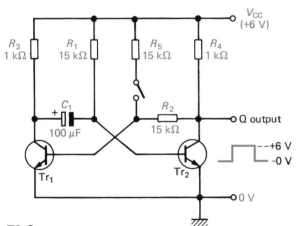

Fig. 71.3

A positive voltage pulse applied via R_5 from the 6 V supply to the base of Tr_1 switches it on. The potential of the right-hand plate of C_1 therefore falls rapidly from +0.6 V to −5.4 V (as explained for the bistable, p. 154) and switches off Tr_2, making Q go 'high' (+6 V). The monostable is now in its second state but only until the right hand plate of C_1 charges up through R_1 to +0.6 V. Then Tr_2 is switched on again and Q goes 'low' (0 V).

The time T of the 'square' output pulse is the time for which Q is 'high' and is given approximately by:

$$T \approx 0.7 \, C_1R_1$$

NAND-GATE ASTABLE AND MONOSTABLE

Astable The circuit using two CMOS inverters A and B (e.g. two one-input NANDs) is shown in Fig. 71.4. Output B is the input to A and, if at switch-on it is 'high',

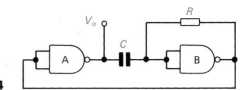

Fig. 71.4

output A, and so also the astable output V_o, will be 'low'. Since input B must be 'low' (if its output is 'high'), C will start to charge up through R.

When the right-hand plate of C (in the diagram) is positive enough, i.e. 'high', output B begins to go 'low' driving input A 'low' and making V_o 'high'. As a result the potential of the right plate (as well as the left) of C rises suddenly because the charge on a capacitor cannot be changed in zero time. The resulting *positive* feedback reinforces the direction input B was going and confirms the 'high' state of V_o.

Because the input to B is now 'high' and its output 'low', C starts to discharge through R and when the input voltage to B becomes 'low', B switches back to its first state with a 'high' output. This in turn makes V_o 'low'. The process is repeated to give a square-wave output V_o, at a frequency f in Hz given approximately by:

$$f \approx \frac{1}{2RC}$$

if R is in ohms and C in farads. For example, if $R = 10 \, k\Omega = 10^4 \, \Omega$ and $C = 0.1 \, \mu F = 10^{-7} \, F$, then:

$$f \approx \frac{1}{2 \times 10^4 \times 10^{-7}} = 500 \, Hz$$

Monostable The circuit of Fig. 71.5 is in its stable state with output B (V_o) 'low' because its input is 'high' via R. Closing S briefly takes output A from 'high' (due to its 'low' input from output B) to 'low' and so forces input B 'low' and V_o 'high'. C then charges through R until input B is 'high' enough to switch its output back to the stable state with V_o 'low'. The time T of the pulse so produced is given in seconds approximately by:

$$T \approx RC$$

where R is in ohms and C in farads.

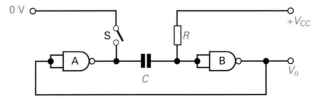

Fig. 71.5

555 TIMER IC

In many electronic systems timing operations are controlled by an astable or a monostable. The popular eight-pin d.i.l. 555 timer IC, Fig. 71.6, can be used as either. It works on any d.c. supply from 3 to 15 V and has the following properties.

(i) When the voltage between 'trigger' (pin 2) and 'ground' (pin 1) falls below $\frac{1}{3}V_{CC}$, the IC is triggered, the 'output' (pin 3) rises to near V_{CC} and 'discharge' (pin 7) is disconnected from 'ground'.

(ii) When the voltage between 'threshold' (pin 6) and ground exceeds $\frac{2}{3}V_{CC}$, the 'output' falls to 0 V and 'discharge' is connected to 'ground'.

Monostable operation The basic connections are shown in Fig. 71.7. R_1 and C_1 are external components whose values determine the time T of the single 'square' output pulse produced when S_1 is switched from X to Y and back to X again. This sends the output 'high' (see (i) above) and allows C_1 to charge up through R_1 (since 'discharge' is now an open circuit).

When the voltage across C_1 reaches $\frac{2}{3}V_{CC}$, the output goes 'low' (see (ii) above) and C_1 discharges (since 'discharge' is now connected to ground), returning the circuit to its stable state to avoid the next trigger pulse. It can be shown that T is given in seconds by:

$$T \approx 1.1\, R_1 C_1$$

if R_1 is in MΩ and C_1 in μF. For example, if $R_1 = 1$ MΩ and $C_1 = 10$ μF then:

$$T \approx 1.1 \times 1 \times 10 = 11 \text{ s}$$

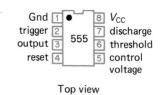

Fig. 71.6

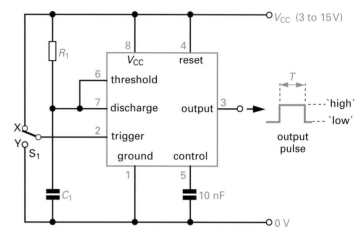

Fig. 71.7

Astable operation In the simplified circuit of Fig. 71.8a, R_1, R_2 and C_1 are external components; their values decide the frequency f of the square-wave oscillations produced automatically at the output. These occur because C_1 charges up through R_1 and R_2 and when the p.d. across it reaches $\frac{2}{3}V_{CC}$, the output goes 'low'. C_1 then starts to discharge (via 'discharge' which is now connected to ground) and when the voltage across it falls below $\frac{1}{3}V_{CC}$, the output goes 'high' and C_1 recharges (since 'discharge' is again an open circuit). This sequence is repeated continuously.

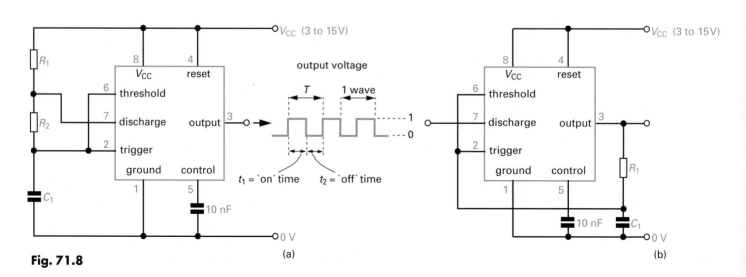

Fig. 71.8

(a)

(b)

The frequency *f* in Hz is given by:

$$f \approx \frac{1.4}{(R_1 + 2R_2)C_1}$$

$$\approx \frac{0.7}{R_2 C_1} \quad \text{if } R_2 \gg R_1$$

The *duty cycle*, or *mark–space ratio* (Chapter 7), is defined by:

$$\text{duty cycle} = \frac{\text{'on' time when output is 1}}{\text{'off' time when output is 0}}$$

$$= \frac{t_1}{t_2} \quad \text{(see Fig. 71.8a)}$$

It can be shown that:

$$t_1 \approx 0.7(R_1 + R_2)C_1$$
$$t_2 \approx 0.7R_2C_1$$

Hence t_1 always exceeds t_2 unless R_1 is very small compared with R_2, in which case $t_1 \approx t_2$ and a true square wave is obtained, i.e. duty cycle = 1. Note that $f = 1/T$, where

$$T = t_1 + t_2 \approx 0.7(R_1 + R_2)C_1 + 0.7R_2C_1$$
$$\approx 0.7(R_1 + 2R_2)C_1$$

Another simple astable circuit using only one resistor and not the discharge terminal is shown in Fig. 71.8b. In this case:

$$t_1 = t_2 \approx 0.7R_1C_1$$

Frequency modulation The control voltage connection (pin 5 on the IC) is usually connected to 0 V via a 0.01 μF capacitor. However, if a voltage (between $\frac{1}{3}$ and

$\frac{2}{3}$ of the supply voltage) is applied to it, the frequency of the astable output can be varied independently of R_1, R_2 and C_1. The process is called *frequency modulation*.

Astable–monostable board See Fig. 71.9. This is a useful board for investigating the properties of the 555.

CRYSTAL-CONTROLLED ASTABLE

Crystal oscillators are commonly used where a fixed, very stable frequency is required. Their action depends on the *piezoelectric* effect (Chapter 17).

When certain crystals, notably quartz, are made to vibrate mechanically, an a.c. voltage develops across their opposite sides (because of the displacement of ions in the lattice) at their *natural frequency* of oscillation. In effect, they behave as very low-loss resonant circuits with a resonant frequency which depends on their size and shape.

Crystals like that shown in Fig. 71.10a (enclosed in a metal can with two leads from its electrodes) with its symbol, can be used in different *feedback* loops to produce oscillators based on a transistor, an op amp or logic gates. The NOT gate circuit of Fig. 71.10b produces a square-wave output suitable for driving logic circuits. Switching on the supply applies a voltage pulse to the crystal, making it vibrate.

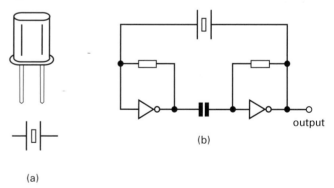

(a)　(b)

Fig. 71.10

OP AMP MONOSTABLE

In the circuit of Fig. 71.11a, when S_1 is pressed and released, one positive pulse is produced whose duration depends on the time constant C_1R_1.

With S_1 open, the + input is grounded via R_1, i.e. $V_2 = 0$ V and the − input is positive owing to the small voltage at the junction of the voltage divider formed by R_2 and R_3. Therefore V_1 is greater than V_2 and the op amp is negatively saturated, i.e. $V_o = -V_s$. Also C_1 is charged with plate X positive relative to plate Y. (In fact, X is at 0 V and Y at $-V_s$ approximately.)

Fig. 71.9

When S_1 is closed briefly, the $-$ input is connected momentarily to $-V_s$, i.e. $V_1 = -V_s$ and so V_2 is greater than V_1. The op amp becomes positively saturated, i.e. $V_o = +V_s$. This makes Y rise to $+V_s$ and X to near $+2V_s$, i.e. $V_2 \approx +2V_s$, thereby holding the op amp in positive saturation even when S_1 opens again. C_1 starts to discharge through R_1, V_2 gradually falls to zero and when it is less than V_1, the op amp reverts to negative saturation again. Fig. 71.11b shows waveforms for V_1, V_2 and V_o.

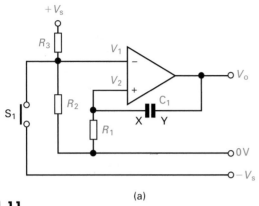

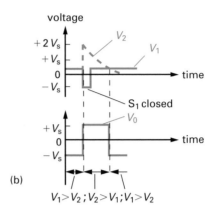

Fig. 71.11

(a)

(b)

$$V_1 > V_2 \; ; V_2 > V_1 ; V_1 > V_2$$

QUESTIONS

1. Draw two circuits to show how you would connect an LED to the output and power supply of a monostable so that the LED lights when the output is (i) 'high', (ii) 'low'.

2. The graph in Fig. 71.12 shows how the output voltage from an astable varies with time.

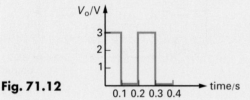

Fig. 71.12

a) What is the period of the output?
b) What is the frequency of the output?
c) What is the value of the duty cycle?

3. The astable in Fig. 71.8a (p. 164) has a frequency of 1 Hz and has a LED with a suitable series resistor connected between it and the 0 V line. R_2 is much greater than R_1.

a) What would you see the LED doing?
b) What happens to the LED if the value of R_2 is halved?
c) What happens to the LED if the value of C_1 is doubled?

72 BINARY COUNTERS

CHECKLIST

After studying this chapter you should be able to:
- recognize and describe the action of a number of toggling bistables connected as a binary up-counter or frequency divider,
- state what is meant by the term 'modulo',
- distinguish between ripple counters and synchronous counters,
- recognize and describe the action of a binary down-counter, and
- draw block diagrams for counters of different modulo.

Counters consist of bistables connected so that they toggle when the pulses to be counted are applied to their clock input. Counting is done in binary code, the bits 1 and 0 being represented by the 'high' and 'low' states of the bistable's Q output.

BINARY UP-COUNTER

A simple three-bit binary up-counter is shown in Fig. 72.1 consisting of cascaded (in series) toggling flip-flops FF_0, FF_1 and FF_2 (i.e. with $\bar{Q}$ joined to D if they are D-types and $J = K = 1$ if they are J–K types) with the Q output of each feeding the clock input (CK) of the next. The total count is given at any time by the states of Q_0 (the l.s.b.), Q_1 and Q_2 (the m.s.b.) and it progresses upwards from 000 to 111 (7 in decimal) as shown in the table, before resetting to 000.

Number of clock pulse	Outputs		
	Q_2	Q_1	Q_0
0	0	0	0
1	0	0	1
2	0	1	0
3	0	1	1
4	1	0	0
5	1	0	1
6	1	1	0
7	1	1	1

Action   Suppose the flip-flops are triggered on the *falling* edge of a pulse and that initially Q_0, Q_1 and Q_2 are all reset to zero.

On the falling edge ab of the first clock pulse, shown in Fig. 72.2, Q_0 switches from 0 to 1. The resulting rising edge AB of Q_0 is applied to CK of FF_1, which does not change state because AB is not a falling edge. Hence the

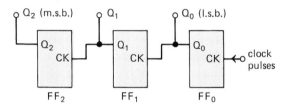

Fig. 72.1

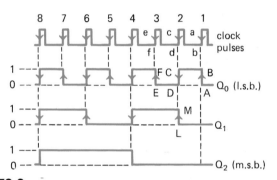

Fig. 72.2

output states are $Q_2 = 0$, $Q_1 = 0$ and $Q_0 = 1$, giving a binary count of 001.

The falling edge cd of the second clock pulse makes FF_0 change state again and Q_0 goes from 1 to 0. The falling edge CD of Q_0 switches FF_1 this time, making $Q_1 = 1$. The rising edge LM of Q_1 leaves FF_2 unchanged. The count is now $Q_2 = 0$, $Q_1 = 1$ and $Q_0 = 0$, i.e. 010.

The falling edge ef of the third clock pulse to FF_0 changes Q_0 from 0 to 1 again but the rising edge EF does not switch FF_1 leaving $Q_1 = 1$, $Q_2 = 0$ and the count at 011. The action thus ripples along the flip-flops, each one waiting for the previous flip-flop to supply a falling edge at its clock input before changing state.

Modulo   The *modulo* of a counter is the number of output states it goes through before resetting to zero. A counter with three flip-flops counts from 0 to $(2^3 - 1) = 8 - 1 = 7$; it has eight different output states representing the decimal numbers 0 to 7 so is a modulo-8 counter.

Dividers In a modulo-8 counter one output pulse appears at Q_2 every eighth clock pulse. That is, if f is the frequency of a regular train of clock pulses, the frequency of the pulses from Q_2 is $f/8$ (i.e. Q_2 is 1 for four clock pulses and 0 for the next four pulses). A modulo-8 counter is a 'divide-by-8' circuit as well. Counters are used as dividers in digital watches (p. 172).

Synchronous counter The counter described above is a *ripple* or *asynchronous* type. In a *synchronous* counter all flip-flops are clocked simultaneously and the *propagation delay time* (i.e. the time taken between the clock pulse being applied and the output of the counter changing) is much less than for a ripple counter with a large number of flip-flops where there is the risk of a 'race condition' arising (Chapter 70: see 'Clocked bistables'). Synchronous counters are therefore used for high-speed counting but their circuits are more complex than those of ripple types.

BINARY DOWN-COUNTER

In a down-counter the count decreases by one for each clock pulse. To convert the up-counter in Fig. 72.1 into a down-counter, the $\bar{Q}$ output (instead of the Q) of each flip-flop is coupled to the CK input of the next, as in Fig. 72.3. The count is still given by the Q outputs.

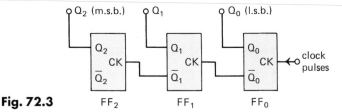

Fig. 72.3

DECADE COUNTER

A four-bit binary up-counter, modified as in Fig. 72.4, acts as a modulo-10 counter and counts up from 0 to 9 before resetting. When the count is 1010 (decimal 10), $Q_3 = 1$, $Q_2 = 0$, $Q_1 = 1$ and $Q_0 = 0$ and since both inputs to the AND gate are 1s (i.e. Q_3 and Q_1), its output is 1 and this resets all the flip-flops to 0 (otherwise it would be a modulo-16 counter counting from 0 to 15).

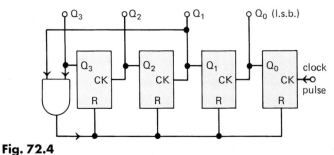

Fig. 72.4

DIVIDING AND COUNTING BOARD

Using the 'debounced switch and pulser' board, Fig. 72.5a, and the 'dividing and counting' board, Fig. 72.5b, the following can be investigated:

(i) the basic action of a D-type bistable,
(ii) the toggling action of a T-type bistable,
(iii) four-bit binary up- and down-counters of various modulo, and
(iv) the action of a shift register (Chapter 73).

(a)

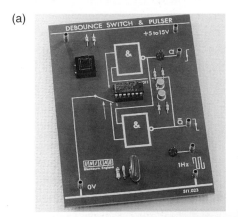

(b)

Fig. 72.5

QUESTIONS

1. a) How many output states are there in a counter which has (i) 1, (ii) 2, (iii) 3, (iv) 4, (v) 5 bistables?
  b) What is the highest decimal number each counter in a) can count before resetting?
2. a) What is meant by the modulo of a counter?
  b) How many flip-flops are needed to build counters of modulo- (i) 2, (ii) 5, (iii) 7, (iv) 10, (v) 30?
  c) If f is the frequency of the clock pulses applied to a modulo-10 counter, what is the frequency of the output pulses from the last flip-flop?
3. Draw the block diagram for a modulo-6 binary up-counter which resets every sixth clock pulse.

73 REGISTERS AND MEMORIES

SHIFT REGISTERS

A shift register is a memory which stores a binary number and shifts it out when required. It consists of several D-type or J–K flip-flops, one for each bit (0 or 1) in the number. The bits may be fed in and out serially, i.e. one after the other, or in parallel, i.e. all together. Shift registers are used, for example, in calculators to store two binary numbers before they are added.

In the four-bit serial-input-serial-output (SISO) type of Fig. 73.1a, clocked D-type flip-flops are used, the Q output of each one being applied to the D input of the next. The bits are loaded one at a time, usually from the left and move one flip-flop to the right every clock pulse. Four pulses are needed to enter a four-bit number such as 0101 and another four to move it out serially.

In the parallel-input-parallel-output (PIPO) type of Fig. 73.1b, all bits enter their D inputs simultaneously and are transferred together to their respective Q outputs (where they are stored) by the same clock pulse. They can then be shifted out in parallel.

In a four-bit serial-input-parallel-output (SIPO) register, the bits are loaded individually in serial form via the D input. They are displayed in parallel form at all four outputs, Fig. 73.1c. The opposite happens with a parallel-input-serial-output (PISO) register.

MEMORIES

Organization A memory stores *data*, i.e. the information to be processed, and the *program of instructions* to be carried out. An IC semiconductor memory consists of an array of memory cells, each storing one bit of data. The array is organized so that the bits are in groups or 'words' of, typically, 1, 4, 8 or 16 bits.

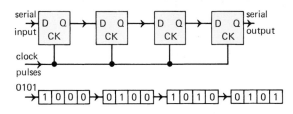

(a) Serial-input-serial-output (SISO)

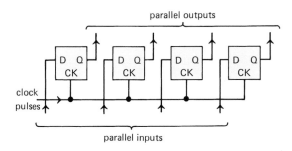

(b) Parallel-input-parallel-output (PIPO)

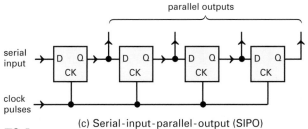

(c) Serial-input-parallel-output (SIPO)

Fig. 73.1

Every word has its own location or *address* in the memory which is identified by a certain binary number. The first word is at address zero, the second at one, the third at two and so on. The table overleaf shows part of the contents of a sixteen-word four-bit memory, i.e. it has sixteen addresses each storing a four-bit word. For example, in the location with address 1110 (decimal 14), the data stored is the four-bit word 0011 (decimal 3).

Address				Data				
Decimal	Binary			Binary				Decimal
	m.s.b.		l.s.b.	m.s.b.			l.s.b.	
0	0	0	0 0	0	1	0	1	5
1	0	0	0 1	1	1	0	0	12
2	0	0	1 0	0	1	1	0	6
3	0	0	1 1	1	0	0	1	9
14	1	1	1 0	0	0	1	1	3
15	1	1	1 1	0	1	1	1	7

In a *random access memory* all words can be located equally quickly, i.e. access is random and it is not necessary to start at address zero.

Types of semiconductor memory

There are two main types—Read Only Memories, i.e. ROMs, and Read and Write Memories which are confusingly called RAMs (because they allow random access, as ROMs also do). A RAM not only lets the data at any address be 'read' but it can also have new data 'written' in. Whereas a RAM normally loses the data stored almost as soon as the power to it is switched off, i.e. it is a *volatile* memory, a ROM does not, i.e. it is *non-volatile*. ROMs are used for permanent storage of fixed data such as the program in a computer. RAMs are required when data has to be changed.

RAMs are of two types—*static* and *dynamic*. In a static RAM (SRAM) each memory cell consists basically of a bistable whose contents are fixed (i.e. a 1 or a 0) until the cell is written into or the power is switched off. The memory cells in a dynamic RAM (DRAM) are tiny capacitors which become charged but due to leakage currents the charged cells have to be 'refreshed' regularly, e.g. every millisecond, from within the chip itself. A charged capacitor represents a 1 and an uncharged one a 0. Non-volatile DRAM ICs are now available with inbuilt low-voltage backup batteries.

For high-capacity memories, DRAMs, despite their need for 'refreshing' circuitry, tend to cost less than SRAMs. This arises from their smaller size due to the use of MOSFETs and the need for only one transistor per memory cell compared with two for a SRAM (which is bistable-based). Power consumption is also lower for MOSFET memory circuits.

ROMs consist of an array of diodes. The programmable ROM or PROM lets the user 'burn' the pattern of bits, i.e. the program, into a ROM by applying a high voltage which fuses a link in the circuit. The disadvantage of this

type is that it does not allow changes or corrections to be made later. When alterations are necessary, for instance during the development of a program, an erasable PROM or EPROM is used in which the program is stored electrically and is erased by exposure to ultraviolet radiation before reprogramming. An EEPROM is an electrically erasable PROM.

Structure The simplified structure of a sixteen-word four-bit RAM is shown by the block diagram of Fig. 73.2; that for a ROM is similar but it has no 'write' provision.

To 'write' a word into a particular address the four-bit binary number of the address is applied to the address inputs and the word (also in binary) is set up at the data inputs. When *write enable* is at the appropriate logic level (say 'high'), the word is stored automatically at the correct address in the memory array, as located by the address decoder.

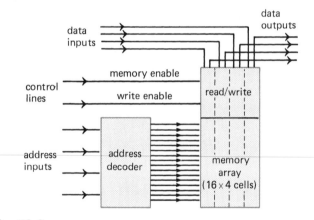

Fig. 73.2

To 'read' a word stored at a certain address, the address code is applied as before and the word appears at the data outputs if write enable is at the required logic level (say 'low').

Storage capacity A sixteen-word four-bit memory has a storage capacity of $16 \times 4 = 64$ bits (it has 64 memory cells) and is limited to four-bit words. An eight-bit word is called a *byte*.

In computer language the symbol K (capital K) is used to represent 1024 (2^{10}). For example, a memory with a capacity of 4 Kbits stores $4 \times 1024 = 4096$ bits, i.e. 512 words if it is organized in bytes or 1024 four-bit words. Do not confuse K with k (small k) which stands for kilo, i.e. 1000.

The number of cells on a memory IC has doubled every year since the early 1970s when the first 1 Kbit DRAM was made. Today semiconductor memories with capacities of 64 million bytes, i.e. 64 megabytes, are common.

MEMORY (RAM) BOARD

The CMOS memory chip on this board, Fig. 73.3, stores up to 16 four-bit binary numbers. In the 'write' mode, data on the four input lines is stored in the address selected as the enable input falls. In the 'read' mode, the data stored is transferred non-destructively to the output sockets and, at the same time, displayed on the output LEDs.

When combined with the 'debounced switch and pulser' and the 'dividing and counting' boards, an automatic sequence of flashing lights can be produced using the memory.

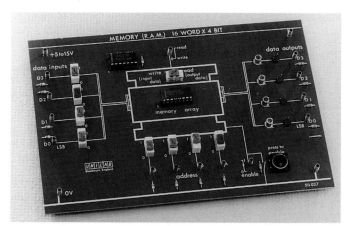

Fig. 73.3

74 SOME DIGITAL SYSTEMS

MICROELECTRONIC OPTIONS

When designing an electronic system using large-scale integrated (LSI) circuits, two broad approaches are possible. In the first, the job done by the system is controlled by its circuit and cannot be changed without altering the circuit. This is the *circuit-controlled* option. In the second, the job done is decided by a program of instructions (software) which tells the system exactly how to perform the task. To change the job done, the program is changed. This is the *program-controlled* option.

The first approach may be best when the system has just one task and the second when it has several. Within the circuit-controlled option there are three choices, to be outlined now. The program-controlled option uses a microprocessor, to be discussed later (p. 230).

Wired logic The system is wired together on a suitable board using standard, off-the-shelf SSI and MSI chips, and perhaps discrete components. The chips might range from logic gates to an ALU (p. 146). This is often the simplest and cheapest method but it is inflexible.

Custom chips In this case all functions demanded of the system are done by one 'dedicated' LSI chip containing the exact number of components for the job. It gives maximum efficiency, minimum size and minimum power consumption.

Uncommitted logic array (ULA) This approach is half-way between that of wired logic and that of custom chips. It uses a chip containing an array of standard logic 'cells', each complete in itself (consisting of logic gates, adders, etc.) but not interconnected into any circuit pattern. Connections are made by the chip manufacturer once a logic diagram has been constructed for the customer's application. ULAs are cheaper than custom chips and can be designed in about one-third of the time. They are also more versatile, within limits.

DIGITAL WATCH

The block diagram of Fig. 74.1 shows the main parts of a digital watch. When activated by a battery, the quartz crystal oscillator produces very stable electrical pulses with a frequency of exactly 32 768 Hz (2^{15} Hz). The dividers reduce this to one pulse per second and after counting and decoding they are applied to the appropriate electrodes of the seven-segment LCD or LED decimal displays to show the time or date.

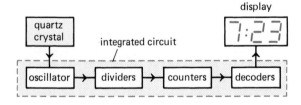

Fig. 74.1

The IC is a custom chip containing over 2000 transistors and incorporates all the system's major circuits.

DIGITAL VOLTMETER

The main parts of a digital voltmeter (p. 91) are shown in the simplified block diagram of Fig. 74.2.

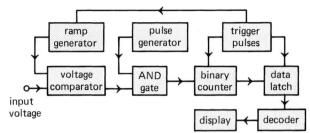

Fig. 74.2

The action is:

(i) a *trigger pulse* sets the *binary counter* to zero and starts the *ramp generator* (p. 123) which produces a repeating sawtooth waveform;

(ii) the *voltage comparator* (p. 121) output changes from 'high' to 'low' when the ramp voltage equals the input voltage;

(iii) the *AND gate* inputs are supplied by the *comparator* and a steady train of pulses from the *pulse generator*; the latter pass through the gate until the *comparator* output goes 'low' and the number which does so depends on the time taken by the ramp voltage to equal the input voltage, i.e. it is proportional to the input voltage (if the ramp is linear);

(iv) the *counter* records the number of output pulses from the *AND gate*;

(v) the *data latch* passes this number to the *decoder* for conversion to decimal before they reach the *display*, where it is held until the next count enters the latch.

The system produces the digital equivalent of the analogue input voltage, i.e. it is an analogue-to-digital (A/D) converter (p. 227) and the waveforms in Fig. 95.3 for a steady d.c. input should also be studied.

ELECTRONIC CALCULATOR

In many calculators a microprocessor is used as the 'brain' but the principles involved can be shown for the addition of two numbers using the very simple system of Fig. 74.3. It has a four-bit adder (p. 138) as its ALU.

Suppose the two numbers to be added are 3 and 5. When switch '3' is pressed on the *keyboard*, the four-bit binary form of decimal 3 is produced by the *encoder* at its four outputs Q_0, Q_1, Q_2 and Q_3. In this case the output values would be Q_3 (m.s.b.) = 0, Q_2 = 0, Q_1 = 1 and Q_0 (l.s.b.) = 1. The binary number 0011 is thus applied to the four inputs of *shift register B*. If the 'store' switch (STO) on the *keyboard* is pressed next, a 'clock' pulse causes *shift register B* to shift 0011 from its inputs to its outputs, where it is stored and becomes the input to *shift register A*.

The second number can now be entered and if switch '5' is pressed, the binary number applied to the inputs of *shift register B* becomes 0101, i.e. $Q_3 = 0$, $Q_2 = 1$, $Q_1 = 0$ and $Q_0 = 1$. On applying a second 'clock' pulse, the first number ($0011 = 3$) is shifted from the inputs of *shift register A* to its outputs where it becomes the $A_3A_2A_1A_0$ input to the *adder*. At the same time, the second number ($0101 = 5$) is shifted from the inputs of *shift register B* to its outputs and becomes the $B_3B_2B_1B_0$ input to the *adder*.

The *adder* adds the two binary numbers $A_3A_2A_1A_0$ and $B_3B_2B_1B_0$ immediately and produces their binary sum $S_3S_2S_1S_0$ at its four outputs. These outputs are applied to the four inputs of the *decoder* which, if it was driving a seven-segment decimal *display*, would create seven outputs each capable of driving one segment. In this example all seven segments would light up and give 8 as the sum of $3 + 5$.

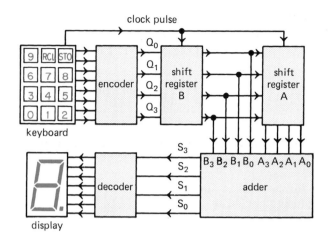

Fig. 74.3

75 PROGRESS QUESTIONS

1. 'One of the characteristics of the bistable circuit is that it has memory'. Explain this statement.

What do you consider are the other important characteristics of the bistable circuit?

How are the properties of the bistable (regarded as a basic module) applied in **(a)** a scaling (i.e. counting) system, **(b)** a shift register? (*O. and C.*)

2. a) Fig. 75.1 represents three bistables connected in series. When a bistable is in the state logical 1 the light-emitting diode (LED) is lit, and when it is in the state logical 0 the LED is off. If a bistable changes state from logical 1 to logical 0, it causes the next bistable along to change state.

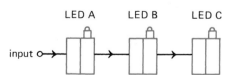

Fig. 75.1

At the start, all the LEDs are off. A series of switching pulses is then fed to the input.

(i) Copy and complete the table to show the states (1 or 0) of the LEDs after each of 7 pulses.

(ii) Explain what happens for the first 3 pulses.

Pulse number	LED A	LED B	LED C
1			
2			
3			
4			
5			
6			
7			

(iii) Describe a use for such an arrangement of bistables.

b) Fig. 75.2 shows a bistable circuit. The lamp A is on after the circuit is first connected up.

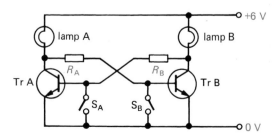

Fig. 75.2

(i) In which transistor is there a large collector current flowing?

(ii) Which switch must be closed momentarily to turn off lamp A? Explain.

(iii) What effect does the closing of this switch have on the base voltage of the transistor Tr B?

(iv) What happens as a result to lamp B? Explain.

(v) How can the initial condition with lamp A be restored? Explain.   (O.L.E.)

3. What is **(a)** a data latch, **(b)** code conversion?

Describe how a digital measuring unit holds, displays and refreshes a reading (as for example in a digital voltmeter). In your answer, be sure to quote all the stages of the process and discuss each as fully as you can.   (O. and C.)

4. a) With the aid of an example, explain the reason for using a Schmitt trigger.

b) A certain Schmitt trigger switches on when the input voltage rises to 2.0 V and then off when it falls to 1.6 V. Draw the output waveform for the input shown in Fig. 75.3, given that the output voltage is 0 V with trigger off and 10 V when it is on.

What is the duration of each output pulse?
                    (O.L.E. part qn.)

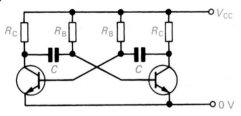

Fig. 75.3

5. Fig. 75.4 shows the circuit of an astable multivibrator.

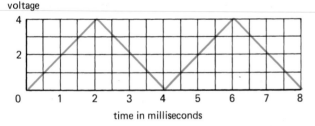

Fig. 75.4

a) Explain the role of the resistors R_B in the operation of the circuit. What factors limit the maximum value which R_B can have? What effect would an increase in the value of R_B have?

b) Explain the role of the capacitors C. What effect would a decrease in the value of C have?

c) Explain the part played by the resistors R_C in the operation of the circuit.

d) Sketch graphs of the variation with time of (i) the p.d. between collector and emitter, and (ii) the p.d. between the base and emitter, for one of the transistors.

e) What effect would a variation in V_{CC} have on the operation of the circuit?

f) How would you change the circuit so that one transistor was on for longer than the other?   (L.)

6. What happens to the pitch of the note from the loudspeaker in the astable circuit of Fig. 75.5 when the amount of light falling on the LDR (i) increases, (ii) decreases?

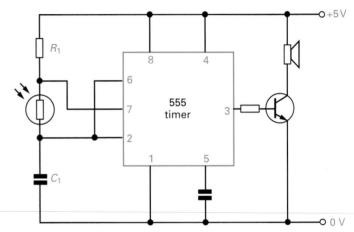

Fig. 75.5

7. The circuit in Fig. 75.6 shows an astable pulse generator made from a 555 timer.

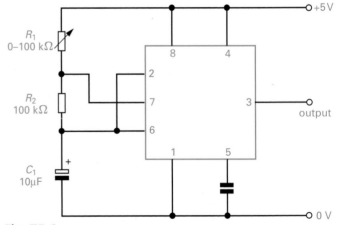

Fig. 75.6

R_1 is a variable resistor. C_1 charges through R_1 and R_2 in series and during this time, t_1, the output is 'high'. C_1 discharges through R_2 only and for this time, t_2, the output is 'low'. The times t_1 and t_2 are given in seconds by:

$$t_1 = 0.7(R_1 + R_2)C_1 \quad \text{and} \quad t_2 = 0.7R_2C_1$$

where R_1 and R_2 are in MΩ and C_1 is in μF (or Ω and F respectively).

a) Copy and complete the table below when R_1 has values of 0, 50 kΩ and 100 kΩ. $T = t_1 + t_2$.

R_1 (kΩ)	R_2 (kΩ)	t_1 (s)	t_2 (s)	T (s)	t_1/t_2
0	100				
50	100				
100	100				

b) What is T called?

c) What is t_1/t_2 called?

d) Which value of t_1/t_2 gives true square waves?

e) Draw a graph of output voltage against time when $R_1 = R_2 = 100$ kΩ; label both axes in the correct units.

8. Figure 75.7 consists of four falling-edge triggered D-type master–slave flip-flops and an AND gate. The flip-flops reset when the R connection is at logic 1.

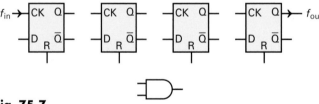

Fig. 75.7

Copy and add appropriate connections to Fig. 75.7 so that:

$$f_{out} = \frac{f_{in}}{10}$$

and has a mark–space ratio of 1 : 1.   (N.)

9. a) Fig. 75.8a shows a falling-edge triggered D-type master–slave flip-flop. Input D is at logic 1 and output Q is initially at logic 0.

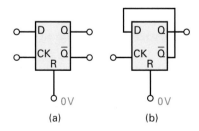

(a)   (b)

Fig. 75.8

Describe the changes in the Q output as the clock input changes from 0 to 1 and back again to 0.

b) In Fig. 75.8b the D input and the inverted output $\bar{Q}$ are connected.

(i) Draw a timing diagram for the circuit showing six clock pulses, with the corresponding Q and $\bar{Q}$ outputs.

(ii) Explain why the circuit reliably changes state after each clock pulse even though the D-input changes almost immediately after a clock pulse.

c) (i) Draw a circuit diagram of a divide-by-eight counter using D-type bistables. Show on the diagram how the counter may be reset to zero.

(ii) What is meant by the term *propagation delay*? Explain the effect it has on the operation of counters working at high speed.

(iii) What is meant by the term *synchronous counter*?

d) (i) Draw a circuit diagram of a four-stage shift register using D-type bistables. Label the reset, clock, data inputs and data outputs.

(ii) State *one* possible use for a shift register.

10. Fig. 75.9 shows a 555 connected as an astable. Time for which V_{out} is high, $t_h = 0.7(R_1 + R_2)C$. Time for which V_{out} is low, $t_l = 0.7R_2C$.

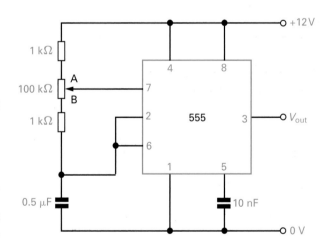

Fig. 75.9

a) When the variable resistor is set to B, calculate:

(i) the time that V_{out} is high,

(ii) the time that V_{out} is low,

(iii) the frequency of the oscillation.

b) The variable resistor is now set to A. Without any further calculations state what will happen to the waveform and the frequency.   (N.)

INFORMATION, ELECTRONICS AND SOCIETY

76 INFORMATION TECHNOLOGY

CHECKLIST

After studying this chapter you should be able to:
• state the relative merits of the two methods of representing information electrically,

• explain the process of pulse code modulation, and
• recall that bit-rate = sampling frequency × no. of bits in code.

Information technology (IT) is concerned with the ability of *computers* to store and process vast amounts of information in split seconds and of *telecommunications* to transmit it almost instantaneously. *Microelectronics* enables the equipment required to be made incredibly small and cheaply. IT is cropping up in every aspect of our lives, bringing changes, as profound as those of the Industrial Revolution, that will affect our homes, shops, offices, factories and schools as well as our health and leisure.

Links are being established allowing information to be sent and received anywhere, in much the same way as telephones enable us to speak to one another. In the broadest sense, *information* includes the messages, programmes and other 'traffic' carried by telephones, radio, television, as well as by computer signals, and the various data-processing devices used by business and government organizations. Today, information is seen, like minerals and energy, as a basic resource which is becoming more easily and widely accessible.

REPRESENTING INFORMATION

Information or data can be represented electrically in two ways.

Digital method In this method electricity is switched on and off and the information is in the form of electrical pulses. For example, in the simple circuit of Fig. 76.1a, data can be sent by the 'dots' and 'dashes' of the Morse code by closing the switch for a short or a longer time. In Fig. 76.1b the letter A (·−) is shown.

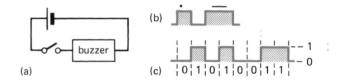

Fig. 76.1

Computers use the simpler binary code with 1 and 0 represented by 'high' and 'low' voltages respectively. They can only handle numbers (i.e. 1 and 0) and so a pattern of 1s and 0s has to be agreed for each of the 26 letters of the alphabet if words are also to be processed.

A five-bit code has 2^5 (32) variations, i.e. from 00000 to 11111, which would be enough. However, many digital systems use eight-bit words, i.e. bytes. The *American Standard Code for Information Interchange* (ASCII) is an eight-bit code that allows $2^8 = 256$ characters to be coded in binary. This is adequate for all letters of the alphabet (capitals and small), the numbers 0 to 9, punctuation marks and other symbols. Fig. 76.1c shows one eight-bit pulse, representing the letter capital S in ASCII code.

Analogue method In this case the flow of electricity is *regulated* (not switched) and a continuous range of voltages (or currents) between 0 and some maximum, is possible, Fig. 76.2a. The actual value at any instant stands for a number.

A crystal microphone (p. 42) produces a voltage which is the analogue of information (in the form of sound). A carbon microphone (p. 43) gives current as the analogue of information.

ADVANTAGES OF DIGITAL SIGNALS

Information in digital form has certain advantages over that in analogue form despite the fact that most transducers produce analogue signals which can be amplified readily. There are two main reasons for this.

First, with digital signals it is only necessary to detect the presence or absence of a pulse (i.e. whether it is 1 or 0). Should they pick up 'noise' (i.e. stray, unwanted voltages or currents) at any stage of processing or transmission, it does not matter as it would with an analogue signal whose waveform would be distorted.

Second, digital signals fit in with modern technology and can be used with both telecommunications and data-processing equipment.

ANALOGUE-TO-DIGITAL CONVERSION

With the increasing popularity of digital signals this operation is frequently necessary. It is performed by an analogue-to-digital converter (p. 227) in a process called *pulse code modulation* (PCM) which involves 'sampling' its value regularly.

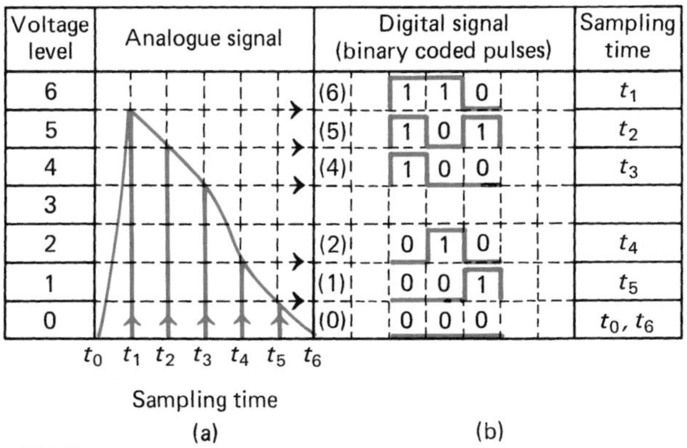

Voltage level	Analogue signal	Digital signal (binary coded pulses)			Sampling time
6		(6)	1 1 0		t_1
5		(5)	1 0 1		t_2
4		(4)	1 0 0		t_3
3					
2		(2)	0 1 0		t_4
1		(1)	0 0 1		t_5
0		(0)	0 0 0		t_0, t_6

t_0 t_1 t_2 t_3 t_4 t_5 t_6

Sampling time
(a)   (b)

Fig. 76.2

Suppose the analogue voltage has the waveform shown in Fig. 76.2a. It is divided into a number of equally spaced voltage levels (six in this case) and measurements taken at equal intervals to find the level at each time. Every level is represented in binary code by a number which has a characteristic series of on–off electrical pulses, i.e. a par-

ticular digital bit-pattern. A three-bit code can represent up to eight levels (0 to 7), as in Fig. 76.2b; four bits will allow sixteen levels to be coded.

The accuracy of the representation increases with the number of voltage levels and the sampling frequency. The latter has to be greater than twice the highest frequency of the analogue signal to be sampled. The highest frequency needed for intelligible speech in a telephone system is about 3500 Hz and a sampling frequency of 8000 Hz is chosen, i.e. samples are taken at 125 μs intervals, each sample lasting for 2 to 3 μs. An eight-bit code (giving $2^8 = 256$ levels) is used and so the number of bits that have to be transmitted is $8000 \times 8 = 64\,000$, i.e. the *bit-rate* is 64 kbit/s and is given by:

$$bit\text{-}rate = sampling\,frequency \times no.\,of\,bits\,in\,code$$

For good quality music where frequencies up to about 16 000 Hz must be transmitted, the sampling frequency is 32 000 Hz and a sixteen-bit code ($2^{16} = 65\,536$ levels) is used. The bit-rate required is $32\,000 \times 16 = 512$ kbit/s. If the music was to be stored on an audio compact disc (Chapter 79) playing for one hour, then:

number of bits stored $= 512 \times 10^3$ bit/s $\times 3600$s

or, since 1 byte = 8 bits,

$$\begin{aligned} number\,of\,bytes\,stored\ &= 512 \times 36 \times 10^5/8 \\ &= 230 \times 10^6\,bytes \\ &= 230\,Mbytes \end{aligned}$$

For television signals, which carry much more information, a bit-rate of $70\,000\,000 = 70$ Mbit/s is needed. (See question 3, p. 215.)

The analogue voltage shown in Fig. 76.2a would be represented in digital form by the train of pulses in Fig. 76.3 using a three-bit code.

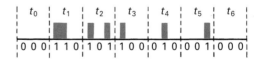

t_0 t_1 t_2 t_3 t_4 t_5 t_6

0 0 0 1 1 0 1 0 1 1 0 0 0 1 0 0 0 1 0 0 0

Fig. 76.3

77 COMMUNICATION SYSTEMS

CHECKLIST

After studying this chapter you should be able to:
- draw a block diagram for a communication system,
- draw graphs to explain AM and FM,
- explain the term 'bandwidth' and state what it is for speech, hi-fi music and television signals,
- account for and calculate the bandwidth of an AM signal, and
- state what is meant by 'multiplexing' and explain how it is done by frequency division and time division.

TRANSMISSION OF INFORMATION

Electrical signals representing 'information' from a microphone, a TV camera, a computer, etc., can be sent from place to place using either cables (electrical or optical) or radio waves. Information in the form of audio frequency (a.f.) signals may be transmitted directly by a cable but in general, and certainly in radio and TV, they require a 'carrier'. This has a higher frequency than the information signal, its amplitude is constant and its waveform sinusoidal.

The general plan of *any* communication system is shown in Fig. 77.1. Signals from the *information source* are added to the carrier in the *modulator* by the process of 'modulation'. The modulated signal is sent along a 'channel' in the 'propagating medium' (i.e. cable or radio wave) by the *transmitter*.

At the receiving end, the *receiver* may have to select (and perhaps amplify) the modulated signal before the *demodulator* extracts from it the information signal for delivery to the *receptor of information*.

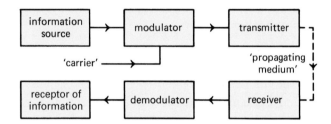

Fig. 77.1

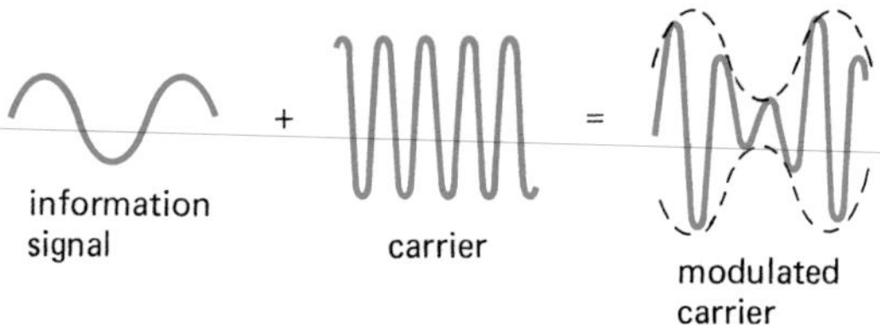

TYPES OF MODULATION

Amplitude modulation (AM) The information signal from, for example, a microphone, is used to vary the *amplitude* of the carrier so that it follows the wave shape of the information signal, Fig. 77.2.

Frequency modulation (FM) In this case the information signal varies the *frequency* of the carrier, which increases if the signal is positive and decreases if it is negative. The effect, much exaggerated, is shown in Fig. 77.3.

Pulse code modulation (PCM) The signal is modulated to form a pattern of *pulses* which, as explained on p. 177, represents in binary code regular samples of the amplitude of the information signal. PCM is at the heart of present-day high-speed digital communications technology.

information signal + carrier = modulated carrier

Fig. 77.2

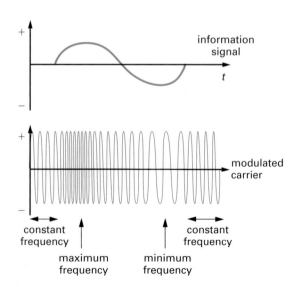

Fig. 77.3

BANDWIDTH

This term is used in two ways.

Bandwidth of a signal

This is the range of frequencies a signal occupies. For intelligible speech it is about 3 kHz (e.g. 300 to 3400 Hz as in the telephone system), for high quality music it is 16 kHz or so and for television signals about 8 MHz.

The bandwidth of a modulated signal is due to the fact that when modulated, other frequencies, called *side frequencies*, are created on either side of the carrier (which is a single frequency). In AM, if the carrier frequency is f_c and a modulating frequency f_m, two new frequencies of $f_c - f_m$ and $f_c + f_m$ are produced, one below f_c and the other above it, Fig. 77.4a.

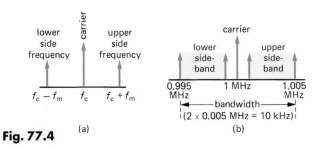

Fig. 77.4

If, as usually occurs in practice, the carrier is modulated by a range of a.fs, each a.f. gives rise to a pair of side frequencies. The result is a band of frequencies, called the *lower* and *upper sidebands*, stretching below and above the carrier by the value of the highest modulating frequency. For example if $f_c = 1$ MHz and the highest $f_m = 5$ kHz $= 0.005$ MHz, then $f_c - f_m = 0.995$ MHz and $f_c + f_m = 1.005$ MHz, Fig. 77.4b. The bandwidth of a carrier modulated by a.f. signals up to 5 kHz is thus 10 kHz. Sidebands also arise in FM.

Bandwidth of a channel

This is the range of frequencies a communication channel can accommodate. High-frequency transmission channels have greater bandwidths than lower frequency ones, i.e. their information-carrying capacity is greater. For instance, the v.h.f. radio band, which extends from 30 MHz to 300 MHz, has 'space' for 2700 signals 10 kHz wide. The medium waveband, 300 kHz to 3 MHz, can carry only 270 such signals.

Cables also have different bandwidths.

MULTIPLEXING

Multiplexing involves sending several different information signals along the same communication channel so that they do not interfere. Two methods are outlined.

Frequency division (for analogue signals) The signals (e.g. speech in analogue form) modulate carriers of different frequencies which are then transmitted together at the same time, a greater bandwidth being required.

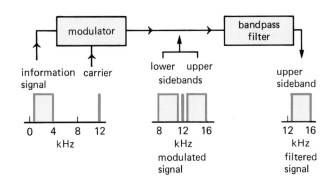

Fig. 77.5

The principle is shown in Fig. 77.5 for cable transmission. The information signal contains frequencies up to 4 kHz and the carrier frequency is 12 kHz. *Each sideband contains all the modulating frequencies*, i.e. all the information and so only one is really necessary. Here, a 'bandpass filter' allows just the upper sideband to pass.

The multiplexing of three 4 kHz wide information signals 1, 2 and 3, using carriers of 12, 16 and 20 kHz is shown in Fig. 77.6.

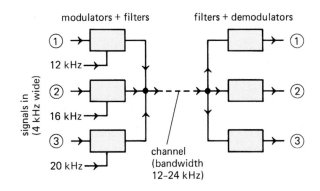

Fig. 77.6

Time division (for digital signals) This method is shown in Fig. 77.7 for three signals. An electronic

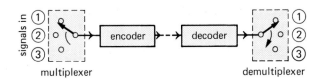

Fig. 77.7

switch, i.e. a multiplexer (p. 147 and here 3 : 1 line), samples each signal in turn (for speech 8000 times per second). The encoder converts the samples into a stream of pulses, representing, in binary, the level of each signal (as in PCM). The transmissions are sent in sequence, each having its own time allocation. The decoder and a demultiplexer (1 : 3 line), which are synchronized with the multiplexer and encoder, reverse the operation at the receiving end.

QUESTIONS

1. What is the bandwidth when frequencies in the range 500 Hz to 3500 Hz amplitude-modulate a carrier?

2. A carrier of frequency 800 kHz is amplitude-modulated by frequencies ranging from 1 kHz to 10 kHz. What frequency range does each sideband cover?

78 ELECTRONICS AND SOCIETY

CHECKLIST

After studying this chapter you should be able to:
- give six reasons why electronics is having a great impact on society,
- give examples of the use of electronics in each of the following areas: home, medical services, industry, offices, banks, shops, communications, weather forecasting, transport, emergency services, scientific research, education, leisure, and
- state some of the positive and negative social and economic consequences of the impact of electronics.

Electronics is having an ever-increasing impact on all our lives. Work and leisure are changing as a result of the social, economic and environmental influence of new technology. In the first industrial revolution, machines replaced muscles. In the second, now upon us, and caused by electronics, brain power is being replaced. Few areas of human activity are likely to escape.

REASONS FOR IMPACT

Why is electronics having such a great impact? Some of the reasons are listed below.

(i) *Mass production* of large quantities of semiconductor devices (e.g. ICs) allows them to be made very cheaply.

(ii) *Miniaturization* of components means that even complex systems can be quite compact.

(iii) *Reliability* of electronic components is a feature of well-designed circuits. There are no moving parts to wear out, no servicing is needed and systems can be robust.

(iv) *Energy consumption* and use of natural resources tends to be much less than for their non-electronic counterparts.

(v) *Speed of operation* can be millions of times greater than for other alternatives (e.g. mechanical devices).

(vi) *Transducers* of many different types are available for transferring information in and out of an electronic system.

To sum up, electronic systems tend to be cheaper, smaller, more reliable, less wasteful, much faster and can respond to a wider range of signals than other systems.

AREAS OF IMPACT

Home   Devices such as washing machines, burglar alarms, telephones and answering machines, electric cookers, microwave ovens and modern sewing machines all contain electronic components. Central heating systems and garage doors may have automatic electronic control. For home entertainment, video cassette recorders

(VCRs), compact disc players, television sets with tele-text operated by remote-control keypads that use infrared, and electronic toys and games are now commonplace. Personal computers (PCs) are finding their way into more and more homes.

Medical services   These have benefited greatly in recent years from the use of electronic instruments and appliances. Electrocardiograph (ECG) recorders for monitoring the heart, whole-body X-ray scanners, ultrasonic scanners for checks during pregnancy, deaf aids, heart pacemakers, artificial kidneys, limbs with electronic control and talking newspapers for the blind are just a few examples. Surgeons are now using a technique called keyhole surgery, using optical fibre technology together with computer monitors (Fig. 78.1a). Also, doctors are computerizing records of patients.

Industry   Microprocessor-controlled equipment (Chapter 97) is taking over in industry. Robots are widely used for car assembly work to do dull, routine, dirty jobs such as welding and paint spraying. In many cases production lines and even whole factories, e.g. sugar refineries and oil refineries, are almost entirely automated.

Computer-aided design (CAD) of widely varying products is increasing, e.g. car components, clothes, furniture. The face of industry is being changed by electronics.

Offices, banks and shops   Word processors are efficient, versatile replacements for typewriters, and computers for filing systems. Mail in the form of text, numbers and pictures can be transmitted by electronic means, e.g. *fax* and *e-mail* (p. 213).

Automatic cash dispensers at banks and building society offices, and at railway stations and supermarkets, are a great convenience for customers. Many now allow money to be deposited and household bills to be paid.

Bar codes (like the one on the back cover of this book) on packaged food and other commodities are used by large shops for stock control in conjunction with a bar code reader (incorporating a laser) and a data recorder connected to a computer. A similar system is operated by libraries to record the issue and return of books.

Communications   Communication satellites enable events on one side of the world to be seen and heard on the other side, as they happen. Electronic telephone exchanges with digital transmission, like *System X* (p. 208), are the order of the day. Mobile telephone systems, such as *Cellnet* (p. 214), permit people in cars, trains and ferries to make calls on the existing telephone network.

The *Internet*, which may be a step on the way to an 'Information Superhighway', is a rapidly growing global network of computer users. It is discussed more fully in Chapter 93.

Weather forecasting   Weather satellites provide forecasters with pictures of large areas of the Earth in visible light and infrared. Radar too is used to detect cloud formations.

Transport   The safety of many forms of transport can be improved by the use of electronic devices to give the user more information about the system, e.g. in a ship (Fig. 78.1b) or a car. Traffic signals and traffic flow are regulated in many large cities by microprocessors. Modern railway signalling systems are much more dependent on electronics. Automatic pilots are common in aircraft, as are simulators for pilot training.

Emergency services   Infrared-operated thermal imagers help rescue workers looking for people trapped in collapsed buildings, after disasters such as earthquakes. Computer-assisted response is now used for 999 calls.

Scientific research   Scientific discoveries aid technology which, in turn, produces better tools and techniques for scientists to do their research. The scanning electron microscope, for example, is used to magnify and 'see' the surface of a material on a cathode ray screen.

Another recently developed tool is the pulsed laser (Fig. 78.1c), now spearheading investigations into very high-speed transmission of digital information using optical signals.

Education   Widespread use of calculators, microcomputers with CD-ROMs (read-only memory on compact discs) and interactive teaching programs, as well as more courses on electronics and computing in schools, colleges and universities, are just some of the innovations in education due to the impact of electronics.

Leisure   For some people, leisure means participating in or attending sporting activities and here the electronic score board may be much in evidence. For the golf enthusiast, electronic machines claim to analyse 'swings' and reduce handicaps. For others, leisure may mean visiting the theatre where the lighting and sound effects in modern musicals are programmed by a computer.

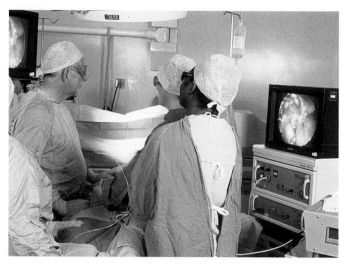

(a) Surgeons removing a gall blader using a minimally invasive technique. A laparoscope (camera) provides a view of the gall bladder on a screen and guides the use of a laser in its removal

(b) A Decca video plotter on a ship

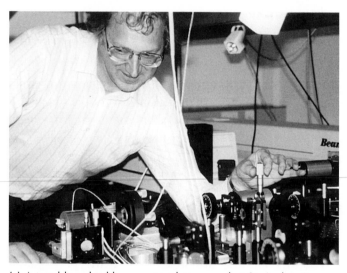

(c) A tunable pulsed laser system being used at St. Andrew's University for research into semiconductors

(d) Children playing an electronic game

Fig. 78.1

Electronic games have become popular with children (Fig. 78.1d). *Multimedia* presentations (p. 220) in which text, colour, graphics, animation, sound and video effects are combined, are much used in computer games.

CONSEQUENCES OF IMPACT

Most of the social and economic consequences of electronics are beneficial but a few cause problems.

Improved quality of life This arguably results from the greater convenience and reliability of electronic systems, increased life expectancy and leisure time, and fewer dull, repetitive jobs.

Better communication The world has become a smaller place due to the speed with which news can be reported to our homes by radio and television. This, and the influence of information systems such as teletext, CD-ROMs and the Internet, should produce a better-informed public.

Databases These are computer programs used to store huge amounts of information which can be searched rapidly and be transmitted rapidly from one place to another. For example, police can obtain by radio, details of a car they are following, in seconds. Databases raise questions, however, about invasion of privacy and security (see Chapter 93).

Early obsolescence of equipment Rapid changes of design due to technological advances mean that equipment can soon become out of date and need replacing.

Employment The demand for new equipment creates new industry and jobs (to make and maintain it), but when electronic systems replace mechanical ones, redundancy and retraining problems arise. Conditions of employment and long-term job prospects can also be affected for many people. For example, one industrial robot may replace four factory workers.

Public attitude Modern electronics is a 'hidden' technology with parts that are enclosed in a tiny package and do not move. It is also a 'throwaway' technology in which the whole lot is discarded and replaced if a part fails—by an expert. For these reasons it may be regarded as mysterious and unfriendly since people feel they do not understand what makes it tick.

THE FUTURE

The only certain prediction about the future is that new technologies will be developed and these, like present ones, will continue to have a considerable influence on our lives.

AUDIO SYSTEMS

79 SOUND RECORDING

CHECKLIST

After studying this chapter you should be able to:
- describe the action of a tape recorder and a CD player, and
- recall the pros and cons of analogue and digital sound recording.

AUDIO TAPE

Recording Sound can be stored by magnetizing plastic tape coated with finely powdered iron oxide or chromium oxide. The *recording head* is an electromagnet having a tiny gap filled with a strip (a 'shim') of non-magnetic material between its poles, Fig. 79.1. When a.c. from the recording microphone passes through the windings, it causes an alternating magnetic field in and just outside the gap. This produces in the tape a chain of permanent magnets, in a magnetic pattern which represents the original sound. This is an analogue method of recording; digital audio tape (DAT) is now being developed.

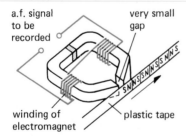

Fig. 79.1

a.f. signal to be recorded
very small gap
winding of electromagnet
plastic tape

The *strength* of the magnet produced at any part of the tape depends on the value of the a.c. when that part passes the gap and this is turn depends on the loudness of sound being recorded. The *length* of a magnet depends on the frequency of the a.c. (which equals that of the sound) and on the speed of the tape. The effect of high and low frequency a.c. is shown in Fig. 79.2.

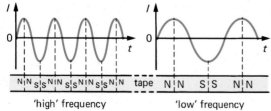

'high' frequency   'low' frequency

Fig. 79.2

High frequencies and low speeds create shorter magnets and very short ones tend to lose their magnetism immediately. Each part of the tape being magnetized must have moved on past the gap in the recording head before the magnetic field reverses. High frequencies and good quality recording require high tape speeds.

Chromium oxide tapes record a wider range of frequencies and their output is greater on playback.

Two-track and stereo tapes are shown in Fig. 79.3. In the former, half the tape width is used each time the tape is turned over. In the latter, the recording head has two electromagnets, each using one-quarter tape-width.

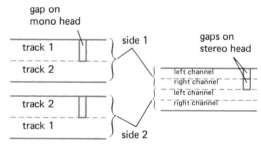

gap on mono head
track 1
track 2
side 1
gaps on stereo head
left channel
right channel
left channel
right channel
track 2
track 1
side 2

Fig. 79.3

Bias Magnetic materials are only magnetized if the magnetizing field exceeds a certain value. When it does, doubling the field does not double the magnetization. That is, the magnetization is not directly proportional to the magnetizing field, as the graph in Fig. 79.4a shows. Therefore, signals fed to the recording head would be distorted when played back (as they also would if the saturation value of the magnetic material was exceeded by too strong a recording signal).

To prevent this, *a.c. bias* is used. A sine wave a.c. with a frequency around 60 kHz is fed into the recording head

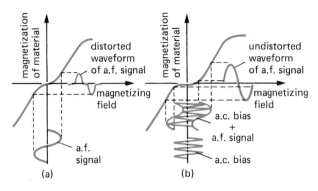

Fig. 79.4

along with the signal. The a.c. bias, being supersonic, is not recorded but it enables the signal to work on linear parts of the graph, Fig. 79.4b.

Playback On playback, the tape runs past another head, the *playback head*, which is similar to the recording head (sometimes the same head records and plays back). As a result, the varying magnetization of the tape induces a small a.f. voltage in the windings of the playback head. After amplification, this voltage is used to produce the original sound in a loudspeaker.

During playback the tape must move at the same constant speed as when recorded. Any variation changes the pitch of the sound produced; it is called *wow* if the frequency wobbles slowly and *flutter* if it is fast. The driving electric motor is usually d.c. and more than one may be used.

Erasing The a.c. bias is also fed during recording to the *erase head*, which is another electromagnet but with a larger gap so that its magnetic field covers more tape. It is placed before the recording head and removes any previous recording by subjecting the passing tape to a decreasing, alternating magnetic field which demagnetizes it.

A simplified block diagram of a tape recorder is given in Fig. 79.5.

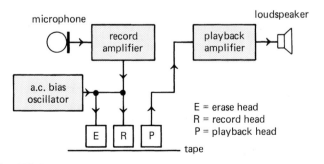

Fig. 79.5

Cassette recorders The audio cassette recorder is now rated as a hi-fi (high-fidelity, or high quality of reproduction) system. At the quite low tape speed of 4.75 cm/s, it can handle frequencies up to 17 kHz. The improvement is due to better head design and tape quality and to the use of noise reduction circuits, e.g. *Dolby*, which improve higher frequency recording by reducing the level of 'hiss' that occurs during quiet playback spells.

The output voltage from a typical cassette tape deck is 300 mV, which is about the same as that from a radio tuner.

The video cassette recorder (VCR) works on the same principle but plays back on a TV receiver, the high-frequency video signals stored on the tape. In this case it is more practical for the playback head to be rotated rapidly (25 revolutions per second) while the tape moves slowly past it (since a much greater relative speed is necessary between tape and head at these frequencies).

COMPACT DISC (CD)

Making a disc A CD is a plastic (polycarbonate) disc, 120 mm (about 5 in) in diameter. It can store about 74 minutes of sound in digital form as a pattern of tiny 'pits' of various length and spacing, which form a spiral track, as shown in the large model of Fig. 79.6. The total track length is several kilometres.

A master disc is made first from a very flat piece of glass by coating it with a thin layer of a photoresist material (Chapter 33). The sound to be recorded is converted from an analogue to a digital electrical signal and used to control a very intense, narrow beam of light from a powerful laser, focused on the photoresist layer on the glass

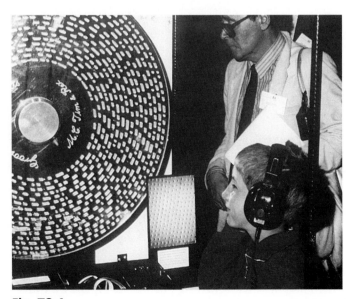

Fig. 79.6

disc as it revolves. When the photoresist is developed and etched, parts that were exposed to light are dissolved away and a spiral track of 'pits' remains, separated by unexposed strips of photoresist. In effect the laser beam 'cuts' the signal into the photoresist in digital form.

The master, too valuable to be used to press commercial discs directly, is strengthened with a metal coating. It is then used in the process which produces a submaster for stamping out, by injection moulding, the plastic discs that are sold to the public. To make them reflective they are coated with aluminium and then lacquered for protection.

Playback The pick-up in a CD player is a laser which sends a fine beam of light to the disc via a partly reflecting mirror and a lens system. If the beam falls on the rough bottom of a pit, Fig. 79.7a, it is scattered, no reflected beam or electrical signal is produced in the photodiode detector and this represents a digital signal 0. If the beam falls on a smooth space between pits, Fig. 79.7b, there is a reflected beam and signal which gives a 1.

The photodiode converts the interruptions of the beam into a digital electrical signal before it undergoes digital-to-analogue conversion and produces a signal of about

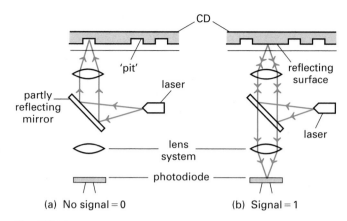

Fig. 79.7

300 mV which operates an amplifier and loudspeaker. The pick-up follows the sound-carrying spiral track by moving across the CD as it revolves.

A CD is not subject to the wear its predecessor, the vinyl record, suffered since there is no physical contact between it and the pick-up.

Recording and retrieving sound digitally is less subject to *distortion* and *noise* than by analogue methods as in tape recording and vinyl records, but is more difficult technically.

80 AUDIO AMPLIFIERS

CHECKLIST

After studying this chapter you should be able to:
- state three essential properties of a preamplifier,
- explain the term 'equalization', how the need for it arises and how it is achieved,
- state that a power amplifier is a current-amplifying buffer that drives a low-impedance loudspeaker,
- state that MOSFETs as power amplifiers have a larger input impedance and a higher power output than bipolar transistors,

- recognize and explain the action of a push–pull follower and state why it can be used as a power amplifier,
- draw a graph to show what is meant by 'crossover distortion' and explain how it is reduced,
- calculate the maximum output of a power amplifier,
- calculate power and voltage ratios on the decibel scale, and
- state what is meant by 'noise', how it arises and how it can be reduced.

Audio frequency (a.f.) amplifiers fall into two broad groups—preamplifiers and power amplifiers.

PREAMPLIFIER

A preamplifier is a voltage (or small-signal) amplifier which amplifies the signal supplied to its input by a transducer, e.g. microphone, pick-up, or tape playback head.

Basically it needs to have:

(i) high gain,
(ii) a high input impedance compared with that of the transducer to ensure maximum voltage transfer,
(iii) an output impedance that matches that of the device receiving its output, e.g. a power amplifier.

Ideally, it should also have:

(iv) a wide frequency response,
(v) low distortion, and
(vi) low noise.

A preamplifier has also to provide *equalization* and *tone control*; both are achieved using frequency-dependent negative feedback (n.f.b.).

Equalization This ensures that the output is a faithful reproduction of the original sound, i.e. that the same frequencies are present in both in the same relative proportions. The need for equalization arises because the input transducer may not respond equally to all frequencies. To compensate for this the preamplifier has to boost some frequencies more than others, which it does by frequency-dependent n.f.b. (see below).

Tone control This permits the overall frequency response of the amplifier to be, to some extent, under the listener's control so that allowance can be made for different loudspeaker performances, room acoustics, etc. Tone control lets the bass and treble frequencies be cut or boosted as required.

Frequency-dependent n.f.b. If the n.f.b. path in an amplifier contains a capacitor C and a resistor R, its impedance varies with frequency f. For example, in the inverting op amp circuit of Fig. 80.1, the reactance X_C of C ($= 1/(2\pi f C)$) is large at low frequencies since f is small, giving a small amount of n.f.b. and therefore a small reduction in the gain A (p. 106).

At high audio frequencies, X_C is small, so n.f.b. is greater and there is a larger fall in A. The amplifier thus, in effect, boosts bass frequencies more than treble frequencies and by suitably choosing component values, gives the equalization required.

Tone control circuits are designed on similar principles, variable resistors being used to allow adjustments to be made.

POWER AMPLIFIER: SINGLE-ENDED

Introduction The small amount of power produced by a preamplifier could not operate a loudspeaker. A power (or large-signal) amplifier, taking the output of a preamplifier as its input, is required. It converts d.c. power from the supply into large swings of output current and voltage, i.e. large a.c. power in the load. Basically it is a robust voltage amplifier (or *buffer*) which uses a transistor (a power transistor) capable of supplying high output currents. However its matching requirement is for maximum power transfer rather than maximum voltage transfer.

The maximum power theorem (p. 14) states that a generator (e.g. a transistor amplifier) delivers maximum power to an external load (e.g. a loudspeaker) when the output impedance Z_1 of the generator equals the input impedance Z_2 of the load. The impedance of a loudspeaker is typically 8 Ω, while the output impedance of a transistor amplifier is several thousand ohms. Direct connection of the speaker as the collector load would give poor results.

Transformer matching One solution to the matching problem is to connect a step-down transformer as in Fig. 80.2 with its primary winding of n_1 turns as the collector load and its secondary winding of n_2 turns supplying the speaker. It can be shown that Z_2 will 'look' equal to Z_1 and the maximum power transferred if the turns ratio n_1/n_2 is chosen so that:

$$\frac{n_1}{n_2} = \sqrt{\frac{Z_1}{Z_2}}$$

For example, if $Z_1 = 8\,\text{k}\Omega$ and $Z_2 = 8\,\Omega$, $n_1/n_2 = \sqrt{8000/8} = \sqrt{1000} \approx 32/1$. A matching transformer with a step-down ratio of about 32/1 would be suitable.

Single-ended power amplifiers (i.e. with one output transistor) are not used much except where low powers

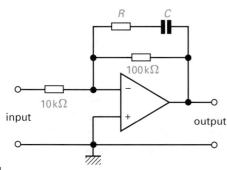

Fig. 80.1

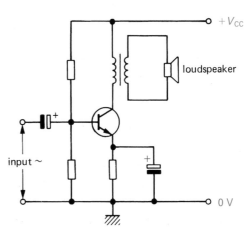

Fig. 80.2

(up to about 1 W) are required, for two reasons. First, transformers are costly, bulky, heavy and cause distortion. Second, the d.c. operating point for the transistor has to be near the middle of the load line to prevent distortion of the output (p. 99); the quiescent collector current is therefore comparatively large and *represents wasted d.c. power which produces unwanted heat, possibly requiring the transistor to have a heat sink (Fig. 35.6). Overall, the efficiency is low.*

POWER AMPLIFIER: PUSH–PULL

Most audio power amplifiers employ the 'push–pull principle'. Two transistors are used and are called a *complementary pair* because one is an n-p-n type and the other a p-n-p type. The basic circuit is given in Fig. 80.3.

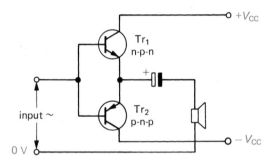

Fig. 80.3

Emitter-follower matching Tr_1 is the n-p-n transistor, its collector is positive with respect to the emitter, the opposite being the case for Tr_2, the p-n-p type. The bases of both transistors are connected to the input terminal. The load (loudspeaker) is joined to both emitters via a d.c. blocking capacitor. The circuit consists of two emitter-followers, which as we saw earlier (p. 110), have a low output impedance. In fact, it is comparable with that of a loudspeaker, so solving the impedance-matching problem.

Push–pull action When an input is applied, the positive half-cycles forward-bias the base–emitter junction of Tr_1, which conducts and 'pushes' a series of positive half-cycles of output current through the loudspeaker. During this time Tr_2 is cut off because its base–emitter junction is reverse biased when the base is positive with respect to the emitter.

Negative half-cycles of the input reverse-bias Tr_1 (i.e. base negative with respect to emitter) and cut it off but forward-bias Tr_2. Tr_2 conducts and 'pulls' a series of negative half-cycles of output current through the speaker. Both halves of the input current are thus 'processed'.

Each transistor only conducts for one half of each input cycle and is cut off during the other half. The quiescent base and collector currents are zero since the base–emitter junctions of neither Tr_1 nor Tr_2 are forward biased. Hence, no d.c. power is wasted when there is no input, which accounts for the high efficiency of a push–pull amplifier.

Tr_1 and Tr_2 should be a 'matched pair' of power transistors, i.e. their characteristics should be identical, otherwise if one has a greater gain than the other, one half of the output waveform will be amplified more and cause distortion. Matched pairs are available for different power outputs, both as n-p-n/p-n-p junction transistors and p-/n-channel enhancement-mode MOSFETs (Chapter 32).

Crossover distortion A disadvantage of the simple circuit in Fig. 80.3 is that each transistor does not turn-on until the input is about 0.6 V. As a result there is a 'dead zone' in the output waveform, as shown in Fig. 80.4, producing *crossover distortion*.

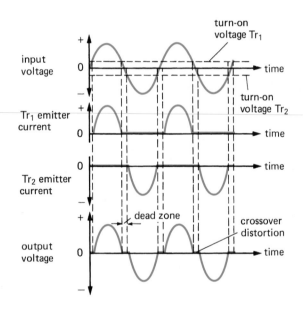

Fig. 80.4

Matters are improved by applying some forward bias to the base–emitter junctions of both transistors so that small quiescent currents flow and the smallest inputs make them conduct.

The bias required can be provided by connecting two resistors R_1 and R_2 as in Fig. 80.5 so that in the quiescent state, the bases of Tr_1 and Tr_2 are, with respect to their emitters, 0.6 V positive and 0.6 V negative respectively. Both transistors are then forward biased.

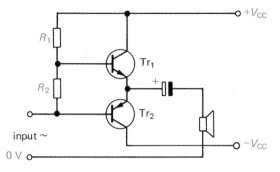

Fig. 80.5

Maximum output power The quiescent voltage at the emitter of either of the matched transistors in Fig. 80.5 is $\frac{1}{2}V_{CC}$ where V_{CC} is the d.c. supply voltage. If the alternating input makes the emitter voltage vary from its maximum positive value of V_{CC} (when Tr_1 saturates), the emitter voltage is then a varying direct voltage. The latter has an alternating component with a peak value of $\frac{1}{2}V_{CC}$ and therefore an r.m.s. value (see Chapter 7) $V_{r.m.s.} = V_{CC}/(2\sqrt{2})$.

If $I_{r.m.s.}$ is the r.m.s. value of the output current, the maximum output power P_{max} of the amplifier is given by:

$$P_{max} = V_{r.m.s.} \times I_{r.m.s.} = (V_{r.m.s.})^2/Z_L$$

where Z_L is the loudspeaker's resistive impedance (since $Z_L = V_{r.m.s.}/I_{r.m.s.}$). Hence:

$$P_{max} = \left(\frac{V_{CC}}{2\sqrt{2}}\right)^2 \times \frac{1}{Z_L} = \frac{V_{CC}^2}{8Z_L}$$

For example, when $V_{CC} = 6$ V and $Z_L = 4\,\Omega$, $P_{max} = 36/32 \approx 1$ W. The transistors would need to be capable of handling the current and power involved without overheating.

If the input is not large enough to drive the transistors into saturation, the output power is less than P_{max}.

DECIBEL SCALE

Power ratios To the human ear the *change* in loudness is the same when the power of a sound increases from 0.1 W to 1 W as when it increases from 1 W to 10 W. The ear responds to the *ratio* of the powers and not to their difference.

The log to base 10 of the ratio of the powers is the same for each of these changes:

$$\log_{10}(1/0.1) = \log_{10} 10 = 1$$
$$\log_{10}(10/1) = \log_{10} 10 = 1$$

This accounts for the way the unit of *change of power*, called the *bel* (B), is defined. If a power changes from P_1 to P_2, then we define:

$$\text{number of bels change} = \log_{10}\frac{P_2}{P_1}$$

In practice the bel is too large and the *decibel* (dB) is used. It is one-tenth of a bel, hence:

$$\text{number of decibels change} = 10\log_{10}\frac{P_2}{P_1}$$

For example, if the power output from an amplifier increases from 100 mW to 200 mW, then:

$$\text{gain in dB} = 10\log_{10}(200/100) = 10\log_{10} 2$$
$$\approx 10 \times 0.30$$
$$= 3 \text{ dB}$$

Hence a 3 dB gain is a *doubling* of power. Similarly it can be shown that a ten-fold change of power is a 10 dB change.

Voltage ratios It is often more convenient to express the power ratio in terms of the ratio of the corresponding r.m.s. voltages, measured across identical resistances. Suppose these are V_1 and V_2 when the powers are P_1 and P_2. Then, since power is proportional to the square of the voltage (p. 14), we can say:

$$\left(\frac{V_2}{V_1}\right)^2 = \frac{P_2}{P_1} \quad \text{or} \quad \frac{V_2}{V_1} = \sqrt{\frac{P_2}{P_1}}$$

Therefore if the power gain is halved, i.e. $P_2/P_1 = \frac{1}{2}$, then $V_2/V_1 = 1/\sqrt{2} \approx 0.7$, so the voltage decreases to 0.7 of its previous value. In Chapter 44 we stated that the bandwidth of a voltage amplifier was given with reference to a 0.7 voltage ratio fall, which we have now shown corresponds to a 3 dB decrease.

We can also write:

$$\text{dB change} = 10\log_{10}\left(\frac{V_2}{V_1}\right)^2$$
$$= 10 \times 2\log_{10}\frac{V_2}{V_1} \quad \text{since } \log_{10} x^2 = 2\log_{10} x$$

so

$$\text{dB change} = 20\log_{10}\frac{V_2}{V_1}$$

NOISE

Noise in an amplifier or any electronic system is an unwanted voltage or current in the output. It can have any frequency and is heard as a 'hiss' on a loudspeaker.

Causes *Internal* noise arises inside the equipment itself, caused for example by the haphazard motion of electrons producing very small random voltages.

External noise is either artificial, e.g. mains hum, sparking at the contacts of electric motors and car ignition systems, or it is natural, e.g. lightning flashes.

Whatever its source it should be reduced as much as possible. It will not improve matters if, in say a radio receiver, we amplify a very weak signal when the noise present with it is strong.

Signal-to-noise ratio S/N is a useful measure of how 'clean' an output is and is defined by:

$$\text{S/N ratio} = \frac{\text{wanted signal power}}{\text{unwanted noise power}} = \frac{P_S}{P_N}$$

It is often stated in the decibel scale as:

$$\text{S/N ratio} = 10 \log_{10} \frac{P_S}{P_N} = 20 \log_{10} \frac{V_S}{V_N}$$

where V_S and V_N are the signal and noise r.m.s. voltages respectively, measured across identical resistances.

An acceptable S/N ratio in an a.f. amplifier is 60 dB at 1 kHz.

Reduction Pick-up of unwanted interference, i.e. noise, from stray electric or magnetic fields can often be reduced by:

(i) keeping all *wires as short as possible*,

(ii) using *pairs of twisted wires* so that the interference produces equal and opposite inputs that cancel out,

(iii) using *coaxial cable* in which the central conductor carrying the wanted signal is shielded by an earthed copper braid screen (Chapter 90) from electric fields,

(iv) using *optical fibre cables* since optical signals are not affected by electric or magnetic fields, and

(v) enclosing vulnerable components (e.g. coils) in a *metal can*.

Noise is not to be confused with *distortion*. Distortion changes the signal and is caused, for example, by non-linearity or 'clipping' in an amplifier (Chapter 44).

AUDIO AMPLIFIER ICs

Preamplifier The LM381 is a dual preamplifier IC designed to amplify small signals from, for example, a pick-up or a tape playback head. A circuit for the latter, giving equalization, is shown in Fig. 80.6. If required the output could be applied to a tone control circuit before the power amplification stage.

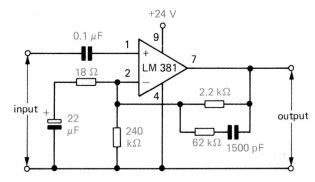

Fig. 80.6

Power amplifier The LM380 is a power amplifier IC with a typical output of 2 W into a 8 Ω speaker on a 20 V supply and using as a heat sink about 6 cm square of copper strip connected to the centre pins on its d.i.l. package. The input can be taken directly from a microphone or from the output of an LM381 preamplifier.

QUESTIONS

1. As applied to a.f. power amplifiers, briefly explain the terms
   a) single-ended,
   b) push–pull,
   c) complementary pair,
   d) heat sink.

2. How is matching achieved to the load in (i) a single-ended, (ii) a push–pull, a.f. power amplifier?

3. A push–pull a.f. power amplifier operates from a 9 V supply. What is the maximum a.c. power it can deliver to an 8 Ω load?

4. a) What are the advantages of stating audio power changes on the dB scale?
   b) If the volume control on a radio receiver is turned up, so increasing the power output from the loudspeaker from 100 mW to 500 mW, what is the power gain in dB? (Take $\log_{10} 5 = 0.7$)

5. Explain the terms
   a) noise,
   b) signal-to-noise ratio.

6. a) Why should screened cable be used to connect a microphone to an amplifier?
   b) State *two* other ways of reducing the problem.

81 COMPLETE HI-FI SYSTEM

CHECKLIST

After studying this chapter you should be able to:
- state the job of a baffle, a woofer and a tweeter, and explain why each is needed,
- state the function of a 'crossover network' and draw a circuit for one, and
- state what high-, low- and band-pass filters do and sketch graphs of output voltage against frequency for them.

MUSIC CENTRE

A modern hi-fi (high-fidelity) music centre and its remote controller are shown in Fig. 81.1a. It comprises:

(i) a CD player,
(ii) a 3-band v.h.f./LW/MW radio tuner (p. 198),
(iii) a tape cassette deck,
(iv) an amplifier (pre- and power) into which the outputs from (i), (ii) and (iii) can be switched separately,
(v) a loudspeaker system.

A block diagram of the components is shown in Fig. 81.1b.

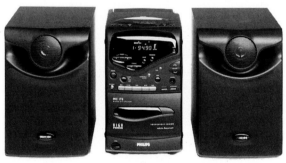

(a)

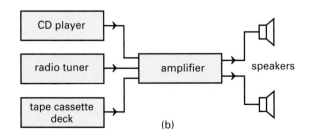

(b)

Fig. 81.1

LOUDSPEAKER SYSTEMS

Baffles and cabinets When the cone of a loudspeaker (p. 44) moves forward it produces a compression (i.e. greater air pressure) in front and a rarefaction (i.e. lower air pressure) behind. If these meet (as they will if the distance from the front to the back of the cone is small), they partly cancel (especially the low frequencies), thereby reducing the sound.

Mounting the speaker on a flat board, called a *baffle*, Fig. 81.2a, helps by increasing the front-to-back path length. A further improvement is to totally enclose the speaker in a cabinet (an infinite baffle), Fig. 81.2b. However this can lead to unwanted resonance of the air in the cabinet and also 'loads' the speaker due to the air being compressed and resisting when the cone moves back into it. Efficient coupling between a speaker and the air in the room requires careful cabinet design.

Woofers and tweeters More than one loudspeaker is needed to handle the full range of audio frequencies efficiently; in practice there are often two or three in the same cabinet. A large-coned one, the *woofer*, cannot move fast enough for high notes and deals with low (bass) frequencies, while a small one, the *tweeter*, cannot move enough air for low notes and handles high (treble) frequencies, Fig. 81.2c. Middle frequencies may also be catered for by a third speaker.

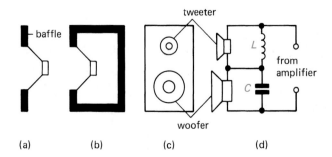

Fig. 81.2

Crossover networks The correct range of frequencies is fed to each speaker by a *crossover network*. In the simple arrangement of Fig. 81.2d, L and C act as a voltage divider with high frequencies developing a voltage across L, for application to the tweeter, and low frequencies creating a voltage across C, for application to the woofer. (Remember that $X_L = 2\pi fL$ and $X_C = 1/(2\pi fC)$.) A typical crossover frequency is about 3 kHz, i.e. frequencies above this go mostly to the tweeter, those below mostly to the woofer.

A resistor–capacitor circuit can be used as a *filter* for a crossover network. In the *low-pass* filter of Fig. 81.3a, high frequencies in the input are short-circuited by C and do not reach the output. (R and C act as a potential divider across the input with C offering a much lower reactance to high frequencies than to low ones and so there is very little high-frequency voltage across C, i.e. in the output.)

In the *high-pass* filter of Fig. 81.3b, C attenuates low frequencies but allows high frequencies to pass to the output.

Typical output voltage–frequency graphs for both types of filter are given in Fig. 81.3c, d. To accommodate the wide range of frequencies a logarithmic scale is used in which equal divisions represent equal changes in the log of the frequency.

For a middle-frequency speaker a *band-pass* filter would be used. It attenuates all frequencies except those between certain values and has an output voltage–frequency graph like that in Fig. 81.4.

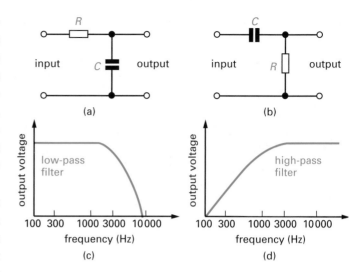

Fig. 81.3

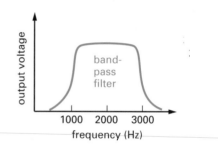

Fig. 81.4

82 PROGRESS QUESTIONS

1. Explain how *either* a CD *or* a magnetic tape contains the audio signal which has been recorded on it.

2. Describe briefly the purpose in a preamplifier of
a) equalization,
b) tone controls.

3. Describe how the following defects of sound reproduction occur:
a) distortion,
b) flutter.

4. The bandwidth of a certain a.f. power amplifier is said to be 10 Hz to 50 kHz at the 3 dB level. What does this mean in terms of (i) power output, (ii) voltage output, at these frequencies?

5. **a)** What is the signal-to-noise ratio in dB of a system in which the signal output power is 1 W and the noise output power is 1 mW?
b) If the input to a radio receiver is 100 µV and the noise level is 1 µV, calculate the signal-to-noise ratio in dB.

6. a) State *three* essential requirements of a preamplifier for use with a microphone.

b) With the aid of a block diagram outline the steps necessary to transfer the signal from a microphone onto a compact disc.

c) A baby alarm is to be built using a microphone, an amplifier and a 40 Ω speaker. The input to the amplifier from the microphone is typically 20 mV r.m.s. and the output power of the amplifier to the speaker is 250 mW. Calculate:

(i) the voltage across the speaker when the power is 250 mW,

(ii) the voltage gain of the amplifier, and

(iii) suggest a suitable frequency response for the amplifier.

d) A certain power amplifier is 60% efficient. State what happens to the 40% of the energy not transferred to the speakers, and indicate how the design of the amplifier allows for it.

(*N. part qn.*)

7. In *amplitude modulation* the carrier wave is modulated by a range of sinusoidal audio frequency waves. The resulting frequency spectrum of the modulated carrier is shown in Fig. 82.1.

(i) Identify the regions A and B.

(ii) If the carrier frequency is 1 MHz and the range of modulating frequencies is 50 Hz to 4.5 kHz give values for the frequency ranges C–D and E–F.

(iii) What is the bandwidth of the transmitted signal?

(*N. part qn.*)

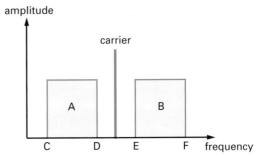

Fig. 82.1

RADIO AND TELEVISION

83 RADIO WAVES

CHECKLIST

After studying this chapter you should be able to:
• recall the grouping of radio waves into frequency bands,
• describe the three ways in which radio waves can travel, and
• describe and state the action of radio transmitting and receiving aerials.

ELECTROMAGNETIC WAVES

Electromagnetic (e.m.) waves are produced when electrons suffer an energy change, e.g. by being accelerated in an aerial. They are energy carriers which are associated with a varying electric field accompanied by a varying magnetic field of the same frequency and phase, the fields being at right angles to each other and to the direction of travel of the wave. A diagrammatic representation of an e.m. wave is given in Fig. 83.1.

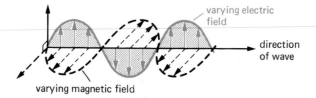

Fig. 83.1

Electromagnetic spectrum A whole family of e.m. waves exists, extending from gamma rays of very short wavelength to very long radio waves. Fig. 83.2 shows the wavelength ranges of the various types but there is no rapid change of properties from one to the next. Although methods of production and detection differ from member to member, all exhibit typical wave properties such as interference and travel in space at a speed $c \approx 3.0 \times 10^8$ m/s, i.e. the speed of light.

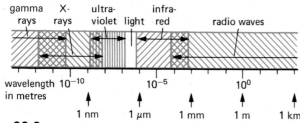

Fig. 83.2

Wave equation For any wave motion, its speed v is related to its wavelength λ and frequency f by:

$$v = f\lambda$$

For example, for violet light of $\lambda = 0.4\ \mu m = 4 \times 10^{-7}$ m, $f = v/\lambda = c/\lambda = 3 \times 10^8/(4 \times 10^{-7}) = 7.5 \times 10^{14}$ Hz. For a radio wave of $\lambda = 300$ m, $f = 10^6$ Hz $= 1$ MHz. The smaller λ is, the larger will be f, since c is fixed. Electromagnetic waves may be distinguished either by their frequency or wavelength but the former is more fundamental since, unlike λ (and c), f does not change when the wave travels from one medium to another.

RADIO FREQUENCY BANDS

Radio frequency (r.f.) waves have frequencies extending from about 30 kHz upwards and are grouped into bands, as in the table below.

Frequency band	Some uses
Low (l.f.) 30 kHz–300 kHz	Long wave radio and communication over large distances
Medium (m.f.) 300 kHz–3 MHz	Medium wave, local and distant radio
High (h.f.) 3 MHz–30 MHz	Short wave radio and communication, amateur and CB radio
Very high (v.h.f.) 30 MHz–300 MHz	FM radio, police, meteorology devices
Ultra high (u.h.f.) 300 MHz–3 GHz	TV (bands 4 and 5) aircraft landing systems
Microwaves (s.h.f.) 3 GHz–30 GHz	Radar, communication satellites, telephone and TV links, cooking

PROPAGATION OF RADIO WAVES

Radio waves from a transmitting aerial can travel in one or more of three different ways, Fig. 83.3.

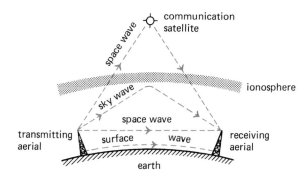

Fig. 83.3

Surface or ground wave This travels along the ground, following the curvature of the earth's surface. Its range is limited mainly by the extent to which energy is absorbed from it by the ground. Poor conductors such as sand absorb more strongly than water and the higher the frequency the greater the absorption. The range may be about 1500 km at low frequencies (long waves) but much less for v.h.f.

Sky wave This travels skywards and, if it is below a certain *critical frequency* (typically 30 MHz), is returned to earth by the *ionosphere*. This consists of layers of air molecules (the D, E and F layers), stretching from about 80 km above the earth to 500 km, which have become positively charged through the removal of electrons by the sun's ultraviolet radiation. On striking the earth the sky wave bounces back to the ionosphere where it is again gradually refracted and returned earthwards as if by 'reflection'. This continues until it is completely attenuated.

The critical frequency varies with the time of day and the seasons. Sky waves of low, medium and high frequencies can travel thousands of kilometres but at v.h.f. and above they usually pass through the ionosphere into outer space.

If both the surface wave and the sky wave from a transmitter are received at the same place, interference can occur if the two waves are out of phase. When the phase difference varies, the signal 'fades', i.e. goes weaker and stronger. If the range of the surface wave for a signal is less than the distance to the point where the sky wave first reaches the earth, there is a zone which receives no signal.

Space wave For v.h.f., u.h.f. and microwave signals, the space wave giving line-of-sight transmission is used.

A range of up to 150 km is possible on earth if the transmitting aerial is on high ground and there are no intervening obstacles such as hills, buildings or trees. These frequencies are also used for satellite transmission.

AERIALS

An aerial radiates or receives radio waves. Any conductor can act as one but proper design is necessary for maximum efficiency.

Transmitting aerials When a.c. from a transmitter flows in a transmitting aerial, radio waves of the same frequency f as the a.c. are emitted if the length of the aerial is comparable with the wavelength λ of the waves. For example, if $f = 100$ MHz $= 10^8$ Hz, $\lambda = c/f = 3 \times 10^8/10^8 = 3$ m; but if $f = 1$ kHz, $\lambda = 300\,000$ m. Therefore if aerials are not to be too long they must be supplied with r.f. currents from the transmitter.

The *dipole* aerial consists of two vertical or horizontal conducting rods or wires, each of length one-quarter of the wavelength of the wave to be emitted, and is centre-fed, Fig. 83.4a. It behaves as a series *LC* circuit (p. 34) whose resonant frequency depends on its length, since this determines its inductance L (which all conductors have) and its capacitance C (which is distributed along it and arises from the capacitor formed by each conductor and the air between them). The radiation pattern of Fig. 83.4b shows that a vertical dipole emits equally in all horizontal directions, but not in all vertical directions, Fig. 83.4c. The electric field of the wave is emitted parallel to the aerial and, if the latter is vertical, the wave is said to be *vertically polarized*.

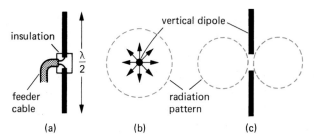

Fig. 83.4

When required (e.g. in radio telephony but not in sound or television broadcasting systems) concentration of much of the radiation in one direction can be achieved by placing another slightly longer conductor, a *reflector*, about $\lambda/4$ behind the dipole and a slightly shorter one, a *director*, a similar distance in front, Fig. 83.5a. The radiation these extra elements emit arises from the voltage and current induced in them by the radiation from the dipole.

It adds to that from the dipole in the wanted direction and subtracts from it in the opposite direction. Fig. 83.5b shows the radiation pattern given by the array, called a *Yagi aerial*. The extra elements reduce the input impedance of the dipole and causes a mismatch with the feeder cable unless a *folded dipole* is used, Fig. 83.5c, to restore it to its original value.

In the microwave band *dish* aerials in the shape of large metal 'saucers' are used. The radio waves fall on the dish from a small dipole at its focus, which is fed from the transmitter, and are reflected as a narrow, highly directed beam, Fig. 83.6.

Dish aerials using frequencies of several gigahertz (1 GHz = 1000 MHz) play an important part in the reception and retransmission of signals between different earth stations and geostationary communication satellites which orbit the earth at 36 000 km in 24 hours and appear at rest. More than half of all intercontinental telephone calls, television broadcasts giving worldwide coverage of major events as well as other information, in particular on the Internet, go by the satellite network.

Receiving aerials Whereas transmitting aerials may handle many kilowatts of r.f. power, receiving aerials, though basically similar, deal with only microwatts due to the voltages (of a few microvolts) and currents induced in them by passing radio waves. Common types are the *dipole*, the *vertical whip* and the *ferrite rod*.

Dipole arrays are used to receive TV (where the reflector is often a slotted metal plate) and v.h.f. radio signals. They give maximum output when they are (i) lined up on

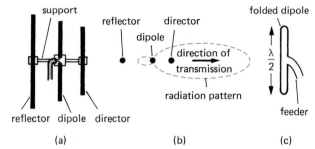

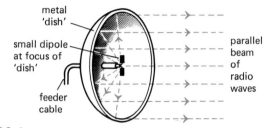

Fig. 83.5

Fig. 83.6

the transmitter and (ii) vertical or horizontal if the transmission is vertically or horizontally polarized.

A vertical whip aerial is a tapering metal rod (often telescopic), used for car radios. In most portable medium/long-wave radio receivers the aerial consists of a coil of wire wound on a ferrite rod, as in Fig. 11.1d (p. 27). Ferrites are non-conducting magnetic substances which 'draw in' nearby radio waves. The coil and rod together act as the inductor L in an LC circuit tuned by a capacitor C. For maximum signal, the aerial, being directional, should be lined up so as to be end-on to the transmitter.

84 RADIO SYSTEMS

TRANSMITTER

An aerial must be fed with r.f. power if it is to emit radio waves effectively (p. 195) but speech and music produce a.f. voltages and currents. The transmission of sound by radio therefore involves *modulating* r.f. so that it 'carries' the a.f. information, as we saw earlier (p. 178). Amplitude modulation (AM) is used in medium-, long- and short-wave broadcasting.

A block diagram for an AM transmitter is shown in Fig. 84.1. In the *modulator* the amplitude of the r.f. carrier from the *r.f. oscillator* (usually crystal-controlled) is varied at the frequency of the a.f. signal from the *microphone*, Fig. 84.2.

The *modulation depth m* is defined by:

$$m = \frac{\text{signal peak}}{\text{carrier peak}} \times 100\%$$

For a signal peak of 1 V, and a carrier peak of 2 V, $m = \frac{1}{2} \times 100 = 50\%$. If m exceeds 100%, distortion occurs, but if it is too low, the quality of the sound at the receiver is poor; a value of 80% is satisfactory.

The modulated signal consists of the carrier and the upper and lower sidebands (p. 179, Fig. 77.4b). The bandwidth it requires to transmit a.fs up to 5 kHz, is 10 kHz. In practice, in the medium waveband, which extends from about 500 kHz to 1.5 MHz, 'space' is limited if interference between stations is to be avoided and the bandwidth is restricted to 9 kHz.

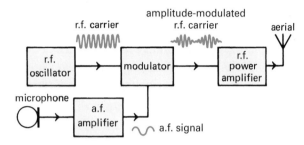

Fig. 84.1

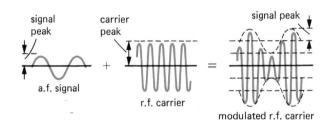

Fig. 84.2

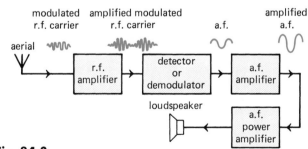

Fig. 84.3

STRAIGHT RECEIVER

Block diagram The various parts of a *straight* or *tuned radio frequency* (TRF) receiver are shown in Fig. 84.3. The wanted signal from the *aerial* is selected and amplified by the *r.f. amplifier*, which is a voltage amplifier with a tuned circuit (p. 34) as its load, and

should have a bandwidth (p. 100, Fig. 44.7) of 9 kHz to accept the sidebands.

The a.f. is next separated from the modulated r.f. by the *detector* or *demodulator* and amplified by the *a.f. pre-* and *power amplifiers* to operate the *loudspeaker*. The bandwidths of the a.f. amplifiers need not exceed 4.5 kHz, i.e. the highest a.f. allowed.

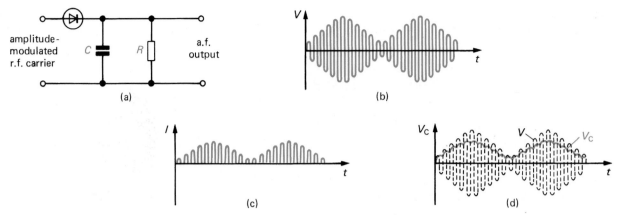

Fig. 84.4

Detection The basic detector circuit is shown in Fig. 84.4a. If the AM signal, voltage V, of Fig. 84.4b is applied to it, the diode produces rectified pulses of r.f. current I, Fig. 84.4c. These charge up C whenever V exceeds the voltage across C, i.e. during part of each positive half-cycle, as well as flowing through R. During the rest of the cycle when the diode is non-conducting, C partly discharges through R. The voltage V_C across C (and R) is a varying d.c. which, apart from the slight r.f. ripple (much exaggerated in Fig. 84.4d), has the same waveform as the modulating a.f.

The time constant CR must be between the time for one cycle of r.f., e.g. 10^{-6} s (1 μs), and one cycle of a.f. (average), e.g. 10^{-3} s (1 ms). If it is too large, C charges and discharges too slowly and V_C does not respond to the a.f.; too small a value allows C to discharge so rapidly that V_C follows the r.f. Typical values are $C = 0.01$ μF and $R = 10$ kΩ giving $CR = 10^{-4}$ s. Point-contact germanium diodes are used as explained before (p. 61).

SUPERHETERODYNE RECEIVER

The 'superhet' is much superior to the straight receiver and is used in commercial AM radios. Fig. 84.5 shows a typical block diagram.

Action The modulated r.f. signal of frequency f_c from the wanted station is fed via an *r.f. tuner* (which may not be an amplifier but simply a tuned circuit that selects a band of r.fs) to the *mixer* along with an r.f. signal from the *local oscillator* which usually has a higher frequency f_o.

The output from the *mixer* contains an r.f. oscillation of frequency $f_o - f_c$, called the *intermediate frequency* (i.f.), having the a.f. modulation. Whatever the value of f_c, f_o is greater than it by the *same* amount, i.e. the i.f. The i.f. ($= f_o - f_c$) is therefore fixed, and in an AM radio is usually 470 kHz.

To achieve this, two variable capacitors, one in each of the tuned circuits of the *r.f. tuner* and the *local oscillator*, are 'ganged' so that their resonant frequencies change in step, i.e. track. Long, medium and short wavebands are obtained by switching a different set of coils into each tuned circuit. Longer waves (smaller frequency) require greater inductance, i.e. more turns on the coil.

The *i.f. amplifiers* are in effect, r.f. amplifiers with tuned circuits having a fixed resonant frequency of 470 kHz. They amplify the modulated i.f. before it goes to the *detector* which, like the *a.f. amplifiers*, acts as in the straight receiver.

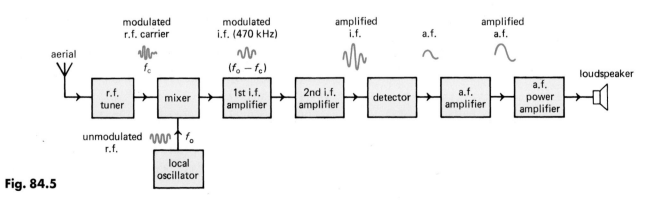

Fig. 84.5

Heterodyning The frequency changing of f_c to $f_o - f_c$ is done in the *mixer* by *heterodyning*. This is similar to the production of beats in sound when two notes of slightly different frequencies f_1 and f_2 ($f_1 > f_2$), create another note, the *beat* note, of frequency $f_1 - f_2$. In radio, the 'beat' note must have a supersonic frequency (or it will interfere with the a.f.), hence the name *super*(sonic)*heterodyne* (or 'superhet') receiver.

Advantages By changing all incoming carrier frequencies to one lower, fixed frequency (the i.f.) *at which most of the gain occurs*, we obtain greater:

> (i) *stability* because positive feedback is less at lower frequencies and so there is less risk of the receiver going into oscillation;
> (ii) *sensitivity* (i.e. the ability to detect weak signals) because more i.f. amplifiers can be used (two are usually enough) as a result of (i), so giving greater gain;
> (iii) *selectivity* (i.e. the ability to distinguish between signals that have frequencies close to one another) because more i.f. tuned circuits are possible without creating the tracking problems that arise when several variable capacitors are ganged.

Further points Instead of the local oscillator generating a frequency f_o that is *higher* than the wanted carrier frequency f_c by 470 kHz, it could produce one that is *lower* than f_c by 470 kHz with the same result. This is due to the fact that the heterodyning process in the mixer creates an i.f. frequency equal to the *difference* between f_o and f_c, i.e. $f_o \sim f_c = 470$ kHz. For example, if the i.f. is 0.5 MHz and the wanted radio signal has a frequency f_c of 2.0 MHz, two suitable values for f_o would be 2.5 MHz and 1.5 MHz.

It is possible for two different carrier frequencies from two different radio stations to be picked up at the same time by a superhet. It would happen if f_c for one signal was 470 kHz *above* f_o and for the other f_c was 470 kHz *below* f_o. For example, if $f_o = 1.5$ MHz and the i.f. is 0.5 MHz, then carrier frequencies of 2.0 MHz and 1.0 MHz will be accepted since f_o heterodynes with both in the mixer to produce the same i.f. of 0.5 MHz. The interference due to one signal or the other could be reduced by improving the selectivity of the r.f. amplifier.

A low i.f. makes it easier to obtain stable amplification.

FM RADIO

Frequency modulation (FM) is used in v.h.f. radio.

Transmitter An FM transmitter is similar to the AM one in Fig. 84.1 but the *frequency* of the r.f. carrier is changed by the a.f. signal (see p. 178, Fig. 77.3). The change or *deviation* is proportional to the amplitude of the a.f. at any instant.

Each a.f. modulating frequency produces a large number of side frequencies (not two as in AM) but their amplitudes decrease the more they differ from the carrier. In theory the bandwidth of an FM system is infinite but in practice the 'outside' side frequencies can be omitted without noticeable distortion. The BBC uses a 250 kHz bandwidth which is readily accommodated in the v.h.f. radio broadcasting band (covering frequencies 88 to 108 MHz) and allows the full range of a.fs needed for 'high quality' sound.

Receiver FM receivers use the superhet principle. To ensure adequate amplification of the v.h.f. carrier, the first stage is an r.f. amplifier with a fairly wide bandwidth. The mixer and local oscillator change all carrier frequencies to an i.f. of 10.7 MHz. These three circuits often form the FM *tuner*.

The detector, called a *ratio detector*, is more complex and quite different from an AM detector. First, it limits any changes in amplitude of the carrier due to unwanted 'noise' being picked up. This makes 'quiet' reception another good feature of FM and produces a signal that varies in frequency only. Second, it contains tuned circuits which create an a.f. voltage that is proportional to the deviation.

DIGITAL RADIO

In the future radio programmes may be transmitted as multiplexed digital signals (p. 179), resulting in many more channels and interference-free reception. However it will require listeners to buy new, self-tuning sets.

85 BLACK-AND-WHITE TELEVISION

CHECKLIST

After studying this chapter you should be able to:
* describe how a two-dimensional scene can be transmitted as a stream of information using a 'raster' pattern,

* draw and label a diagram to show the composition of a TV signal, giving values, and
* describe using a labelled block diagram the processing of the signal in a TV receiver.

TELEVISION CAMERA

A black-and-white (monochrome) TV camera changes light into electrical signals.

Action The *vidicon* tube, shown simplified in Fig. 85.1, consists of an electron gun which emits a narrow beam of electrons, and a target of photoconductive material (lead monoxide) on which a lens system focuses an optical image.

The target in effect behaves as if the resistance between any point on its back surface and a transparent aluminium film at the front depends on the brightness of the image at the point. The brighter it is, the lower is the resistance. The electron beam is made to scan across the target and the resulting beam current (i.e. the electron flow in the circuit consisting of the beam, the target, the load resistor R, the power supply and the gun) varies with the resistance at the spot where it hits the target. The beam current thus follows the brightness of the image and R turns its variations into identical variations of voltage shown in Fig. 85.2, for subsequent transmission as the video signal.

Scanning The scanning of the target by the electron beam is similar to the way we read a page of print, i.e. from left to right and top to bottom. In effect the picture is changed into a set of parallel lines, called a *raster*.

Two systems are needed to deflect the beam horizontally and vertically. The one which moves it steadily from left to right and makes it 'flyback' rapidly, ready for the next line, is the *line scan*. The other, the *field scan*, operates simultaneously and draws the beam at a much slower rate down to the bottom of the target and then restores it suddenly to the top. Magnetic deflection is used in which relaxation oscillators (p. 125) act as *time bases* and generate currents with sawtooth waveforms, Fig. 85.3a, at the line and field frequencies. These are passed through two pairs of coils mounted round the camera tube.

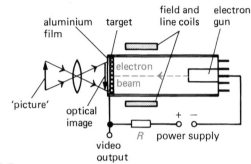

Fig. 85.1

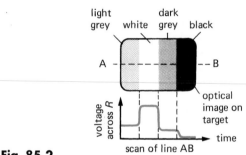

Fig. 85.2

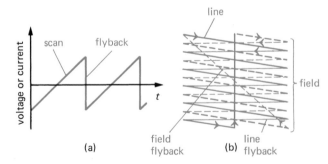

Fig. 85.3

For the video signal to produce an acceptable picture at the receiver, the raster must have at least 500 scanning lines (or it will seem 'grainy') and the total scan should occur at least 40 times a second (or the impression of continuity between successive scans due to persistence of vision of the eye will cause 'flicker'). The European TV system has 625 lines and a scan rate of 50 Hz.

It can be shown that for such a system, the video signal would need a bandwidth of about 11 MHz, owing to the large amount of information that has to be gathered in a short time. This high value would make extreme demands on circuit design and for a broadcasting system would require too much radio wave 'space'.

However, it can be halved by using *interlaced* scanning in which the beam scans alternate lines, producing half a picture (312.5 lines) every 1/50 s, and then returns to scan the intervening lines, Fig. 85.3b. The complete 625 lines or *frame* is formed in 1/25 s, a time well inside that allowed by persistence of vision to prevent 'flicker'.

TRANSMISSION

Synchronization   To ensure that the scanning of a particular line and field starts at the same time in the TV receiver as in the camera, synchronizing pulses are also sent. These are added to the video signal during the fly-

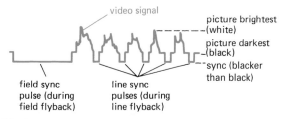

Fig. 85.4

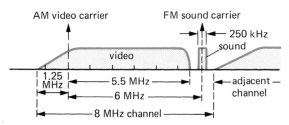

Fig. 85.5

back times when the beam is blanked out. The field pulses are longer and less frequent than the line pulses (50 Hz compared with $312.5 \times 50 = 15\,625$ Hz). Fig. 85.4 shows a simplified video waveform with line and field sync pulses.

Bandwidth   In broadcast television the video signal is transmitted by amplitude modulation of a carrier in the u.h.f. band. Since the video signal has a bandwidth of 5.5 MHz, the bandwidth required for the transmission would be at least 11 MHz owing to the two sidebands on either side of the carrier.

In practice a satisfactory picture is received if only one sideband and a part (a vestige) of the other is transmitted. This is called *vestigial sideband* transmission. The part used contains the lowest modulating frequencies closest to the carrier frequency, for it is the video information they carry that is most essential. The video signal is therefore given 5.5 MHz for one sideband and 1.25 MHz for the other, Fig. 85.5.

The accompanying audio signal is frequency modulated on another carrier, spaced 6 MHz away from the video carrier. The audio carrier bandwidth is about 250 kHz and is adequate for good quality FM sound. The complete video and sound signal lies within an 8 MHz wide channel.

TELEVISION RECEIVER

In a black-and-white TV receiver the incoming video signal controls the number of electrons travelling from the electron gun of a cathode ray tube (p. 49) to its screen. The greater the number, the brighter the picture produced by interlaced scanning as in the camera.

The block diagram in Fig. 85.6 for a broadcast receiver shows that the early stages are similar to those in a radio

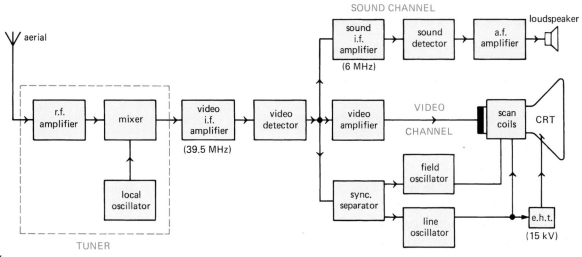

Fig. 85.6

superhet, but bandwidths and frequencies are higher (e.g. the i.f. is 39.5 MHz). The later stages have to demodulate the video and audio signals as well as separate them from each other and from the line and sync pulses.

Separation occurs at the output of the *video detector*. There, the now demodulated AM video signal is amplified by the *video amplifier* and applied to the modulator (grid) of the *CRT* to control the electron beam. The still-modulated FM sound signal, having been heterodyned (p. 199) in the *video detector* with the video signal to produce a sound i.f. of 6 MHz (i.e. the frequency difference between the sound and video carriers, shown in Fig. 85.5), is fed into the sound channel where, after amplification and FM detection, it drives the *loudspeaker*.

The mixed sync pulses are processed in the time base channel by the *sync separator* which produces two sets of different pulses at its two outputs. One set is derived from the line pulses and triggers the *line oscillator*. The other set is obtained from the field pulses and synchronizes the *field oscillator*. The oscillators produce the deflecting sawtooth waveforms for the *scan coils*. The *line oscillator* also generates the extra high voltage or tension (e.h.t.) of about 15 kV required by the final anode of the *CRT*.

SOME FACTS AND FIGURES

Television channels and frequencies (MHz)

Band	Channel	Vision	Sound	Receiving aerial group
IV	21	471.25	477.25	A ⎫
	22	479.25	485.25	⎬ W = wideband
	34	575.25	581.25	E ⎭
V	39	615.25	621.25	
	68	847.25	853.25	

Transmitting (T) and relay (R) stations

Station	Channel				Polarization	Power (kW)
	BBC1	BBC2	ITV	Ch4		
Crystal Palace (T)	26	33	23	30	Horizontal	1000
Woolwich (R)	57	63	60	67	Vertical	0.63
Winter Hill (T)	55	62	59	65	Horizontal	500
Windermere (R)	51	44	41	47	Vertical	0.5

86 COLOUR TELEVISION

CHECKLIST

After studying this chapter you should be able to:
- describe how colours are reproduced on the screen of a colour television, and
- outline how teletext works.

TRANSMISSION

Colour television uses the fact that any colour of light can be obtained by mixing the three primary colours (for light) of red, green and blue in the correct proportions. For example, all three together give white light; red and green give yellow light.

The principles of transmission (and reception) are similar to those for black-and-white television but the circuits are more complex. A practical requirement is that the colour signal must produce a black-and-white picture on a monochrome receiver; this is called *compatibility*.

Therefore in a broadcast system the bandwidth must not exceed 8 MHz despite the extra information to be carried.

A monochrome picture has only brightness or *luminance* variations ranging from black through grey to white. In a colour picture there are also variations of colour or *chrominance*. The signals for both are combined without affecting each other or requiring extra bandwidth.

In a colour TV camera, three vidicon tubes are required, each viewing the picture through a different primary colour filter. The 'red', 'green' and 'blue' signals so obtained provide the chrominance information, which is modulated by encoding circuits on a carrier. If added together correctly, they give the luminance as well, Fig. 86.1.

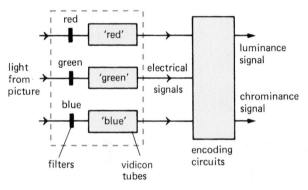

Fig. 86.1

RECEIVER

In a colour TV receiver, decoding circuits are needed to convert the luminance and chrominance signals back into 'red', 'green' and 'blue' signals and a special CRT is required for the display.

One common type of display is the *shadow mask tube* which has three electron guns, each producing an electron beam controlled by one of the primary colour signals. The principle of its operation is shown in Fig. 86.2. The

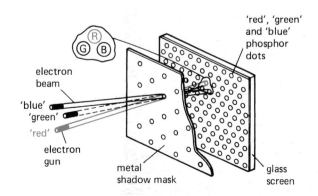

Fig. 86.2

inside of the screen is coated with many thousands of tiny dots of red, green and blue phosphors, arranged in triangles containing a dot of each colour. Between the guns and the screen is the shadow mask consisting of a metal sheet with about half a million holes (for a 26 inch diagonal tube).

As the three electron beams scan the screen under the action of the same deflection coils, the shadow mask ensures that each beam strikes only dots of one phosphor, e.g. electrons from the 'red' gun strike only red dots. When a particular triangle of dots is struck, it may be that the red and green electron beams are intense but not so the blue. In this case the triangle would emit red and green light strongly and so appear yellowish. The triangles of dots are excited in turn and since the dots are so small and the scanning so fast, we see a continuous colour picture.

The holes in the mask occupy only 15% of the total mask area; 85% of the electrons emitted by the three guns are stopped by the mask. The beam current in a colour tube therefore has to be much greater than in a monochrome tube for a similar picture brightness. Also, the final anode voltage in the tube is higher, about 25 kV.

More recent colour TV tubes are the *Precision-in-line* (PI) tube with elongated holes in the shadow mask and the *Trinitron* tube that has an 'aperture grille' with vertical slits matching up to vertical phosphor stripes on the screen, Fig. 86.3. Due to the positioning of the grille each electron beam can only energize a 'spot' on its own colour phosphor stripe.

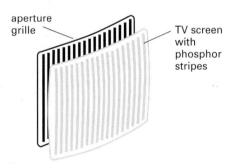

Fig. 86.3

TELETEXT

This is a system which, aided by digital techniques, displays on the screen of a modified domestic TV receiver, up-to-the-minute facts and figures on news, weather, sport, travel, entertainment and many other topics. It is transmitted along with ordinary broadcast TV signals, being called *Ceefax* ('see facts') by the BBC and *Oracle* by ITV.

During scanning, at the end of each field (i.e. 312.5 lines), the electron beam has to return to the top of the screen. Some TV lines have to be left blank to allow time for this and it is on two or three of these previously blank lines in each field (i.e. four or six per frame of 625 lines) that teletext signals are transmitted in digital form.

One line can carry enough digital signals for a row of up to 40 characters in a teletext page. Each page can have up to 24 rows and takes about $\frac{1}{4}$ second (i.e. $12 \times 1/50$ s) to transmit. The pages are sent one after the other until, after about 25 seconds, a complete magazine of 100 pages has been transmitted before the whole process starts again.

The teletext decoder in the TV receiver picks out the page you asked for (by pressing numbered switches on the remote control keypad) and stores it in a memory. It then translates the digital signals into the sharp, brightly coloured words, figures and symbols that are displayed a page at a time on the screen.

DIGITAL TELEVISION

Like radio, in the future, TV may also use digital transmission. It will provide more channels, high-definition wide-screen pictures and CD quality sound, as well as interactive services to which the viewer can respond.

In the system adopted by one station (Channel 5), signals are sent via satellite links to about 30 transmitter sites around the country. There the signals are converted from digital to analogue form for our TV receivers.

87 CABLE AND SATELLITE TV

CHECKLIST

After studying this chapter you should be able to:
- describe how interactive cable TV could promote the information revolution, and
- describe how geostationary communication satellites provide 24-hour TV.

CABLE TELEVISION

In cable television, pictures are sent to the homes of viewers via a cable, as distinct from over-the-air transmissions picked up by rooftop aerials.

In the *star* distribution system, Fig. 87.1, every home is linked directly to a local cable station from which many channels could be available to choose from, including foreign broadcasts beamed via satellites to a dish aerial at the cable station. The subscriber's telephone can also be linked in. The cable company controls a local telephone network and links the user to the national trunk network. Such an *interactive* system opens the way, via a computer modem (Chaper 91), for additional facilities that include:

(i) shopping by ordering directly from home after the goods have been viewed on the TV screen,

(ii) home banking in which bank accounts could be debited automatically at the push of a button,

(iii) two-way teaching,

(iv) booking holidays via the 'box' at home, after browsing through travel brochures seen on it,

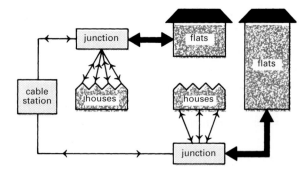

Fig. 87.1

(v) advertising articles for sale electronically,

(vi) taking public opinion poll samples very quickly and even replacing the polling booth by the TV set,

(vii) providing a burglar or fire alarm system if linked to detection sensors in the home and the police or fire station, and

(viii) reading of domestic meters (e.g. electricity) remotely.

Cable systems using modern coaxial cables (p. 209) or optical fibres (p. 209) can carry simultaneously several TV channels, ordinary telephone links and other telecommunication services because of their wide bandwidth (p. 179). The cables are laid in ducts (pipes) under the street. For economic reasons, cable TV is available only in major population centres.

SATELLITE TELEVISION

British television viewers saw their first live broadcast from America on 11 June 1962. The microwave signals were beamed from a large steerable dish aerial on the east coast of the U.S.A. to the satellite *Telstar* which amplified and retransmitted them back down to a 26 m diameter dish at Goonhilly Downs earth station in Cornwall. *Telstar* orbited the earth in $2\frac{1}{2}$ hours at a height varying from 320 to 480 km (200 to 300 miles) but signals were received for only about 20 minutes when it could be 'seen' by the aerials on both sides of the Atlantic.

Geostationary (synchronous) satellites
The idea proposed by the space science writer Arthur C. Clarke in 1945 has been realized because these days rockets are powerful enough to place communications satellites in *geostationary* orbit 36 000 km (22 500 miles) above the equator where they circle the earth in 24 hours and appear at rest. *Early Bird*, launched in 1965 over the Atlantic, was the first geostationary satellite to give round-the-clock use.

There are now over 200 major earth stations throughout the world like the one at Goonhilly (in the UK there are others at Madley, near Hereford, and in London's Docklands).

The global system of communications satellites is managed by the International Telecommunications Satellite Organization (Intelsat), which currently has 25 satellites in orbit serving over 200 countries/territories around the world. Microwave frequencies (in the GHz range) are used for signalling. Satellites are launched by American, European, Russian or Chinese rockets or by American space shuttles. *Intelsat VIII* satellites, Fig. 87.2, are due to be launched in 1997. The European organization Eutelsat operates satellites for intra-European television and telecommunications.

There are about 90 communications satellites in geostationary orbit at any given time. The current practical limit is near 180, up from 120 a few years ago. The limit arises from earth-station aerials being unable to separate (i.e. resolve) signals from two neighbouring satellites.

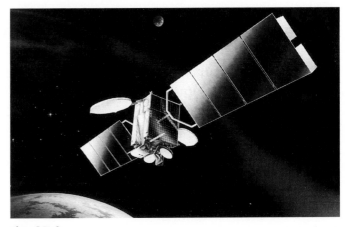

Fig. 87.2

Communications satellites have a limited life, about ten to fifteen years, after which they run out of power and a small booster rocket sends them into a higher orbit to make way for their replacements.

Direct broadcasting by satellite (DBS)
This enables homes in any part of a country to become low-cost 'earth stations' and receive TV programmes from a geostationary satellite if they have a small rooftop dish aerial (Fig. 87.3), 0.9 m in diameter, pointing towards the satellite. In this way, one satellite, having had its programme beamed from an earth station, will give country-wide coverage.

To receive satellite signals on a domestic TV receiver extra circuits are needed as well as a dish aerial. First a frequency converter in the base of the aerial has to convert the 12 GHz signals from the satellite down to a lower frequency before they go indoors to the receiver. Here a second converter must reduce the frequency further to u.h.f. Existing terrestrial TV transmissions are AM but satellite transmitters use FM to carry the programmes (because of the limited power available). The conversion circuits in the receiver must therefore change the signal from FM to AM. Finally, as the sound signals accompanying the TV signals are digitally encoded, conversion to analogue form is also required.

Fig. 87.3

88 PROGRESS QUESTIONS

1. Draw a diagram illustrating the various regions of the electromagnetic spectrum in order of increasing wavelength.

To which regions of the spectrum do the following radiations belong?

(i) Radiation of 632.8 nm wavelength from a helium-neon laser.

(ii) 3 cm waves from a klystron transmitter.

(iii) Radiation of 300 MHz.

(iv) Radiation of 0.1 nm wavelength produced in the electron bombardment of a metal surface.

(C. part qn.)

2. a) What do radio waves consist of?

b) Describe briefly the ways in which radio waves travel.

c) Models are often controlled by radio signals of approximately 27 MHz which are received by an aerial of one-eighth of the wavelength. The speed of radio waves is 3×10^8 m/s. Calculate the length of the aerial.

3. a) Explain why modulation is necessary in radio broadcasting.

b) Draw a simple sketch to show the nature of an amplitude-modulated signal. Why must a definite bandwidth be allowed for such a signal?

c) Explain how the signal is demodulated at the receiver. *(A.E.B.)*

4. a) If a radio set was tuned to receive a radio signal of constant amplitude and frequency, state and explain what sound, if any, would be heard from the loudspeaker.

b) What is done to a radio wave in order to *amplitude-modulate* it with a sound wave?

c) What is meant by *frequency modulation*?

(O.L.E. part qn.)

5. A circuit for a very simple diode radio receiver is shown in Fig. 88.1.

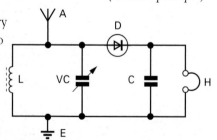

Fig. 88.1

Which components enable the receiver to (i) pick up a weak signal, (ii) select the signal required, (iii) produce a.f. signals from the modulated r.f. signals?

If the headphones H have a resistance of 1 kΩ estimate, by calculation, a suitable value for C, taking 1 MHz and 1 kHz as typical frequencies for r.f. and a.f. signals respectively.

6. a) State *three* advantages of the superheterodyne (superhet) radio receiver.

b) Draw a block diagram of a superhet radio receiver.

c) If a superhet has an intermediate frequency of 10 MHz, what *two* local oscillator frequencies would receive a signal of 30 MHz?

7. State the advantages and disadvantages of AM and FM radio systems.

8. a) Explain the terms (i) raster, (ii) line timebase, (iii) field timebase, in relation to a television picture.

b) Draw a block diagram of a black-and-white *television system*, and give a brief account of the basic elements included in your system.

c) A television picture with 625 lines uses interlaced scanning in which half the lines are scanned every 1/50th second. What is (i) the line frequency, (ii) the field frequency? *(C.)*

9. a) Draw a block diagram of a superhet receiver which includes the following: mixer, local oscillator, RF amplifier, Intermediate Frequency (IF) amplifier, AF amplifier, detector.

b) State *two* functions of the IF amplifier.

c) A superhet receiver has an IF of 10.7 MHz and a local oscillator that is at a higher frequency than the received signal. When tuned to a weak radio signal of frequency 94.2 MHz, a signal on 115.6 MHz can be heard at the same time.

(i) Explain why this second higher frequency radio signal is being received at the same time.

(ii) Explain what improvements could be made to the superhet receiver to eliminate this problem.

d) The local oscillator of the superhet receiver has a coil of inductance 0.1 μH. Calculate the capacitance that would be needed so that a signal of 94.2 MHz can be received.

e) Explain how the IF amplifier of a television differs from that of a VHF radio.

f) Starting with the video signal and the line and frame synchronization pulses, explain how a picture is formed on a black-and-white CRT. *(N.)*

TELEPHONY

89 TELEPHONE SYSTEM

CHECKLIST

After studying this chapter you should be able to:
• outline the telephone exchange network, and

• describe briefly System X including the facilities it offers.

TELEPHONE RECEIVER

The telephone receiver consists of a handset with a *microphone* (Chapter 17) at the lower end, an *earpiece* at the upper end and, in a modern press-button telephone, a *keypad* for dialling.

Lifting the handset connects the telephone to the 'line' (i.e. cable) going to the telephone exchange. When a number is called, it is stored in an IC in code or as tones of different frequencies (called 'multi-frequency signalling') depending on the type of keypad. A second IC then arranges for the correct number of electrical pulses to be sent to the exchange, e.g. two if 2 is keyed.

TELEPHONE EXCHANGES

The exchange network In a telephone system involving many subscribers, connection of each one directly to every other subscriber would be extremely complex and costly. The number of connections can be greatly reduced if the subscribers in one area are each connected to a *local exchange*, located centrally. Any subscriber in that area can then be connected to any other subscriber in the area or to subscribers in other areas via other exchanges. Power supplies, other essential equipment and staff can also be conveniently based at the exchange.

In the same way, and for similar reasons, local exchanges are connected to *trunk exchanges*, or *Digital Main Switching Units* (DMSUs), Fig. 89.1. There are about 60 DMSUs located throughout the country.

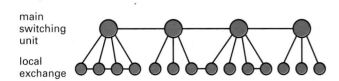

main
switching
unit

local
exchange

Fig. 89.1

Switching To reach its destination a telephone call must be directed or *routed*. This involves *switching* at the exchange to ensure the correct lines are connected.

(a)   (b)

(c)

Fig. 89.2

In the earliest exchanges switching was done manually by operators, Fig. 89.2a, and in later cordless switchboards, Fig. 89.2b, by pushing buttons and switches. The first practical automatic exchanges (called Strowger exchanges after their inventor, an undertaker from Kansas City, USA) used electromechanical switches, known as selectors, Fig. 89.2c. This system was in use in the UK from 1926 until 1990. However, the selectors contained moving parts that wore out and the exchanges were costly to maintain. They were also liable to faults such as crossed lines, buzzes, crackles and wrong numbers. During the 1980s they were replaced by electronic exchanges.

DIGITAL EXCHANGES

System X   Developed by British Telecom in co-operation with industry, System X is a family of electronic exchanges ranging in size from small local exchanges to large international exchanges that is more compact and cheaper than existing exchanges. It uses pulse code modulated (PCM: p. 178) *digital signals*, *electronic switching* and *computers* to control the routing of calls.

The first System X exchange (Fig. 89.3) was opened in the City of London in 1980 and by the early 1990s the whole of the trunk network and about half of local telephone calls were via digital exchanges.

In addition to clearer speech, less noise, faster connection, fewer wrong numbers, System X offers numerous new facilities to subscribers, including the following:

(i) *Call waiting*. A bleep lets a phone user know that another caller is trying to get through, and the caller is asked to wait to be connected.

(ii) *Call diversion*. Incoming calls can be transferred to another number when required.

Fig. 89.3

(iii) *Three-way calling*. While holding one call, another can be made, then a three-way conversation held.

(iv) *Reminder call*. The exchange can be programmed by the subscriber to ring back at a certain time.

(v) *Call return*. On dialling a four-figure number the subscriber will be told the telephone number of the last caller and the time and date of the call.

(vi) *Charge advice*. At the end of a call the exchange calls back giving the cost.

90 TELEPHONE LINKS

CHECKLIST

After studying this chapter you should be able to
• recall the construction, capacity and use in telecommunications of copper coaxial cables and optical fibres,

• state the advantages of optical fibres, and
• recall the use of microwave links in telecommunications.

COPPER CABLES

The earliest telephone links were copper wires carried overhead on wooden poles. Today most cables are underground and a typical one may contain hundreds of pairs of wires, each covered with plastic insulation and all bunched together inside a thicker plastic covering called the cable sheath, Fig. 90.1. The sheath provides protection and excludes moisture. The cables are laid in earthenware or plastic pipes, called 'ducts', and are joined either in 'joint boxes' just under footpaths or in 'manholes' (as large as small rooms) often under the roadway.

On local links, the cables carry a.f. currents; on trunk lines amplifiers, called *repeaters*, are required to boost the signals which may be multiplexed (p. 179). That is, the a.f. speech currents are modulated on different high-frequency carriers, enabling many telephone calls to be carried by one circuit.

Multiplexing requires cables to transmit r.f. currents with minimum loss, i.e. to have a wide bandwidth. Low-loss *coaxial* cables, Fig. 90.2, can handle frequencies in the u.h.f. range. The central conductor is solid copper and carries the signal. It is separated from the outer conductor of copper braid which is earthed and also acts as a screen against stray signals, i.e. 'noise'.

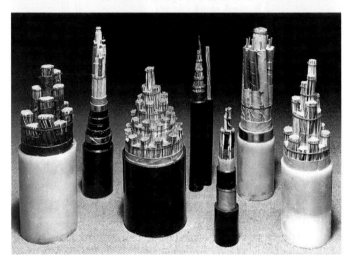

Fig. 90.1

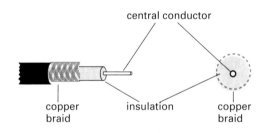

Fig. 90.2

OPTICAL FIBRES

Principle   The suggestion that information could be carried by light and sent over long distances in thin fibres of very pure (optical) glass, was first made in the 1960s. In just eleven years, the world's first optical fibre telephone link was working in Britain. Now there are over 3 million km of optical fibre cabling installed in the UK network, carrying over 84% of trunk traffic. Undersea optical fibre links exist in the English Channel, the North Sea and the Atlantic, as well as elsewhere in the world.

Light, like radio waves, is electromagnetic radiation.

When modulated and guided by glass fibre cables installed in (existing) cable ducts, it *escapes the severe attenuation* (weakening) it would suffer from rain and fog if sent through the air.

Light signals in optical fibres have a considerably greater information-carrying capacity than electrical signals in copper cables. They are also *free from 'noise'* due to electrical interference and distances of over 50 km can be worked *without regenerators*, compared with 2–3 km for coaxial cables.

Compared with copper cables, optical fibre cables are lighter, smaller and easier to handle, Fig. 90.3. They are also cheaper to produce, do not suffer from corrosion problems and are generally easier to maintain.

91) and computer data to be sent over the same transmission path. An A4 page of type or a colour photograph can be transmitted in 1 second, the contents of the *Concise Oxford Dictionary* in 32 seconds and of the entire *Encyclopaedia Britannica* in 30 minutes.

Inmarsat (International Maritime Satellite Organization) uses satellites to provide direct communication with ships at sea.

SUMMARY

Some of the main properties of telephone links are summarized in the table below.

Link	Typical bit-rate	No. of voice lines	Regenerator spacing
Pair of twisted copper wires	144 kbit/s	2	2–3 km
Microwaves	70 Mbit/s	1000	15–50 km
Coaxial cable	100 Mbit/s	1500	2–3 km
Optical fibre	560 Mbit/s	9000	50 km

Note. The bit-rate required for good reproduction is:
(i) 64 kbit/s for a telephone channel,
(ii) 512 kbit/s for a music channel,
(iii) 70 Mbit/s for a TV channel.

Fig. 90.9

91 OTHER TELEPHONE SERVICES

CHECKLIST

After studying this chapter you should be able to:
- describe briefly facsimile (fax), e-mail, telex, videophones, Videoconferencing, radiopaging, cordless phones and cellphones,
- state three reasons why modems are required when ordinary telephone lines are used to process digital signals, and
- perform bit-rate calculations and recall the meaning of the term 'baud'.

FACSIMILE (FAX)

Facsimile (meaning 'exact copy') or *fax* allows a document to be 'sent' over the telephone system and is a kind of electronic mail. When the document is inserted in the sender's fax machine, the 'optical eye' scans it and produces a string of electrical pulses which, on reaching the receiver's machine, makes it print out the same document which is then available for discussion by both parties if required. An A4 page of text can be transmitted and printed in 1 minute. Fax machines often incorporate an answering machine and phone, like the one in Fig. 91.1.

Fig. 91.1

Fig. 91.2

Fig. 91.3

E-MAIL

Electronic mail or *e-mail* allows users instant communication, using the ordinary telephone network or special ISDN lines (see p. 214) to carry text from the sender's microcomputer and be stored in an electronic 'mailbox' in the service provider's mainframe computer. The receiver can open the mailbox using a personal password and either file the message, send an immediate reply using his or her microcomputer or make a hard copy using a printer.

TELEX

This is of interest mainly to firms with overseas connections. The message to be transmitted is keyed by an operator at the sender's end on the keyboard of a teleprinter which translates letters, figures, signs and punctuation marks into code. In the modern telex terminal in Fig. 91.2, the VDU (visual display unit) permits editing and a printer gives a permanent copy of the message if it is required.

The receiving teleprinter decodes the message and prints it automatically on to a sheet of paper without anyone needing to be present.

Connection between telex subscribers is by a network of *tele*printer exchanges.

VIDEOPHONES

These are now available, Fig. 91.3. They have a built-in camera and screen which allow users to see as well as hear each other. They cost more than ordinary phones but use the same network.

VIDEOCONFERENCING

Using this British Telecom service, several groups of people in different parts of the country can see and speak to each other at the same time without travelling, i.e. they can have a 'conference', Fig. 91.4.

Fig. 91.4

RADIOPAGING

Radiopaging allows people who are out and about to be contacted. By dialling a ten-digit number on any phone, a low-power radio transmitter in one of forty zones covering the UK is triggered. The signal it emits causes the small radio receiver carried by the person whose code has been dialled, to bleep. Contact is then made with the caller from the nearest telephone.

MOBILE TELEPHONES

Using mobile phones people can stay in touch and work while moving.

Cordless phones   In these a *base unit* plugs into a telephone socket and an electricity mains socket. The handset is portable and is powered by batteries that recharge automatically when the handset is resting on the base. Calls are made and received up to 100 metres from the base unit over a radio channel.

Cellphones   In a cellular network, such as *Cellnet*, the country is divided into small areas or 'cells'. Each cell has its own low-powered microwave transmitter, called a *base station*, and is linked to a regional mobile exchange. These exchanges are in turn linked to the existing telephone system in which the full range of services is available. When a caller travels from one cell to another (by car, train, ferry or on foot), a computer automatically switches the caller to another cell so that the call can be continued.

There are several types of cellphone.

 (i) *Car phones* operate from the car battery and modern versions permit 'hands-free' calls while driving.
 (ii) *Transportable types* are for outdoor use, as on a building site, over a period of time. They have a built-in battery.
 (iii) *Hand portables* fit into the pocket but have to be recharged regularly.

Mobile phones that use digital transmission are now taking over from analogue types. They are more secure, suffer less interference and give longer battery life.

TRANSMISSION OF DIGITAL DATA

Modems   Ordinary copper telephone lines are designed to carry audio frequency signals in the range 300 Hz to 3500 Hz. They are not meant to handle long strings of fast-rising and -falling d.c. pulses which would be affected by capacitive and inductive effects.

However, digital data can be sent over the telephone network so long as there is a device called a *modem* (*mod*-ulator–*dem*odulator) at each end. The modem has to do three things.

 (i) At the sending end it changes digital data into audio tones; in one system it encodes a 1 as a burst of high tone (2400 Hz) and a 0 as a burst of low tone (1200 Hz).
 (ii) At the receiving end it decodes the tones into digital data.
 (iii) At both ends it electrically isolates the low voltage levels of the digital equipment (e.g. a computer) from the high voltage levels of the telephone system and so prevents damage to either.

Fax and e-mail are among the services that require a modem when transmission is via the ordinary telephone network. However, *ISDN* (*Integrated Services Digital Network*) is a development which enables all digital communications—speech, text, data, drawings, photographs, music—to be carried speedily from computer to computer using a *terminal adaptor* instead of a modem, on high-quality copper cables.

Bit-rate   Digital data is sent in bits, i.e. 1s or 0s, and the rate at which transmission occurs is measured in *bauds* where:

$$1 \text{ baud} = 1 \text{ bit per second}$$

The greater the bandwidth of the telephone line, i.e. the greater the range of frequencies it can transmit satisfactorily, the greater can be the bit-rate. This is because each bit of data requires as a *minimum*, one cycle of audio tone. Therefore the higher the tone frequency the more bursts can be 'fitted in' but this depends on the bandwidth of the line.

A typical modem might have a capacity of 33 600 baud.

Example   Consider the transmission of data from a database to a computer. If a page (screen) of data consists of 20 lines, each with 50 characters, then:

$$\text{number of characters per page} = 20 \times 50 = 1000$$

Suppose each character of data comprises 1 start bit, 8 data bits and 2 stop bits (Chapter 93), then:

$$\text{total number of bits to be transmitted} = 1000 \times 11$$

If the sending rate is 1100 baud, then:

$$\text{time to transmit one page} = \frac{1000 \times 11}{1100} = 10 \text{ s}$$

If several sets of *separate* signals had to be sent to different destinations on the *same* telephone line, *time-division multiplexing* (Chapter 77) would be used, each terminal receiving the data being allotted a time slot. Frequency-division multiplexing would be restricted by the bandwidth of the line.

92 PROGRESS QUESTIONS

1. Write brief notes on the use of the following in telecommunications:
a) coaxial cables,
b) optical fibres,
c) microwaves,
d) satellites.

2. Some telecommunication services are: telex, fax, Videoconferencing, radiopaging.
a) Describe how *two* of these services are useful to business organizations.
b) Describe how a *third* service could have a considerable effect on a person's everyday life.

3. If a TV picture is made up of 350 000 dots (picture elements or pixels) and an eight-bit binary code is needed to describe their many different colours and brightness levels, how many bits per second must be sent when digital transmission is used? (Note: 25 complete pictures occur on the screen per second, p. 200.)

4. a) Many long-distance telephone links have been replaced by optical fibres. State *three* advantages of optical fibres over other types of cable.
b) To enable more cost effective use of telephone cables, many telephone calls are *multiplexed* together, either in *time domain* or *frequency domain*. Explain what is meant by (i) time-domain multiplexing, (ii) frequency-domain multiplexing.
c) To send computer data along a telephone link, a modem must be used. State *three* functions of a modem. (*N. part qn.*)

5. a) What is the maximum baud rate of the data transmitted by a *modem* which sends a logic 1 as a 2400 Hz tone and a logic 0 as a 1200 Hz tone? Explain your answer.
b) How long will it take to send 4 Kbytes of data if each byte has 1 start bit, 8 data bits and 2 stop bits? (1 Kbyte = 2^{10} or 1024 bytes.)

COMPUTERS AND MICROPROCESSORS

93 DIGITAL COMPUTERS

CHECKLIST

After studying this chapter you should be able to:
- draw a block diagram for a computer system and state the function of each part,
- state the function of tri-state gates,
- recall how and when data is transmitted in serial and in parallel form,
- state examples of computers being used for data processing, process control, robotics and computer-integrated manufacture (CIM),
- name major computer users,
- state how computers are used in the Internet, in multimedia including digital interactive TV, and in generating virtual reality, and
- explain how the Data Protection Act attempts to protect individuals from the misuse of computer-stored data.

Computers are playing an increasing role in our everyday lives at home, at work and in our leisure. This is due to the speed with which they can undertake almost any task involving the processing of information (data).

The power of a computer depends on its speed of working and the amount of data it can handle at the same time. There are three very broad classes.

(i) *Mainframe* computers are usually the most powerful and contain several large units, housed in an air-conditioned room and operated by a team of people. They are used by big organizations to prepare bills and payrolls and for large calculations, e.g. in weather forecasting.

(ii) *Minicomputers* are made from smaller units, only require one or two operators and are used, for example, by government departments, hospitals and businesses and to control manufacturing operations, such as steel-making (i.e. in automation). They were developed for the space industry.

(iii) *Microcomputers* are the desktop personal computers (PCs) found today in homes, schools and offices. They have been made possible by the development of microprocessors and consist of just one or two connected units. The smallest, portable types go under such names as 'laptops', 'palmtops', 'think-pads', 'notebooks' and 'pocket books'.

COMPUTER BUILDING BLOCKS

All computers, whatever their size and power, are basically similar. They contain the building blocks, called *hardware*, shown in Fig. 93.1.

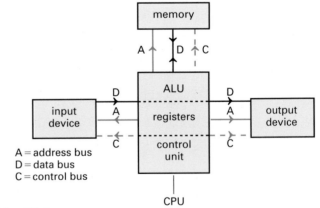

A = address bus
D = data bus
C = control bus

Fig. 93.1

Central processing unit (CPU) The CPU performs, organizes and controls all the operations the computer can carry out. It is sometimes called the 'brain' of the computer and what it can do depends on the set of instructions (typically fifty or more), called the *instruction set*, which it is designed to interpret and obey.

It consists of an *arithmetic and logic unit* (ALU, Chapter 66), a number of *shift registers* (Chapter 73) and a *control unit.* The ALU performs arithmetic calculations and logical operations. The registers are temporary stores; one of them, the *accumulator*, contains the data actually being processed. Two important components of the control unit are the *clock* and the *program counter*. The clock is a crystal-controlled oscillator (Chapter 51) which generates timing pulses at a fixed frequency (typically 33 MHz to over 200 MHz) to synchronize the computer's operations and ensure they all occur at the right time. The higher the 'clock speed', the faster and smoother is the computer's performance. The program counter counts the timing pulses in binary as they initiate the next step in the program and points to the address of the next instruction.

The operation of the CPU and some of its other components will be described more fully when the microprocessor, which is a CPU on a single silicon chip, is considered (Chapter 97).

Memory   A computer needs a memory to store the program of instructions (the *software*) it has to execute. If the program is fixed (as in a computer-controlled washing machine or electronic game), the memory has only to be 'read' and a ROM (Chapter 73), programmed by the manufacturer, is used since its contents are retained when the power is switched off. A PROM, EPROM or EEPROM would also be suitable if the user wanted to construct the program and keep it permanently in the memory.

If the program has to be changed, the 'read' and 'write' facilities of a RAM are required to allow instructions to be written in, read out and altered at will. A RAM is also needed to store the *data* for processing because it too may change. Sometimes in small computers a RAM is used for both the program of instructions and the data and so all is normally lost at switch off. As we saw earlier (Chapter 73) each location in a memory has its own *address* which allows us to get directly to any instruction or item of data.

To sum up, the computer's memory (usually RAM) is the 'work space' it uses to perform the calculations and operations it makes when running a program. The larger its capacity in megabytes (MB), the larger the program that it can handle or the more programs it will run at the same time. A typical desktop PC today might require 4 MB for simple tasks such as word processing and 16 MB for multimedia applications (p. 220).

In addition to these internal *semiconductor* memories, there is usually provision for *permanent* storage of programs and data not in current use in *external* magnetic or optical memories (Chapter 94).

Input and output devices   The CPU accepts digital signals from the input device, e.g. a keyboard, a mouse or a scanner, via its *input port*. After processing these are fed out via its *output port* to the output device, e.g. a *visual display unit* (VDU; usually called a *monitor*) or a *printer*, in a form we can understand. A mouse would be connected by a *serial* port and a printer by a *parallel* port, see below.

Buses   The CPU is connected to other parts by three sets of wires, called *buses* because they 'transport' information in the form of digital electrical signals. The *data bus* carries data for processing and is a two-way system with 4, 8 or 16 lines, each carrying one bit at a time. The one-way *address bus* conveys signals from the CPU which enable it to find data stored in a particular location of the memory. It has anything from 4 to 32 lines depending on the number of memory addresses there are (8 lines give $2^8 = 256$ addresses). The *control bus* transmits timing and control signals and could have 3 to 10 lines. In diagrams each bus is represented by one line.

To stop interference between parts that are not sending signals to a bus and parts that are, all parts have in their output a circuit called a *tri-state gate*. Each gate is enabled or disabled by the control unit. When enabled, the output is 'high' or 'low' and pulses can be sent to the bus or received from it. When disabled, the output has such a large impedance that it is in effect disconnected. Then it can neither send nor receive signals and does not upset those passing along the bus between other parts.

Serial and parallel transmission   When data has to be sent from one computer to another it can either be in serial or parallel form.

In serial form, which has advantages for long distances, the bits are sent in order one after the other with extra bits to denote starts and stops.

In parallel form, which is faster, the data is sent simultaneously a byte at a time using *handshaking* control signals which ensure synchronization with the output peripherals.

SOME FURTHER POINTS

Operating system   This is the software which makes the computer work and controls such operations as how, for example, it reads programs. It is the language the computer 'speaks'.

Two popular systems are *Windows* (by Microsoft) and *Macintosh* (by Apple); PCs designed to use one system will not run on the software of the other. Recently a new system called *Java* has been developed which, it is claimed, can be run on almost any computer.

Cache The very rapid increase in the clock speed of processors during the last 25 years (from 0.1 MHz to 200 MHz, due to miniaturization) has not been matched by a decrease in the time it takes to access instructions from memory. As a result, processors are often kept waiting by a slow memory.

A *cache* is a small, high-speed memory (typically 256 kilobytes) incorporated in the process chip which reduces the problem by storing and allowing faster access to information that is used frequently. The slower main memory is then accessed only for information not in the cache.

Viruses A computer virus is a program which reproduces itself and can become attached to other programs. There, it can cause the unwelcome changes planned by its originator. It is spread for instance via the Internet but may be countered using protective software.

USES OF COMPUTERS

The impact on society of electronics in general was considered in Chapter 78. Here we will discuss very briefly some of the vast number of ever-increasing applications of, in particular, computers, some of which are shown in Fig. 93.2.

Data processing This accounts for about three-quarters of all computer use today. It includes:

(i) *payroll preparation* for firms with a large number of employees and requires a knowledge of the hours worked, rate of pay, income tax code, national insurance and other deductions to be made,

(ii) *stock control* in large shops, supermarkets and factories so that an up-to-date record of present stock is kept and orders are placed for new stock when it runs low,

(iii) *airline ticket booking* for travel agents, giving details of flights and seats available,

(iv) *preparation of invoices* (bills) to accompany goods that will not be paid for until after delivery,

(v) *processing insurance claims*, and

(vi) *production of gas and electricity bills* from quarterly meter readings.

Fig. 93.2

Top left Office workers using desktop computers

Top right Primary school pupils being taught with the aid of an interactive computer program

Bottom left A computer-controlled laser device used for cutting fabric at a clothing manufacturer

Bottom right A student using a portable laptop computer

Process control (automation) Large-scale control applications using mainframe or minicomputers are common in process industries where raw materials are converted into useful goods. Some examples are:

- **(i)** steel making,
- **(ii)** oil refining,
- **(iii)** brick making,
- **(iv)** chemical manufacture,
- **(v)** paper making,
- **(vi)** sugar refining, and
- **(vii)** glass production.

Hundreds of *sensors* are needed to control temperatures, pressures, flow-rates and other factors, and *activators* (switches, valves, motors) are also needed to keep the process going. Apart from achieving automation, the aim is to maintain the quality of the end-product with maximum economy and output. By monitoring the process at all stages, warnings can be given when things go wrong and modifications can be made automatically by a computer-controlled feedback loop or, if necessary, the operation can be shut down.

Robotics An industrial robot is like a human arm with several separately controllable joints. In small arms the joints are worked by *stepper motors*. These rotate through exact angles when an electrical pulse is received from a computer that has the required movement program stored, often in a ROM chip. The 'hand' at the end of the arm may be a claw-like device which grips things or a special tool such as a paint sprayer or a spot welder.

Sometimes the control program is worked out by monitoring and recording the actions of a human operator taking the robot arm through the job.

Larger arms use hydraulic or pneumatic systems to operate the joints. In some factories many assembly jobs are done entirely by robots.

Computer-integrated manufacture (CIM) At present this is being researched and hailed as the next great leap forward for industry. It is attempting to bring under computer control all the various stages of manufacturing, from designing the product to ordering the materials, from quality control to issuing invoices, as well as actually making the product.

Traditional technologies such as production, mechanical and electrical engineering will be married with computer and information technology in the hope of creating a single completely integrated operation. Warehousing, accounting procedures and management information will be fully computerized and by using robots, automation will be introduced into the actual manufacturing process itself.

The potential for progress through CIM appears to be enormous and many industrialized nations are now involved in the race to achieve what will undoubtedly be another technological revolution. It is a development that countries which rely on exporting manufactured goods for their livelihood cannot afford to ignore.

Other computer users These include:

- **(i)** the *police* for storing details of criminals, fingerprints, lists of missing persons and information about major crime investigations that would previously have been stored in a card index system;
- **(ii)** *banks* for handling cheques, keeping customers' accounts, dealing with standing orders and direct debits, processing cash dispenser transactions and Electronic Funds Transfer at Point of Sale (EFT-POS) in which the cost of the shoppers' purchases is automatically deducted from their bank account and credited to the shop's account;
- **(iii)** the *Driver and Vehicle Licensing Agency* (*DVLA*), Swansea, for storing information (registration number, make, model, colour, engine size, owner's name and address) about every vehicle in the country;
- **(iv)** *newspaper publishers* who use desktop publishing packages to speed up the editing and production (with fewer employees) of their papers and enable some or all of them to be transmitted as digital signals to regional centres for printing;
- **(v)** *scientists, engineers, geologists and weather forecasters* who can make models or simulations of real-life situations using a computer and predict likely outcomes, e.g. engineers can predict how structures they design will behave without building prototypes, so saving time and money and allowing alternative designs to be studied;
- **(vi)** *doctors and dentists* for storing patients' records;
- **(vii)** *solicitors and estate agents* for whom computers with word-processing facilities are more efficient than conventional typewriters;
- **(viii)** *farmers* for planning the use of agricultural chemicals, keeping records of livestock and recording production of milk, cereals, etc;
- **(ix)** schools and colleges for educational and administrative purposes;
- **(x)** film producers, especially for special effects in science fiction production.

FURTHER APPLICATIONS

The Internet The use of computers in this communication network was mentioned briefly earlier (Chapter 78). It originated in 1969 when the US Defence Department created a computer network designed to maintain essential worldwide communications in the event of nuclear war, using ordinary telephone lines. Later, many other computer networks were established (e.g. by universities and businesses), all based on the same networking language. This has enabled computers on different networks to communicate directly with one another.

In the 1990s commercial companies gave individual home computer users access to this *inter*connected collection of *networks*—now called the *Internet*, via a telephone line and a modem. These *Internet Service Providers* (*ISPs*) use fast, powerful computers, known as servers, to give their subscribers (for a monthly fee) a temporary link to the Internet (after a password is exchanged) via their fast permanent Internet link.

Among the services available is the *World Wide Web* (*WWW*), known simply as the *Web*, which is rapidly becoming the most popular way to distribute information on the Internet. The information consists of millions of pages of text, graphics, photographs, sound and video residing on thousands of Web server computers distributed across the Internet. The pages are combined as an on-screen page and presented to the user (called a *browser*, who has sent his request using Web browser software) as formatted pages, known as *hypertext* documents. Each page on the Web has its own address, often starting with 'http://' (standing for *hypertext transfer protocol*). In addition, Web pages also contain link words, usually shown in a different style to the other text on the page, that when 'clicked' call up pages from other parts of the Web.

Another service available to Internet subscribers is *e-mail*, which was outlined earlier (Chapter 91).

Multimedia A textbook like this one is a dual-media system which uses two media, viz. text and illustrations. *Multimedia* is the integration of text, colour graphics, animation, photographs, sound and video using a computer, for education and home entertainment, Fig. 93.3. The software containing the large amounts of data required is stored in digital form mostly on CD-ROMs but also on floppy and hard disks (Chapter 94). When a disk is inserted into the appropriate disk drive slot in the computer, data transfer occurs.

Encyclopaedias, atlases, instruction manuals and educational programs can be viewed and the user allowed to

Fig. 93.3

interact, unlike normal TV. Computer games accompanied by sound effects, speech and music become more realistic.

Digital interactive TV It may be that in the future, the *home computer*, *TV set* and *telephone* will be combined. Computing, broadcasting and telecommunications will be brought together in what is called *digital interactive TV* (*iTV*) using one piece of equipment and digital signals. Separate TV sets, video cassette recorders (VCRs), CD players, telephones and computers will be unnecessary.

The process has already started with several electronics companies having launched systems that combine the personal computer (*PC*) and the television receiver (*TV*) into a *PCTV*, using the TV as a full-screen monitor. Audio CDs, photo CDs (that store the data for up to 100 colour snaps) and CD-ROMs can all be played by the equipment. The Internet is also accessible as are facsimile (fax) and answerphone facilities. Letters can be written and sent (i.e. *e-mail*) over telephone lines using the system's *modem* from up to 3 metres away by a remote wireless keyboard connected to the system by an infrared link.

It is envisaged that a fully integrated multimedia system, controlled by a *Set Top Box* (a switching and memory device plugged into a TV set at one end and into a telephone line at the other) would provide, possibly via the so-called Information Superhighway, hundreds of TV channels, films, videos and computer games on demand as well as voice and visual telephone communication, computer data links and a host of interactive activities like home banking, tele-shopping, etc. (Chapter 87).

The basic idea behind PCTV is that the TV receiver should become the central device for getting all kinds of information into the home. One problem the approach has highlighted is the different demands made of displays by computers and TV sets. A computer screen is made for close viewing by an individual and is comparatively dim but has the high resolution (Chapter 94) for showing text. On the other hand, TV sets have much brighter screens intended to be viewed at a distance by several people and their resolution is lower.

Virtual reality This is an artificial world seen when computer-generated graphics are viewed on a monitor. Using the effect, an architect can 'walk' through a building that hasn't yet been built to detect any problems before construction starts. Doctors are experimenting with it to teach surgical techniques. In Fig. 93.4 'painless' surgery is being carried out on simulated organs inside a 'virtual body'. A special ultrasonic hand controller is used by the 'surgeon' in conjunction with the computer to perform the 'operation'. At present the greatest use of virtual reality is in video games where special goggles and headsets are worn to produce a cartoon-like world of wizards and monsters.

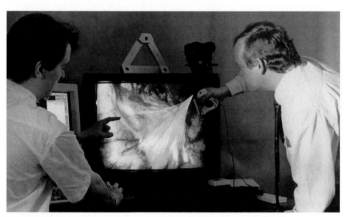

Fig. 93.4

BRIEF HISTORY OF COMPUTERS

Charles Babbage, a mathematics professor at Cambridge University in the 1830s, is generally regarded as the father of the computer. He designed, but never completed, a mechanical device that could perform calculations. He called it an 'Analytical Engine' and believed it could process information that was represented by numbers. Furthermore, he saw his 'engine' as being able to follow different sets of instructions and so perform various tasks. That is, it would respond to what we now call 'software' and be a general purpose device rather than one dedicated to a particular job.

In the 1930s and 40s *Alan Turing* and *Claude Shannon* did important theoretical work on calculating machines, logic circuits and information representation. Also, *John von Neumann* suggested that the architecture of general purpose machines could be simplified if they had a memory to store different instructions – which is the principle of modern computers.

During *World War II* electronic computers were built in the UK, USA and Australia using thousands of thermionic valves (vacuum tubes) and relays as switches. These monsters occupied rooms 10 m x 5 m, weighed several tonnes, consumed vast amounts of power (e.g. 100 kW), had only a few kilobytes of memory and performed a mere 1000 operations per second. Computer electronics was revolutionised with the arrival of integrated circuits in the early 1960s and of the microprocessor in the early 1970s. They heralded the birth of the personal computer.

DATA PROTECTION AND THE PUBLIC

In Britain most law-abiding adults have data stored about them in computers operated by:

  (i) their employer,
 (ii) their bank and/or building society,
(iii) a local authority,
 (iv) the Department of Social Security,
  (v) the National Health Service,
 (vi) the DVLA,
(vii) the Inland Revenue, and
(viii) credit providers.

To help to allay public fears about privacy the government passed the *Data Protection Act* in 1984 and created a Data Protection Registrar (i) to enforce the provisions of the Act which included 'obligations for data users', and (ii) to set up a public register of all data users that was available to anyone who wished to know who had data about him or her.

Exemptions are allowed for matters involving some police work, tax collection and national security.

COMPUTER ACRONYMS

These are common in technical and sales literature.

MMX: MultiMedia eXtensions. An advanced technology used in microprocessor architecture.

MPEG: Motion Picture Experts Group. A compression technology which allows a PC to give full quality TV from a CD-ROM

PCMCIA: Personal Computer Memory Card Industry Association. A peripheral card of electronic circuits which provides additional features when it is plugged into an expansion slot in a PC. A *sound card* allows CD sound to be played. A *graphics card* or *accelerator* and a *video card* are needed for advanced computer games and for playing CD video. A *TV card* turns a PC into a TV. A modem connects a PC to telephone line facilities.

SVGA: Super Video Graphics Adaptor. A standard with which good quality monitor screens should be compatible for good colour pictures.

EDO RAM: Extra Data Output – a fast RAM.

94 COMPUTER PERIPHERALS

CHECKLIST

After studying this chapter you should be able to:
- describe the action of input devices such as the keyboard, the mouse, the joystick, the touch screen, optical readers, sensors and scanners,
- describe the VDU (monitor) as an output device explaining what is meant by the terms 'pixels' and 'resolution',

- recall and compare the three main types of present-day printers,
- recall, describe and compare (e.g. in terms of storage capacity, access time) floppy disks, hard disks and CD-ROMs as back-up memories, and
- state why interfacing chips are required between the CPU and its peripherals.

The input, output and external memory devices of a computer are called *peripherals*, Fig. 94.1. They enable the computer to communicate with the outside world.

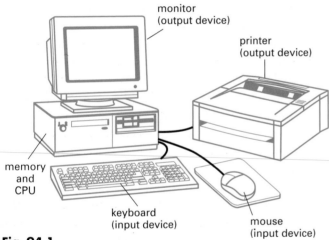

Fig. 94.1

INPUT DEVICES

Keyboard This is similar to the one on a typewriter and contains the usual range of alphanumeric characters (A to Z and 0 to 9) as well as 'command' keys for giving instructions to the computer.

Mouse Controlling and giving commands to a computer can also be done by a small box called a *mouse*, which makes complex programs with a large number of options easier to use. It also enables (together with the appropriate software) the computer user to draw on the screen.

The mouse is moved by hand over a special mat or a desk top. As it does so, it rolls on a ball that causes electrical pulses to be sent to the computer. These indicate the mouse's location and the computer reacts by moving a *pointer* (→) on the screen of the monitor in the direction indicated by the mouse's movement. If the pointer points to one of the symbols, i.e. *icons*, on the screen, the computer can be given commands by 'clicking' a switch on the mouse. The command represented by the icon is then executed, e.g. to draw.

When the ball rolls, it turns two slotted discs at right angles to one another. Each has two photodiodes and two LEDs and electrical pulses are produced in the photodiodes by light from the LEDs passing through slots. These pulses give the horizontal and vertical coordinates of the pointer on the screen.

A *trackball* works like a mouse but stays at rest when the ball is rotated by the fingers.

Joystick This allows the position of a cursor or graphics character to be controlled when the joystick handle is moved. Some are just four-position switches (left, right, up, down). Others are basically variable resistors that give instructions by applying different voltages. They are used mostly for fast response in computer games.

Touch screen Different systems are used but all consist of a clear panel (of glass or plastic) placed in front of, or incorporated into, the screen of a monitor. When the panel is touched, the x and y coordinates of the location of the finger are worked out, decoded and used to control the computer as would a mouse or keyboard. The system is useful for menu choice, i.e. obtaining and choosing from a list of optional facilities available to the user.

The touch-sensitive panel contains a matrix of cells comprising devices which respond when the finger (i) interrupts infrared beams that cross the panel, or (ii) causes a pressure change where the panel is touched, or (iii) alters the capacitance formed by tiny metal dots on the panel and the finger which connects it to earth, thereby causing charge flow.

Optical readers These are input devices used in data processing which read the data directly from the source of information.

The *bar code reader* (which may be a laser) used in large shops for stock control was mentioned earlier (Chapter 78). It works on the principle that a narrow beam of light from the reader is reflected back from the pattern of black lines of various widths on the bar code, and changed into a series of electrical pulses. These enable the computer to identify the item.

An *optical mark reader* (OMR) works on the same principle and detects marks on specially prepared documents, e.g. answer sheets for multiple-choice questions in an examination!

An *optical character reader* (OCR) reads letters and figures, enabling documents to be 'scanned in', thus saving rekeying. They are used in the preparation of gas and electricity bills.

Magnetic ink character readers (MICR) are used by banks for handling cheques. Some of the characters on cheques are printed in a special magnetic ink.

Sensors In industrial process control, in science, in medicine and in other areas, inputs come from transducers or sensors that produce analogue voltages. These may represent continuously varying quantities such as temperature, light intensity, sound level, pressure, fluid flow rate or movement, and must be changed to digital signals by an analogue-to-digital converter before being input to the computer (Chapter 95).

Scanners These are able to transfer full-colour images from documents into a digital form which can be used by the computer in many different ways.

Other input devices Other methods of control that are claimed to be easier to use are emerging. One of the most natural but difficult of all is spoken commands of the *voice*.

Another method is based on *pens* that 'write' on a sheet of paper next to the computer and simultaneously transfer the writing to the screen.

OUTPUT DEVICES

Monitors or visual display units (VDUs) In general the monitor of a desktop PC is a colour *cathode ray tube* (CRT), similar to that used in a domestic TV receiver but giving higher resolution (see below). A portable computer has a *liquid crystal display* (LCD,

Chapter 20) as a monitor because it is flatter, uses less power and is much less heavy than a CRT monitor. It also gives a sharper image with more colours and does not emit radiation. However, it costs more to make and at present its size is limited to 15 or 16 inches (measured diagonally).

A monitor screen, whether CRT or LCD, is divided into picture elements, called *pixels*, each being the smallest size of dot (typically 0.3 mm) that can be controlled independently by the computer; it is the building block from which graphics are constructed. The original drawing, picture or photograph has to be 'digitized' (i.e. represented in binary code) so that it can be treated as a two-dimensional array of data and reproduced as a series of small dots—just like a newspaper photograph.

The *resolution* of a system measures the fineness of detail that it shows, i.e. the degree of sharpness of the image or printed character on the screen. It is usually stated in terms of the number of pixels. For example, a resolution of 320×200 ($= 64\,000$) means that there are 320 pixels across the screen and 200 down, i.e. $64\,000$ altogether. The higher the resolution, the smaller the pixels and the more of them that will be displayed at one time. High resolution is considered to be 1024×768 pixels. There are many ways of controlling pixels by a program.

Monitors using *non-interlaced scanning* produce displays with less flicker than those that use interlacing (Chapter 85) because of the higher rate at which they refresh the display.

In LCD monitors the crystals are arranged in a matrix of cells which represent pixels. There are two types, called *active matrix* (or TFT, standing for Thin Film Transistor) and *passive matrix*. Both need illumination from behind. In the active type every LCD in the screen has its own transistor switch to turn it on and off (when a voltage is applied) and so block the light or let it pass. In the passive type all LCDs in a certain row or column share the same transistor, giving a cheaper device but not such good colour.

Another flat-screen technology under development is the *gas plasma* display in which each pixel is a tiny flourescent tube. When a current passes through one 'column' electrode and one 'row' electrode, the tube at their intersection glows. A plasma monitor with a width of 106 cm (measured diagonally) and a depth of 8 cm has been demonstrated. Other promising new technologies include the *field emission* display, the *polymer* display and the *laser semiconductor* display

The bulky CRT which for many years has produced high-quality pictures at low cost, clearly has rivals that may in the future hang like pictures on the wall.

Printers A printer is used when a permanent record or *hard copy* of the output is required. There are several types.

(i) *Dot-matrix printers* form letters and numbers from tiny dots at a typical rate of 300 characters per second. The computer sends the code for each letter or other character to a chip in the printer head which converts the code into a binary signal that turns electromagnets in the head on and off. The head contains a row of pins, some or all of which (depending on the character) are struck by a hammer when their corresponding electromagnets are switched on. As a result the pin strikes an inked ribbon which makes a dot on the paper. Good quality printers have a row of 24 pins in the head and produce 360 dots per inch (d.p.i.). Stepper motors (Chapter 23) move head and paper to the correct positions.

(ii) *Ink-jet and bubble-jet printers* give better quality printing than dot-matrix types but cost more. They work by firing small droplets of ink onto the paper, typically 600 d.p.i., from over 1000 jets, each half the thickness of a human hair, at the rate of 6000 drops per second from each.

(iii) *Laser printers* combine extremely high-quality black and colour printing with flexibility but are expensive. They make up characters with dots, like a dot-matrix printer, but the dots are much smaller, giving very sharp printing at 600 d.p.i. They work by melting fine plastic powder onto the page to give a permanent result.

Multifunction peripherals are also available, which combine printing, scanning and faxing facilities in one machine.

Activators Outputs from the computer may be used to control activators, often electric motors, which in turn operate the switches, gears, valves, etc. that in turn control the machine or an industrial process. If an activator requires an analogue voltage to operate, the digital output from the computer has to be converted by a digital-to-analogue converter (Chapter 95).

EXTERNAL MEMORIES

In addition to internal memories such as semiconductor RAMs, external memories are required to provide *permanent* storage space (i.e. act as a library) for programs and data which can be transferred to the computer's internal memory when necessary. Three types are common.

Floppy disk This is a flexible plastic disc, usually 3.5 inches in diameter coated with a thin magnetic film, which stores data *magnetically*. When it is inserted into a floppy disk drive (FDD) in 'write' mode, digital electrical pulses (1s and 0s) from the computer are changed by the *write/read head* into small magnetized bands on the disk in either of two directions, one way for a 1, the other way for a 0. In 'read' mode, digital electrical signals are induced in the head (by the magnetized bands on the disk) for transfer to the computer.

The disk is divided into concentric circular *tracks* (e.g. 40) and *sectors* (e.g. 10), Fig. 94.2a. Each sector stores a certain number (e.g. 256) of bytes, one byte representing one stored character. The first two tracks near the edge of the disk store its 'catalogue'. This is the disk's filing system which contains the location (i.e. track and sector) of the required data. Setting up the catalogue, a process called *formatting* the disk, has to be done before the disk can be used. The necessary procedure is usually supplied with the computer.

A precision stepper motor (Chapter 23) moves the write/read head across the disk and locates the track while another motor simultaneously rotates the disk at high speed to locate the sector by timing from a reference point in the rotation, Fig. 94.2b. In this way rapid, random access to data at any location on the disk is obtained.

Floppy disks are used in PCs and typically have a capacity of 1.44 megabytes (1.44 MB). They are widely used as portable *external* memories for data storage.

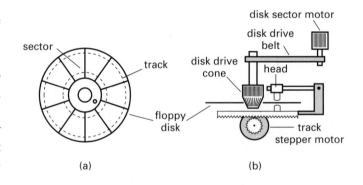

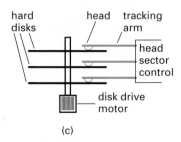

Fig. 94.2

Hard disk This is a rigid magnetic disk, similar to a floppy disk but with a much greater storage capacity, e.g. 1600 megabytes (1.6 gigabytes). It is driven by a hard disk drive (HDD) which often has several disks on the same shaft, Fig. 94.2c.

Compared with semiconductor memories the *access time* (i.e. the time it takes to find and deliver a piece of data) for hard and floppy disks is slow, especially for the latter. Storage is permanent, i.e. non-volatile, unless the disk is demagnetized by a magnetic field or by excessive temperature or by physical deformation.

Many systems have hard disks that can be changed by pulling out a cartridge, so giving greater security and flexibility. *Zip*, *Jaz* and *Syjet* are examples of removable storage systems.

CD-ROM Based on the technology of the audio compact disc (Chapter 79), this is a 5 inch plastic disc that can store graphics, pictures, information and computer programs as well as sound. It is also called an *optical*, *video* or *laser* disc but more commonly a CD-ROM because initially they could be 'written' on only once, but 'read' many times.

Huge amounts of data can be stored. For example, one CD-ROM can store 650 megabytes, enough for the contents of all 40 000 pages of the *Encyclopaedia Britannica*.

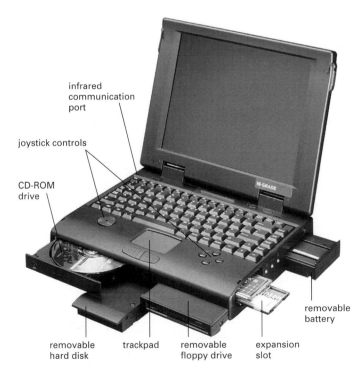

infrared
communication
port

joystick controls

CD-ROM
drive

removable   trackpad   removable   expansion
hard disk   floppy drive   slot

removable
battery

Fig. 94.3

Also, the rate of transfer of data from disc to computer by a CD-ROM is high. A portable computer having a CD-ROM drive as well as a floppy disk drive and a hard disk drive is shown in Fig. 94.3.

As in audio, photo and video CDs, in CD-ROMs digital data is stored in a series of tiny 'pits' or 'craters' of different lengths and spacing on the disc and its reflecting surface. A narrow laser beam 'reads' the data as explained previously (Chapter 79).

The next generation of compact discs will probably be the *Digital Versatile Disc* (DVD), a development of the CD-ROM. It will have a capacity of up to 17.9 gigabytes, about 30 times greater than that of a normal CD-ROM. It will be used to store either programs for a multimedia system (with much-improved graphics) or full-length films for playback on a digital video disc player.

Two other high-capacity backup storage systems that are now available are *rewritable optical discs* and high-speed magnetic *digital audio tapes* (DATs).

INTERFACING CHIPS

Data in its original form cannot usually be fed directly from an input device into the CPU. It must be presented in digital form. Similarly digital signals from the CPU may not be acceptable to an output device. Very often interfacing chips are required between the CPU and its peripherals. They may also have to compensate for differences of, for example, voltage levels (Chapter 62), operating speeds and codes.

Analogue-to-digital (A/D) and digital-to-analogue (D/A) converters
The need for chips to perform these operations for certain input and output devices has been mentioned. Their action will be considered in the next chapter.

Encoders and decoders
To interface a keyboard to a computer, an encoder is required to produce a different pattern of binary bits, according to the ASCII code (Chapter 76), for every key operated. The action of one kind of decoder was considered earlier (Chapter 66).

Modems
A computer connected to a telephone line requires a modem at each end to convert digital signals to and from analogue signals, since the telephone system is designed to handle analogue inputs only. A 1 is encoded in one system as a high audio tone and a 0 as a low tone (Chapter 91).

95 ANALOGUE/DIGITAL CONVERTERS

CHECKLIST

After studying this chapter you should be able to:
- recognize the circuit for a 'binary weighted resistor' digital-to-analogue (D/A) converter and outline how it works, and
- recognize the block diagram of an analogue-to-digital (A/D) converter and outline how it works.

DIGITAL-TO-ANALOGUE (D/A) CONVERSION

There are many occasions when digital signals have to be converted to analogue ones. For example, a digital computer is often required to produce a graphical display on the screen of a monitor. This involves using a D/A converter to change the two-level digital output voltage from the computer, into a continuously varying analogue voltage for the input to the CRT, so that it can deflect the electron beam and make it 'draw pictures' at high speed.

The principle of the *binary weighted resistor* D/A converter is shown in Fig. 95.1 for a four-bit input. It is so called because the values of the resistors, R, $2R$, $4R$ and $8R$ in this case, increase according to the binary scale, the l.s.b. of the digital input having the largest value resistor. The circuit uses an op amp as a *summing amplifier* (p. 120) with a feedback resistor R_f. S_1, S_2, S_3 and S_4 are digitally controlled electronic SPDT switches.

Each switch connects the resistor in series with it to a fixed reference voltage V_{REF} when the input bit controlling it is a 1 and to ground (0 V) when it is a 0. The input voltages V_1, V_2, V_3 and V_4 applied to the op amp by the four-bit input (via the resistors) therefore have one of two values, either V_{REF} or 0 V.

Using the summing amplifier formula (p. 120), the analogue output voltage V_o from the op amp is:

$$V_o = -\left(\frac{R_f}{R} \cdot V_1 + \frac{R_f}{2R} \cdot V_2 + \frac{R_f}{4R} \cdot V_3 + \frac{R_f}{8R} \cdot V_4\right)$$

If $R_f = 1\ k\Omega = R$, then:

$$V_o = -(V_1 + \tfrac{1}{2}V_2 + \tfrac{1}{4}V_3 + \tfrac{1}{8}V_4)$$

With a four-bit input of 0001 (decimal 1), S_4 connects $8R$ to V_{REF}, making $V_4 = V_{REF}$. S_1, S_2 and S_3 connect $2R$, $4R$ and $8R$ respectively to 0 V, making $V_1 = V_2 = V_3 = 0$. And so, if $V_{REF} = -8$ V, we get:

$$V_o = -\left(0 + 0 + 0 + \frac{(-8)}{8}\right) = -\frac{(-8)}{8} = +1\ V$$

With a four-bit input of 0110 (decimal 6), S_2 and S_3 con-

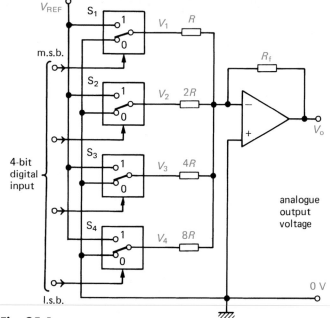

Fig. 95.1

nect $2R$ and $4R$ to V_{REF}, making $V_2 = V_3 = V_{REF} = -8$ V; S_1 and S_4 connect R and $8R$ to 0 V, making $V_1 = V_4 = 0$ V. Therefore:

$$V_o = -\left(0 + \frac{(-8)}{2} + \frac{(-8)}{4} + 0\right)$$
$$= -(-4 - 2) = +6\ V$$

From these two examples we see that the analogue output voltage V_o is directly proportional to the digital input. V_o has a 'stepped' waveform with a 'shape' that depends on the binary input pattern, Fig. 95.2, but it does vary.

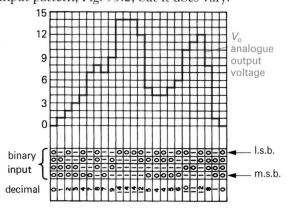

Fig. 95.2

ANALOGUE-TO-DIGITAL (A/D) CONVERSION

In many modern measuring instruments (e.g. a digital voltmeter), the reading is frequently displayed digitally, but the input is in analogue form. An A/D converter is then needed.

The block diagram for a four-bit *counter* type A/D conversion circuit is shown in Fig. 95.3a, with waveforms to help explain the action. An op amp is again used but in this case as a *voltage comparator* (p. 121). The analogue input voltage V_2 (shown here as a steady d.c. voltage) is applied to the non-inverting (+) input; the inverting (−) input is supplied by a ramp generator with a repeating sawtooth waveform voltage V_1, Fig. 95.3b.

The output from the comparator is applied to one input of an AND gate and is 'high' (a 1) until V_1 equals (or exceeds) V_2, when it goes 'low' (a 0), as in Fig. 95.3c. The other input of the AND gate is fed by a steady train of pulses from a pulse generator, as shown in Fig. 95.3d. When both these inputs are 'high', the gate 'opens' and gives a 'high' output, i.e. a pulse.

From Fig. 95.3e, you can see that the number of pulses so obtained from the AND gate depends on the 'length' of the comparator output pulse, i.e. on the time taken by V_1 to reach V_2. This time is proportional to the analogue voltage if the ramp is linear. The output pulses from the AND gate are recorded by a binary counter and, as shown in Fig. 95.3f, are the digital equivalent of the analogue input voltage V_2.

In practice the ramp generator is a D/A converter which takes its digital input from the binary counter, shown by the dashed lines in Fig. 95.3a. As the counter advances through its normal binary sequence, a staircase waveform with equal steps (i.e. a ramp) is built up at the output of the D/A converter, like that shown by the first four steps in Fig. 95.2.

Similar conversion techniques are used in digital voltmeters (p. 91).

A to D and D to A converter board

This board, shown in Fig. 95.4, is a dual-function unit designed to show in 'slow motion' as well as at 'speed' how analogue quantities are converted into digital signals, and vice versa.

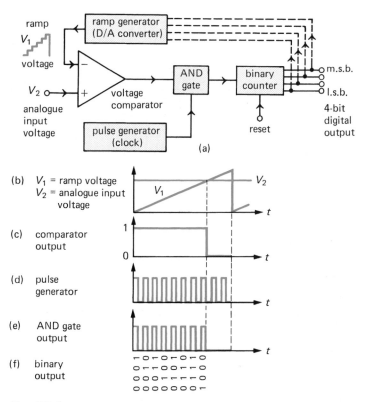

Fig. 95.3

Fig. 95.4

96 PROGRAMS AND FLOWCHARTS

CHECKLIST

After studying this chapter you should be able to:
- state that a computer works in machine code,
- distinguish between low-level and high-level languages using the terms 'assembler' and 'compiler',
- state what a flowchart does, and
- construct and interpret flowcharts using flowchart symbols.

PROGRAM LANGUAGES

A computer must be programmed so that it knows what to do. A *program*, referred to as *software* (or *firmware* if it is stored in a ROM) consists of a series of instructions (from the CPU's instruction set), each followed by an address for data and involves the computer in a 'fetch-and-execute' process.

Programs can be written in *machine code*, i.e. the 0s and 1s of the binary system in which the computer must work. But it is a tedious, time-consuming process, liable to error. Program 'languages' have therefore been developed to make the job easier.

Low-level or assembly languages These are close to the binary system. Programs are written in mnemonics (i.e. memory aids) and by referring to the instruction set for the CPU we might find, for example, that the mnemonic for 'load data from memory into accumulator' is LDA. It would also give the binary code for this operation, say 1001 1110. To enable the CPU to understand the instruction LDA, i.e. to convert the input into machine code, a special program called an *assembler* is stored in a ROM.

High-level languages These are more like everyday English and are easier to understand and work with since they use terms such as LOAD, ADD, FETCH, PRINT, STOP. However they need more memory space and computer time because each statement converts into several machine code instructions, not just one as is usual in a low-level program. A *compiler* or an *interpreter* (the equivalents of an assembler and, like it, consisting of a program) does the translation into machine code. BASIC (Beginner's All-purpose Symbolic Instruction Code) is a popular high-level language. COBOL is designed for business use, while FORTRAN is suitable for scientific and mathematical work.

FLOWCHARTS

All programs, whether high- or low-level, must give absolutely clear instructions to the computer. Writing a program is easier if a flowchart is drawn up initially to show the sequence of events required to complete the task. A flowchart consists of a number of box-like symbols joined together by arrowed lines. The shape of a symbol depends on what it is used to indicate; some are shown in Fig. 96.1.

Symbol	Used to show
 ⬭	when flow of data *starts* or *stops*
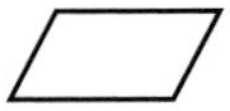 ▱	when data is *input* or *output*
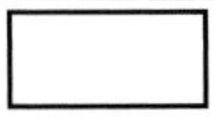 ▭	when data is *processed*
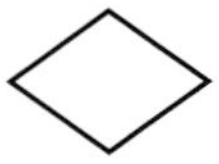 ◇	when a *decision* is made

Fig. 96.1

SOME SIMPLE EXAMPLES OF FLOWCHARTS

Voltage calculation The voltage V across a resistor R carrying a current I is to be calculated by a computer. This is most unlikely, but the flowchart in Fig. 96.2 shows the steps which have to be taken to perform the task, however it is done.

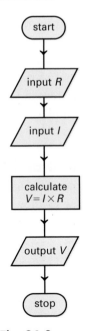

Fig. 96.2

Rain alarm A computer is to be programmed to cause an alarm to be sounded when a rain sensor detects rain by supplying a logic 1 signal to input A of the computer. The flowchart for the system is given in Fig. 96.3. Note the loop to ensure processing continues until the condition 'input A = 1' is satisfied.

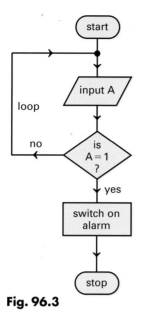

Fig. 96.3

House search problem An estate agent wishes to search a list of properties he has for sale to discover for a prospective buyer which have three bedrooms and a double garage. If he decided to use a computer and write his own program, he might first construct a flowchart like that in Fig. 96.4.

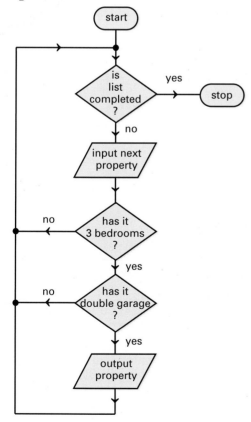

Fig. 96.4

LOOPS

Counter-controlled

When a fixed number of items is to be processed in an operation, a 'counter-controlled' loop is included in the flowchart. For example, if we wish a computer to stop when it has produced a list of the bowling averages of five bowlers in a cricket team, a flowchart like that in Fig. 96.5 would be constructed.

Condition-controlled

If the number of times a loop has to be repeated is not known beforehand, the operation is stopped by inserting a rogue value at the end of the data. This is chosen to be so different from the other data that it is easily identified. The loop, called a 'condition-controlled' loop, then repeats until the rogue value is met and is the signal for data entry to end.

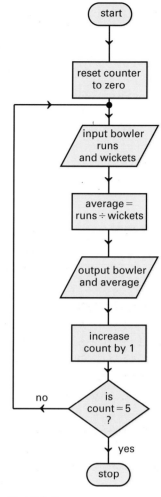

Fig. 96.5

The flowchart for a system to print a long list containing an unknown number of names is shown in Fig. 96.6. It uses the rogue value XXX.

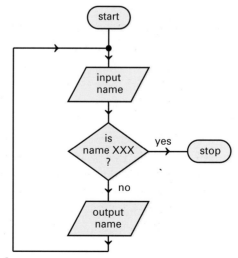

Fig. 96.6

97 MICROPROCESSORS

INTRODUCTION

A microprocessor (MPU or µP) is a miniature version of the CPU of a digital computer, i.e. ALU, registers and control unit. It is an LSI chip containing thousands of transistors, Fig. 97.1, developed in the early 1970s when, as ICs became more complex and specialized, the need was felt for a general-purpose device suitable for a wide range of jobs.

Its versatility is due to the fact that it is *program-controlled*. Simply by changing the program it can be used as the 'brain' not only of a microcomputer but of a calculator, a cash register, a washing machine, a juke box or a petrol pump. Alternatively, it will control traffic lights or an industrial robot. The market for such a flexible device is much greater than for a 'dedicated' chip doing just one job.

Smart cards, which are taking us closer to the 'cashless' society, contain a microprocessor (and memory). They are being tried out by banks as a more secure alternative to magnetic strip cards used at present at automatic teller machines (ATMs). In some areas they are now being used as travel cards whereby passengers 'pay their fare' by wiping the card across a special reader on entering the bus. Telephone companies are also giving them a trial, Fig. 97.2.

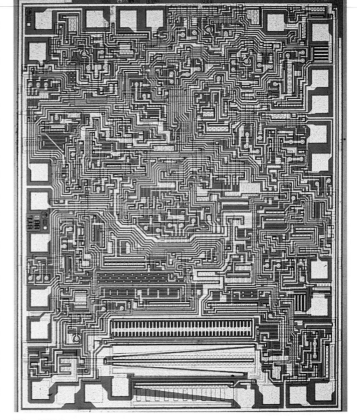

Fig. 97.1

Fig. 97.2

Microprocessors can now be designed only by using computers which probably contain microprocessors. Note that an MPU is not a computer itself—to be a micro*computer* it needs memories and input/output units. However the trend is to incorporate as many as possible of the peripheral support chips into the MPU package.

There are many MPUs on the market, with instruction sets for various tasks and which operate on words of different lengths (usually four-, eight- or sixteen-bit). Some cost just a few pounds and are often the cheapest part of a system. They are often housed in 40 pin 0.6 inch wide d.i.l. packages and operate from 5 V and/or 12 V power supplies.

The first MPU to be useful in a computer was *Intel's* 8080, introduced in 1974 and the direct ancestor of today's family of *Pentium* processors used in about 80% of modern PCs (i.e. *IBM* models and their clones). At the same time *Motorola* were promoting their 68 000 family and at present their *Power PC* processor is used in *Apple* computers. The latest MPUs have over 5 million transistors on one tiny silicon chip, a number which, according to *Moores's Law*, *doubles every two years or so* – as does the MPU clock speed. The microprocessor is unquestionably one of man's greatest intellectual and technological accomplishments.

Current research suggests that MPU development in the near future will be in three main directions. *First*, they will have faster clock-speeds (over 500 MHz), *second*, they will offer extra on-chip multimedia functions more cheaply than today's arrangement of plug-in expansion cards and peripherals (giving better graphics, audio, video and telephone line access) and *third*, there will be chips that are used in network devices and are Internet-based.

ARCHITECTURE AND ACTION OF AN MPU

The simplified block diagram in Fig. 97.3 of a typical microprocessor chip can be used to outline how it works. Assume it is programmed with the necessary instructions and data to add two numbers and that it has been 'reset', either manually by a switch or automatically when the power is applied, by a signal to its reset input. The *program counter* then reads zero, all *registers* contain 0s and the *clock* has stopped.

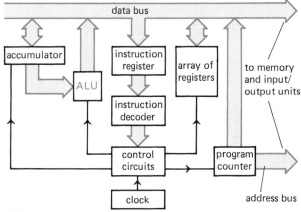

Fig. 97.3

Action The *program counter* is connected by the *address bus* to the *memory*, e.g. a ROM or PROM, in which the instructions are stored. The binary output (count) from the *counter* is the address input to the *memory* and is initially all 0s (e.g. 0000 0000 for an MPU with an eight-bit *address bus*). The instruction in this first address is thus 'read' out of the *memory* into the MPU via the *data bus*.

The instruction is held by the *instruction register* (until another is received) whose outputs are changed by the *instruction decoder* (p. 144) into a signal that goes to the *control circuits*. By opening and closing logic gates these set up the routes that enable the MPU to perform the operation required by the first instruction. Suppose it is LOAD.

If the *clock* is now started and advances the *program counter* by one count, the program advances by one step (line) and the data stored at the second address (e.g. 0000 0001) is 'read' out of the *memory* and loaded ('copied' is more exact) via the *data bus* into the *accumulator*. (The data will have been entered previously by the programmer and transferred into a RAM).

To obtain addition of the data (number) in the *accumulator* and the data (number) stored at another address in the RAM, the program must give the necessary instructions on succeeding clock pulses to enable the first data to be shifted from the *accumulator* to one of the internal *registers* in the MPU and for the second data to be copied into the *accumulator*.

If the ADD instruction is then given, the *instruction decoder* arranges for the *ALU* to perform the addition (i.e. act as a full-adder) and to store the result in the *accumulator* for subsequent transfer to the *output unit*.

Register stack and subroutines Sometimes instead of changing the program counter by one, it is useful to 'jump' from the step-by-step sequence of the main program to what is known as a *subroutine*. For example, multiplication involves a fairly long process of 'shift-and-add'. To save writing this out every time it is used in a program, it can be written just once as a subroutine, stored and recalled when needed, Fig. 97.4a.

The store used is called a *register stack* because the data is 'stacked' on top of each other in order and then recovered from the top in reverse order, i.e. last-in, first-out, Fig. 97.4b. Only one address is needed, that of the top and if the stack is an external RAM, a *stack pointer* is used to give the first location in the RAM chosen by the programmer as stack.

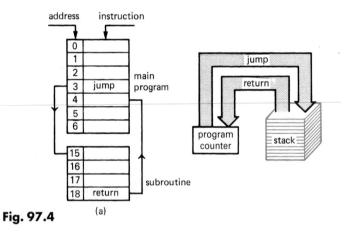

Fig. 97.4 (a)

Flags A flag is a flip-flop which is set to the 'high' (1) state to show that a particular operation has been completed. For instance, in some MPUs the *carry flag* is set to

1 if there is a carry bit in the accumulator and the *zero flag* to 1 if an instruction puts a 0 in it.

AUTOMATIC CONTROL BY AN MPU

Many systems and manufacturing processes can be controlled automatically by suitable transducers supplying (via appropriate interface circuits, p. 225) inputs to an MPU whose output takes corrective action when the quantity being monitored goes outside the permitted range.

Action The block diagram of Fig. 97.5a shows how the temperature of, for example, an electrically heated furnace can be kept constant by an MPU. An analogue input voltage is supplied by a temperature sensor to an A/D converter which sends data continuously (eight bits at a time) to the MPU from D_0 to D_7. If the required temperature (which may have been set on a keyboard) is not maintained, the MPU sends a signal to a circuit like that in Fig. 97.5b which switches the large heater on or off.

If the required temperature is say 250 °C, and the output voltage from the sensor is proportional to temperature in the range 0 to 250 °C and at 250 °C is 1.00 V, then, since an eight-bit A/D converter has $2^8 = 256$ levels, the smallest temperature change that can be detected is $1.00 \times 250/256 \approx 0.98$ °C.

Practical points When the $\overline{CS}$ input on the A/D converter is activated by the MPU, the Conversion Starts; it ends when $\overline{EOC}$ (*End Of Conversion*) is active. Activation of $\overline{OE}$ (*Output Enable*) enables the output and allows a set of data to be sent from the A/D converter to the MPU when it is ready to receive it. (The bars over $\overline{CS}$, $\overline{EOC}$ and $\overline{OE}$ indicate that these inputs are active when 'low', i.e. 0.)

In the heater circuit, operated separately on the a.c. mains supply, the relay is driven by a power MOSFET because of its very high power gain and large input impedance.

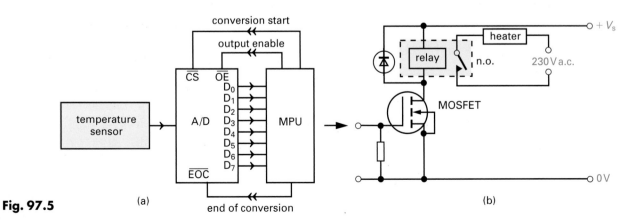

Fig. 97.5 (a)   (b)

98 PROGRESS QUESTIONS

1. a) Draw the block diagram for a simple digital computer showing its *four* main parts. Say what each part does.

b) Distinguish between mainframe, mini- and micro-computers.

2. Explain in a sentence the function of each of the following in a CPU:

a) clock,

b) program counter,

c) instruction accumulator,

d) instruction decoder,

e) ALU,

f) register,

g) data bus,

h) address bus,

i) control bus.

3. a) State how RAMs, ROMs and PROMs are used in computer systems.

b) Briefly describe three kinds of external memory.

4. Distinguish between the following:

a) hardware and software,

b) high-level and low-level languages, and

c) microprocessor and microcomputer.

5. A computer controls an industrial process in which the raw materials have to be kept at a certain temperature. It is programmed to bring on a heater if the temperature falls below 200 °C and turns it off if the temperature is above 210 °C.

a) Construct a flowchart describing the sequence of events in the process.

b) Draw a circuit diagram of a temperature sensor using a thermistor which gives an increasing output voltage when the temperature rises.

c) What kind of interfacing circuit could be needed to use the voltage in **b)** as an input to the computer?

6. With reference to a microprocessor system:

a) (i) What is the purpose of the address bus?

(ii) What is the purpose of the data bus?

(iii) State *one* difference between these two buses.

b) (i) State the possible outputs of a *tri-state device*.

(ii) Why is it essential for RAM to have tri-state data output lines?

c) State and explain *two* differences between a PROM and an EPROM.

d) (i) What is the function of an A/D converter?

(ii) Explain why the output from an A/D converter may be inaccurate if the input signal changes quickly.

(iii) State how this problem can be overcome. (*N.*)

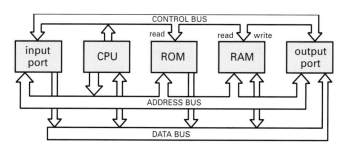

Fig. 98.1

7. The diagram in Fig. 98.1 shows the main parts of a typical computer and the directions in which signals pass between them via the buses.

a) Which buses are (i) one-way, (ii) two-way?

b) In which part do the control signals arise?

c) Why does RAM receive both 'read' and 'write' control signals but ROM only receives 'read' signals?

d) State, giving reasons, *two* situations where it would be better to use an EPROM rather than a ROM.

e) Why is it possible for parts not involved in signal exchange at a certain time, to be connected to the bus systems without interfering with parts that are?

f) Data might be transferred to a peripheral via a *parallel output port* using *handshaking* control lines. Explain the meaning of the two terms in italics.

8. If Moore's Law holds and in 1996 certain microprocessors contained 5 million transistors, how many might we expect the microprocessors of 2010 to have?

PRACTICAL ELECTRONICS

Practical work in electronics usually has two broad aims:

- **(i)** to aid the *understanding of basic circuits and theory*, and
- **(ii)** to enable *projects to be tackled* involving the design and construction of simple electronic systems.

The 'hardware' available takes many forms.

ELECTRONIC KITS

Three kits are briefly described as typical examples.

Locktronics[1] Circuits are built by plugging components (discrete and ICs) and connecting links into sockets on a baseboard, Fig. A1, to give a layout that is easy to follow. It is particularly suitable for experiments on *basic electronics*, being easily assembled, and flexible.

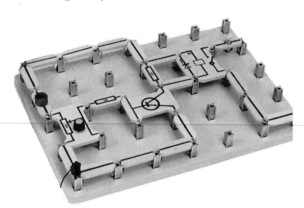

Fig. A1

Alpha system[2] This consists of a number of separate modules, Fig. A2, each with a certain function and a clearly marked circuit. Their properties can be studied individually or *systems* can be constructed by joining two or more together, mechanically and electrically, using 'alphalinks'.

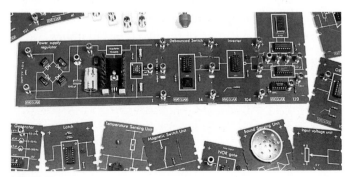

Fig. A2

'Blue Chip' boards[2] These circuit boards have been shown at various points in the book. They lend themselves to in-depth investigations of the behaviour of certain circuits in *digital electronics*.

PROTOTYPE BOARDS

These are solderless circuit boards ('breadboards') on which connections are made by pushing wires and component leads into holes. They are useful for building temporary circuits and have a bracket for mounting switches etc. Several can be interlocked. There are two types.

S-DeC This is designed for discrete components only (not ICs) and has 70 contact points arranged in two sections, each having 7 parallel rows of 5 connected contacts, Fig. A3.

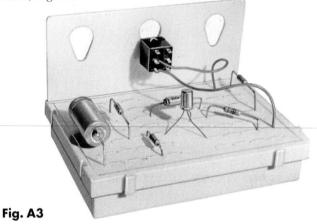

Fig. A3

Professional This accepts discrete components and ICs. The large number of contact points (550 on a 6 inch long board, Fig. A4) form a 0.1 inch grid, arranged as in S-DeC but also having continuous contact rows at the top and bottom for use as power supply rails. With the 'holes' being so close together, care is needed to ensure the correct connection is being made.

Fig. A4

1 Locktronics Ltd, Avon BS18 7TZ
2 Unilab, Blackburn BB1 3BT

SOLDERED BOARDS

'Permanent' circuits require components to be soldered together on some type of insulating (resin-bonded) board pierced with regularly spaced holes (typically 0.1 inch apart).

Matrix board   See Fig. A5a. Single- or double-sided press-fit terminal pins, Fig. A5b, are pushed through the appropriate holes and components soldered to them.

(a)

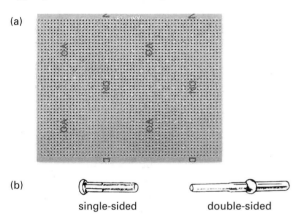

(b)

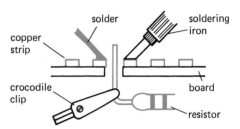

single-sided   double-sided

Fig. A5

Stripboard   See Fig. A6a. This has pierced copper strips bonded to the board on one side to form the connections between the components. Leads from the latter are inserted from the other side of the board and soldered in place on the coppered side (see Fig. A7). When the copper strip has to be broken (to prevent unwanted connections), a stripboard cutter like that in Fig. A6b is inserted into the hole where the break is required and rotated clockwise a few times. Alternatively, an ordinary twist drill held between the fingers and rotated will do.

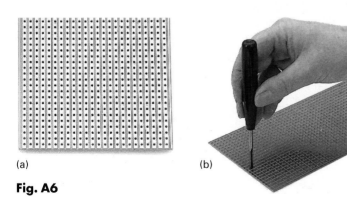

(a)   (b)

Fig. A6

SOLDERING

A badly soldered joint can cause trouble which may be hard to trace. It has a dull surface and forms a 'blob', a good joint is shiny with only a little solder in the joint. A 15 W iron with a 1.5 mm or 2 mm tip is suitable for most purposes but with ICs a 1 mm tip is better.

The steps in soldering a resistor to stripboard are shown in Fig. A7. The crocodile clip acts as a 'heat sink' (to prevent heat damage to the resistor and is essential when soldering diodes and transistors). ICs are best mounted on holders soldered to the board.

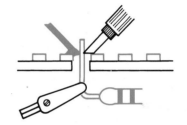

① Tip of iron held on copper strip to heat it

② Tip of iron moved to touch copper strip and resistor till molten solder fills joint

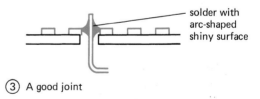

③ A good joint

Fig. A7

Parts to be soldered should be clean (rubbing with emery cloth will ensure this) and grease-free and the joint should not be moved before the solder solidifies.

ANSWERS

1 ELECTRIC CURRENT
2. (i) 5 C   (ii) 50 C   (iii) 1500 C
3. (i) 5 A   (ii) 0.5 A   (iii) 2 A
4. (a)   (i) 1000 mA   (ii) 500 mA   (iii) 20 mA
   (b)   (i) 2000 µA   (ii) 400 µA   (iii) 5 µA
5. $A_2 = A_3 = A_4 = A_5 = 0.5\,A; A_6 = 1\,A$

2 E.M.F., P.D. AND VOLTAGE
1. (i) 12 J   (ii) 36 J
3. (a) 1.5 V   (b) 1.2 V   (c) 0.3 V
4. (a) 4.5 V   (b) 1.5 V
5. 18 V, 2 V, 8 V
6. (i) A = +6 V, B = +3 V, C = 0 V
   (ii) A = 0 V, B = −3 V, C = −6 V
   (iii) A = +3 V, B = 0 V, C = −3 V
7. (a)   (i) 1000 mV   (ii) 700 mV   (iii) 20 mV
   (b)   (i) 1.6 V   (ii) 0.4 V   (iii) 0.05 V

3 RESISTANCE AND OHM'S LAW
1. (a) 1.8 kΩ   (b) 5 V   (c) 2 MΩ
2. (i) 3 kΩ   (ii) 1.5 kΩ   (iii) 4.5 kΩ
3. (a) 75 Ω   (b) 2 V
4. $I = 3\,A, I_1 = 2\,A, I_2 = 1\,A$
5. (i) A, 5 V; B, 5 V   (ii) A, 5 V; B, 0 V

4 METERS AND MEASUREMENT
1. (a) Shunt 0.05 Ω   (b) Multiplier 9950 Ω
2. (i) 1000 Ω   (ii) 5000 Ω. The 5 V range because its resistance is higher
3. (i) 0.5 mA   (ii) 50 µA
4. (i) 4 V   (ii) 3 V

5 POTENTIAL DIVIDER
1. (a) 1 V   (b) 2 V   (c) 2 V   (d) 6 V
2. 6 V
3. 4 V

6 ELECTRIC POWER
1. (i) 1/10 A   (ii) 2/10 A   (iii) 1/60 A
2. (a) 3 kΩ   (b) $27/10^3\,W = 27\,mW$
3. (a) 4.0 V, 2.5 Ω   (b) 2.5 Ω, 1.6 W

7 ALTERNATING CURRENT
1. (a) 2 ms   (b) 500 Hz   (c) 3 V   (d) 2.1 V
2. (a) 12 V   (b) 17 V   (c) 24 W
3. (a) 3 : 1   (b) 1 : 4
4. 10 V steady d.c. + 5 V peak a.c.

8 PROGRESS QUESTIONS
2. (i) Resistance of metal wire increases, therefore current decreases
   (ii) Resistance of semiconductor decreases, therefore current increases
3. 22.5 Ω; 22.2 cm
4. Multiplier of 4995 Ω in series with meter
5. (i) Fig. 8.2a   (ii) Fig. 8.2b
6. (i) 12 V   (ii) 8 V
7. (i) 9 V   (ii) 3 V   (iii) 0 V
8. (a)   (i) 6 V, 4 Ω   (ii) 0.25 A   (iii) 1.25 W
   (b)   (i) 4 Ω   (ii) 2.25 W
9. (a) 4 Ω   (b) 0.125 A (1/8)   (c) 1.75 V
10. With voltmeter (8/13) × 9 = 5.54 V
    Without voltmeter 6 V
11. (c) $2E, E$ (d)   (i) 2.4 V   (ii) 1.6 V
12. 7 V, 50 Hz

9 RESISTORS
1. (a) $R_1 = 1\,k\Omega \pm 10\%; R_2 = 47\,k\Omega \pm 5\%; R_3 = 560\,k\Omega \pm 20\%$
   (b) $R_4 = 1\,k\Omega \pm 1\%; R_5 = 330\,\Omega \pm 5\%; R_6 = 51\,k\Omega \pm 2\%$
2. (a)   (i) brown green brown silver   (ii) brown black black gold
       (iii) orange white red silver   (iv) brown black orange brown
       (v) orange orange yellow red   (vi) brown black green silver
   (b)   (i) brown blue black black red   (ii) red yellow black brown gold   (iii) violet green black orange brown
3. (i) 2.2 kΩ ± 20%   (ii) 270 kΩ ± 5%   (iii) 1 MΩ ± 10%   (iv) 15 Ω ± 1%
4. (i) 100RJ   (ii) 4K7G   (iii) 100KK   (iv) 56KM
5. (i) 1.2 kΩ   (ii) 4.7 kΩ   (iii) 68 kΩ   (iv) 330 kΩ

10 CAPACITORS
1. (a) 1 C   (b) 2 µF   (c) 5 V
2. (a) $1000\,\mu C = 10^{-3}\,C$ (b) 0.25 J
3. (i) 6.9 µF   (ii) 50 µF
4. (i) 159 Ω   (ii) 0.159 Ω
5. (b) 7.5 mA
7. (a)   (i) 0.1 nF   (ii) 8.2 nF   (iii) 10 nF
   (b)   (i) 0.33 µF   (ii) 1.0 µF   (iii) 0.047 µF
8. (a)   (i) 820 pF   (ii) 1000 pF   (iii) 56 000 pF   (d) 100 000 pF
   (b)   (i) 221   (ii) 823   (iii) 104

11 INDUCTORS
1. Resistance
2. R same; X_C decreases; X_L increases
3. 100 mH
4. (a) 9.4 kΩ   (b) 6.3 kΩ
5. (a) 3 Ω   (b) 90 Ω   (c) 90 Ω   (d) 0.29 H (290 mH)

12 CR AND LR CIRCUITS
1. 12 V
2. (i) 1 s   (ii) 5 s
3. (a)   (i) 4 V   (ii) $5\frac{1}{3}$ V   (iii) 6 V
   (b)   (i) 2 V   (ii) 2/3 V   (iii) 0 V
4. (a)   (i) 3 V   (ii) 1 V   (iii) 0 V
   (b)   (i) −3 V   (ii) −1 V   (iii) 0 V
5. (a) 10 mA   (b) 1500 µC   (c) 0.15 s
6. (i) −6 V   (ii) +6 V   (iii) −12 V

13 LCR CIRCUITS
1. 10 Ω
2. 159 kHz

14 TRANSFORMERS
1. (a) 20 : 1   (b) 1600   (c) $V_p \times I_p = V_s \times I_s, \therefore 240 \times I_p = 12 \times 2$, i.e. $I_p = 24/240 = 0.1\,A$

16 PROGRESS QUESTIONS
1. 22.4 mA
2. (a) brown, black, red
   (b) (i)   (ii)

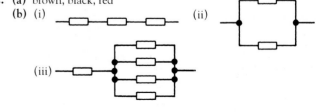

   (iii)

   (c)

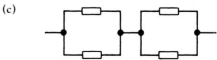

3. 2 μC
4. (a) 200 μC = 2×10^{-4} C (b) 10^{-2} J
5. (a) 4 μF (b) 200 μC (c) 40 V
6. (a) 17.5 mA (b) 20 Ω (c) 0.25 A
7. 4.5 mA; 100 Hz
8. (a) 1 mA (b) 10 μC (c) 10^{-2} s
9. (b) 9 V (c) 3 V
12. 1.27 μF
13. 15.9 MHz to 7.1 MHz
14. (a) (i) 6.3 μA (ii) 100 μA
    (b) (i) current leads by about 90° since X_C ($10^6/2\pi\Omega$) is large compared with R (10^4 Ω) and circuit as a whole is capacitive; (ii) current and voltage are almost in phase since X_C ($1/2\pi\Omega$) is small compared with R (10^4 Ω) and circuit as a whole is resistive
15. 240/20 = 12/1
16. (a) 360 s

19 HEAT, LIGHT AND STRAIN SENSORS
4. 0.3 kΩ (300 Ω)

22 RELAYS AND REED SWITCHES
1. 10 V
2. 8.6 mA

24 PROGRESS QUESTIONS
3. (b) 45 °C
4. (b) 40 mA (c) 320 mW

26 JUNCTION DIODE
1. (a) L_1 bright, L_2 off
   (b) L_1 bright, L_2 off because the diode is forward biased and offers an easy path for the current to bypass L_2
   (c) L_1 and L_2 both dim because the reverse-biased diode forces current through L_2 and there is 4.5 V across each lamp—they need 6 V (or more) to be bright
2. (a) 5 Ω (b) 1 W

27 OTHER DIODES
1. From 3 V to 6 V the brightness gradually increases and is then 'tied' at that brightness by the Zener diode from 6 V to 9 V
2. 0.5 A (500 mA)
3. 680 Ω

28 PROGRESS QUESTIONS
2. (b) 4.2 V (c) 68 Ω
3. (b) 0.04 A (40 mA) (i) current increases, p.d. = 6 V
   (ii) current zero, p.d. = 4 V
4. (iii) 5 V
5. (a) C (c) $I = (9 - 2)/680 = 7/680$ A ≈ 0.01 A = 10 mA

29 TRANSISTOR AS A CURRENT AMPLIFIER
1. (a) ASRBE (b) ALCE
2. (b) and (d)
3. (a) 50 (b) 102 mA
4. 10 mA
5. $80 \times 100 = 8000$

30 TRANSISTOR AS A SWITCH
1. (a) 0.5 mA (b) 40 mA
   (c) No, because at saturation, p.d. across transistor is 0 V and 5 V across 100 Ω, therefore saturation collector current = 5 V/100 Ω = 0.05 A = 50 mA
2. (i) 0 V (ii) 6 V
3. (a) (i) 0 V (ii) 6 V
   (b) (i) 6 V (ii) 0 V

(c) (i) large (ii) small
(d) (i) 0 to +0.6 V (ii) ≈ +0.6 V
(e) (i) To prevent excessive base currents destroying the transistor if R_1 is made zero (ii) To act as a 'load' which causes the collector–emitter (output) voltage to change from 6 V to 0 V when the transistor switches fully on

31 MORE ABOUT THE JUNCTION TRANSISTOR
1. (i) 100 (ii) 2 kΩ (iii) 50 kΩ (iv) 5 V
2. It is the maximum p.d. between the collector and emitter when the transistor is saturated by a collector current of 150 mA and a base current of 15 mA. Important in switching circuits where V_{CE} should be as small as possible (ideally 0 V) to reduce power loss

34 PROGRESS QUESTIONS
1. (a) ≈ 0
   (b) 0.1 A = 100 mA
   (c) 0 (since Tr short-circuits load)
   (d) 0.02 A = 20 mA
2. $I_B = 10$ V/500 kΩ = 1/50 mA ∴$I_C = 100 \times 1/50 = 2$ mA
   $V_{CE} = 10$ V − 2 mA × 4.5 kΩ = 10 V − 9 V = 1 V
3. Max $I_C = 9$ V/1 kΩ = 9 mA = $9/10^3$ A
   (a) (max $I_C/2)^2 \times 1$ kΩ W = $(9/(2 \times 10^3))^2 \times 10^3$ W = $(81 \times 10^3)/(4 \times 10^6)$ W = $81/4 \times 10^{-3}$ W = 81/4 mW ≈ 20 mW
   (b) total power supplied − power in resistor ≈ (9 V × 4.5 mA − 20 mW) ≈ 20 mW
4. (a)

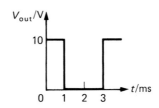

(b) 10 kΩ
5. (a) 0.4 mA (400 μA) (b) 30 kΩ (c) See p. 70
   (d) See Fig. 30.6: replace bell by lamp and make R a variable resistor
6. See p. 70
8. (i) 40 (ii) 50
9. (a) When S is closed the transistor (and L) is on since it can receive base current via S and R_B. When S is open, initially the p.d. V_C across C is zero and the p.d. V_R across R, is 6 V—the latter keeps the transistor on. As C charges up through R, V_C rises and V_R falls (see Chapter 12). Eventually V_R is too small to make the base–emitter p.d. V_{BE} equal to 0.6 V (for a silicon transistor) and the transistor switches off—the delay being determined by the time constant CR. Finally $V_C = 6$ V and $V_R = 0$.
   (b) See p. 70
10. (a) (i) near 0 V (ii) 0.6 V

35 RECTIFIER CIRCUITS: BATTERIES
2. During each half-cycle of V_i the current supplied has to pass through two conducting diodes (either D_2 and D_4 or D_1 and D_3) across each of which there is a voltage drop (of about 1 V per diode for silicon)
   When the load current is zero (i.e. R infinite) $V_o = V_i$ since then no current flows through the diodes and so there is no voltage drop across them
3. Half-wave circuit: 50 Hz
   Full-wave circuit: 100 Hz

36 SMOOTHING CIRCUITS

2. The peak value of the a.c. input, i.e. $\sqrt{2} \times$ r.m.s. value
3. (i) 10 V peak (ii) 20 V peak

37 STABILIZING CIRCUITS

2. (a) (i) 2 V (ii) 3 V (iii) 3 V
   (b) (i) 0 (ii) 10 mA (iii) 30 mA
3. (a) 100 mA, 10 mA, 90 mA
   (b) 100 mA, 30 mA, 70 mA
   (c) 30 Ω
4. (a) 12.7 V (b) $12.7 - 2 \times 0.85 = 11.0$ V
   (c) (i) $11.0 - 5.0 = 6.0$ V (ii) 3.0 W

39 PROGRESS QUESTIONS

1. (a) (i) 5 V (less p.d. across one diode) (ii) 10 V approx.
   (b) For 50 Hz supply, period $T = 1/50 = 0.02$ s. The time constant CR should be large compared with T for good smoothing. Take $CR = 0.1$ s. Hence $C = 0.1/100$ F $= 0.1 \times 10^6/10^2$ μF $= 1000$ μF If R on open circuit p.d. between P and Q = 5 V peak
2. (i) 8.8 mA (ii) Current through load = 5.6 mA and diode current falls to $(8.8 - 5.6) = 3.2$ mA (iii) If load too small, current through diode falls below value for it to work on breakdown part of its characteristic
3. (a) 60 mA (i) current in Z rises to 80 mA (ii) current through load = 40 mA and current through Z falls from 60 mA to 20 mA Minimum resistance across AB = 4 V/0.06 A = 66.7 Ω since the current through Z = 0
   (b) See p. 86
4. (b) 16 to 1 (c) (ii) $2\sqrt{2} \times 15 \approx 43$ V; voltage rating 50 V
   (d) $\approx 10\ \Omega$ (e) 10 000 μF; $\sqrt{2} \times 15 \approx 21$ V; voltage rating 25 V

40 MULTIMETERS

2. $2/3 \times 10$ V $= 6.7$ V; 5 V

41 OSCILLOSCOPES

1. (i) $2 \times 10/2 = 10$ V (ii) $0.7 \times 10 = 7$ V
2. 50 Hz

43 PROGRESS QUESTIONS

1. 4.5 kΩ
3. Faulty: D (short-circuits R_2). Not faulty: R_1
5. Sine waveform of varying d.c. with amplitude 0.2 V and frequency 2.5 kHz (period 400 μs)
7. (b) (i) 300 Hz (ii) 100 Hz
9. (a)

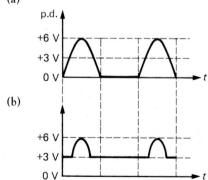

(b)

44 TRANSISTOR VOLTAGE AMPLIFIERS (I)

2. (i) 5 V (ii) 0.02 mA (20 μA) (iii) 420 kΩ

45 TRANSISTOR VOLTAGE AMPLIFIERS (II)

1. (a) $V_{CC} = 10$ V, $R_L = 2$ kΩ

(b) $V_{CE} = 5$ V, $I_C = 2.5$ mA, $I_B = 20$ μA
(c) $(V_{CE} - V_{BE})/I_B = (5 - 0.6)$ V/20 μA $= 4.4$ V/0.02 mA $= 220$ kΩ
2. (a) 9 V/1.8 kΩ = 5 mA
   (b) $I_C = 2.5$ mA, $I_B = 30$ μA, $V_{CE} = 4.5$ V
   (c) Power $= I_C \times V_{CE} = 2.5 \times 4.5 = 11.3$ mW
   (d) (i) 2.5 to 6.5 V $= 4.5$ V $\pm$ 2.0 V (ii) ± 2.0 V peak
   (e) ± 20 mV
   (f) 2.0 V/20 mV = 100
   (g) $R_B = \dfrac{V_{CE} - V_{BE}}{I_B} = \dfrac{(4.5 - 0.6)\ \text{V}}{30\ \mu\text{A}} = 130$ kΩ

48 AMPLIFIERS AND FEEDBACK

3. 40
4. (a) 1/100 (b) 1/10

49 AMPLIFIERS AND MATCHING

2. (a) 80 mV (b) 4 V (c) 2 V (d) 1 W

50 IMPEDANCE-MATCHING CIRCUITS

2. 5/2005 = 1/401; step-down, turns ratio of $\sqrt{2000/5} = \sqrt{400/1} = 20/1$

51 TRANSISTOR OSCILLATOR

1. 7.12 MHz, 2.25 MHz
2. 500 pF, 125 pF

52 PROGRESS QUESTIONS

1. (a) n-p-n (b) A (+) and D (−) (c) E and F
2. (a) 2 kΩ (b) 4.5 mW (c) 4.5 mW; decreased
3. (i) 0.4 V (ii) 4.6 V (iii) 0 V
6. 9 (if the open-loop gain is much greater than the closed-loop gain)
8. 150 mV; 5.6 mW
9. (a) 2.5 V (b) 5/8 mW (c) 2 kΩ
11. (a) 3 V (b) 2.4 V (c) 4.8 mA
12. 0.01 μF

53 OPERATIONAL AMPLIFIER

2. $\pm 15/10^5$ V $= \pm 150$ μV

54 OP AMP VOLTAGE AMPLIFIERS

1. (a) −2 V (b) −9 V (i.e. maximum of supply negative)
2. (a) +3 V (b) +9 V (i.e. maximum of supply positive)

55 OP AMP SUMMING AMPLIFIER

1. (a) −10 V (b) +6 V

56 OP AMP VOLTAGE COMPARATOR

2. (a) +3 V
   (b)

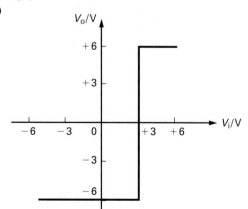

(c)

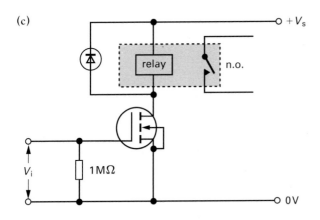

(d) Higher input impedance, larger power gain

57 OP AMP INTEGRATOR

1. (a) fall; -6 V/s   (b) nearly -18 V, after about 3 s

60 PROGRESS QUESTIONS

1. (a) (i) -3 V (ii) -9 V   (b) Resistor $R_i = 2$ kΩ
3. -10 V
8.

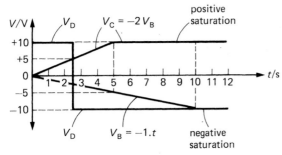

9. (a) $V_1 > V_2$
   (b) (i) resistance large   (ii) $V_2 > V_1$ (iii) goes positive
   (iv) switches on   (v) switches on
   (c) interchange R_1 and LDR
10. (a) $+4$ V   (b) $+12$ V

61 LOGIC GATES

1. (a) AND   (b) OR
2.

A	B	C	D	F
0	0	0	0	0
1	0	0	0	0
0	1	0	0	0
1	1	0	1	1
0	0	1	0	1
1	0	1	0	1
0	1	1	0	1
1	1	1	1	1

3. (i)   (ii)

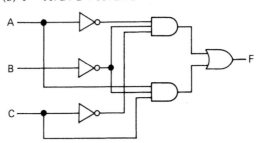

62 LOGIC FAMILIES

1. (a) (i) and (iii)
   (b) (iii) because (i) requires the output of the TTL gate to act as
   a 10 mA source (i.e. provide the 10 mA required to light the LED);
   (iii) uses the TTL gate to *sink* the 10 mA which lights the LED
   and this it can easily do.
   (c) $16/1.6 = 400/40 = 10$
   (d) (i) and (iii)

63 BINARY ADDERS

1. (a) 100, 1101, 10101, 100110, 1000000
   (b) 7, 25, 42, 50
   (c) 1001, 0001, 0111, 0010, 1000, 0011 0111 0000, 0110 0100
   0101
   (d) 3, 10, 13, 30, 421   (e) 6, B, F, 1F, 53, 12C
   (f) 113_8, 305_8 (g) 137, 153

64 LOGIC CIRCUIT DESIGN (I)

1. (c)

A	B	P	Q
0	0	1	0
0	1	1	1
1	0	0	0
1	1	1	0

2.

A	B	C	D	E	F	G
0	0	1	1	0	0	1
0	1	1	0	0	1	0
1	0	0	1	1	0	0
1	1	0	0	0	0	1

3. (a) AND gate   (b) OR gate

65 LOGIC CIRCUIT DESIGN (II)

1. (a) $F = \bar{A}.\bar{B}$

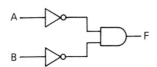

(b) $F = \bar{A}.\bar{B}.\bar{C} + A.\bar{B}.C$

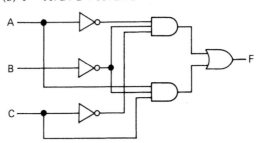

2. (a)

Number	Binary input			Output
	A	B	C	F
0	0	0	0	0
1	0	0	1	1
2	0	1	0	0
3	0	1	1	1
4	1	0	0	0
5	1	0	1	0
6	1	1	0	0
7	1	1	1	0

$$F = \bar{A}.\bar{B}.C + \bar{A}.B.C$$
$$= \bar{A}.C(\bar{B} + B) = \bar{A}.C$$

(b)

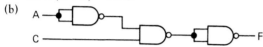

66 LOGIC CIRCUIT DESIGN (III)

1. $F = A.\bar{B}$

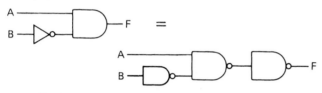

2. $F = A.\bar{B}.\bar{C} + A.B.\bar{C} + \bar{A}.B.C + A.B.C$
$= A.\bar{C}.(\bar{B} + B) + B.C(\bar{A} + A)$
$= A.\bar{C} + B.C$ since $\bar{B} + B = \bar{A} + A = 1$

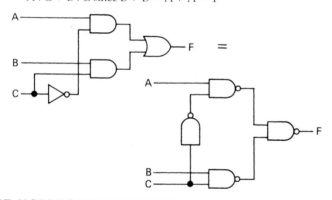

67 MORE BOOLEAN ALGEBRA

2. (i) A (ii) A (iii) 1 (iv) A (v) 1 (vi) A (vii) 0 (viii) 0
(ix) 1 (x) 0 (xi) 1 (xii) 1
4. (a) $A.(A + B) = A.A + A.B = A + A.B$ (since $A.A = A$)
$= A(1 + B) = A$ (since $1 + B = 1$)
(b) $(A + B).(B + C) = A.B + B.B + A.C + B.C$
$= A.B + B + A.C + B.C$ (since $B.B = B$)
$= B.(A + 1) + C.(A + B)$ (since $A + 1 = 1$)
$= B + C.(A + B)$
$= B + A.C + B.C$
$= B(1 + C) + A.C$
$= B + A.C$ (since $1 + C = 1$)
(c) $\overline{A + B + C} = \overline{A + B}.\bar{C}$ (1st theorem)
$= \bar{A}.\bar{B}.\bar{C}$ (1st theorem)

(d) $\bar{A}.\bar{B}.C + \bar{A}.B.C + A.C = \bar{A}.C(\bar{B} + B) + A.C$
$= \bar{A}.C + A.C$
(since $\bar{B} + B = 1$)
$= C(\bar{A} + A)$
$= C$ (since $\bar{A} + A = 1$)
(e) $\overline{(A + B).(B + C)} = \overline{(A + B)} + \overline{(B + C)}$ (1st theorem)
$= \bar{A}.\bar{B} + \bar{B}.\bar{C}$ (1st theorem)
$= \bar{B}.(\bar{A} + \bar{C})$
(f) $\overline{A.B + B.\bar{C}} = \overline{A.B}.\overline{B.\bar{C}}$ (1st theorem)
$= (\bar{A} + \bar{B}).(\bar{B} + \bar{\bar{C}})$ (1st theorem)
$= (\bar{A} + \bar{B}).(\bar{B} + C)$ (since $\bar{\bar{A}} = A$ and $\bar{\bar{C}} = C$)
$= \bar{A}.\bar{B} + \bar{B}.\bar{B} + \bar{A}.C + \bar{B}.C$
$= \bar{A}.\bar{B} + \bar{B} + \bar{A}.C + \bar{B}.C$ (since $\bar{B}.\bar{B} = \bar{B}$)
$= \bar{B}.(\bar{A} + 1) + C.(\bar{A} + \bar{B})$
$= \bar{B} + C.\bar{A} + C.\bar{B}$ (since $A + 1 = 1$)
$= \bar{B}(1 + C) + C.\bar{A}$
$= \bar{B} + \bar{A}.C$ (since $1 + C = 1$)

5. (a) (i) $\overline{\bar{A}.\bar{B}} = A.B$ (i.e. an AND gate)
Q = 1 when A = 1 and B = 1
(ii) $\overline{A + B}$ (i.e. a NOR gate)
Q = 1 when A and B are both 0
(iii) $\overline{A.B + C} = \overline{A.B}.\bar{C} = (\bar{A} + \bar{B}).\bar{C} = \bar{A}.\bar{C} + \bar{B}.\bar{C}$
Q = 1 when A and C = 0 or B and C = 0
(iv) $\bar{A}.\bar{B}.C$
Q = 1 when A = 0, B = 0 and C = 1
(v) $\overline{\overline{\bar{A}.\bar{B}} + \bar{C}} = \overline{\overline{\bar{A}.\bar{B}}}.\bar{\bar{C}} = \bar{A}.\bar{B}.C$
Q = 1 when A = 1, B = 0 and C = 1
(vi) $\bar{A}.B + \bar{C}$
Q = 1 when A = 0 and B = 1 or when C = 0
(b) (iii)

A	B	C	A.B	Q	
0	0	0	0	1	← $\bar{A}.\bar{C} + \bar{B}.\bar{C}$ (A and
0	0	1	0	0	C = 0 or B and
0	1	0	0	1	C = 0) ← $\bar{A}.\bar{C}$ (A and C = 0)
0	1	1	0	0	
1	0	0	0	1	← $\bar{B}.\bar{C}$ (B and C = 0)
1	0	1	0	0	
1	1	0	1	0	
1	1	1	1	0	

(iv)

A	B	C	$\bar{A}.\bar{B}$	Q	
0	0	0	1	0	
0	0	1	1	1	← $\bar{A}.\bar{B}.C$ (A = 0,
0	1	0	0	0	B = 0 and C = 1)
0	1	1	0	0	
1	0	0	0	0	
1	0	1	0	0	
1	1	0	0	0	
1	1	1	0	0	

6. (a) To get an excl-NOR gate from an excl-OR gate the output from the former has to be inverted, hence if $Q = \bar{A}.B + A.\bar{B}$ is for an excl-OR gate, for an excl-NOR gate we can say, inverting the expression, that:

$$Q = \overline{\bar{A}.B + A.\bar{B}} = \overline{\bar{A}.B}.\overline{A.\bar{B}}$$
$$= (\bar{\bar{A}} + \bar{B}).(\bar{A} + \bar{\bar{B}}) = (A + \bar{B}).(\bar{A} + B)$$
$$= A.\bar{A} + \bar{A}.\bar{B} + A.B + \bar{B}.B$$
$$= \bar{A}.\bar{B} + A.B \quad (A.\bar{A} = B.\bar{B} = 0)$$

(b) In effect gates 1 and 2 behave as an NOR gate (OR followed by NOT) to inputs A and B with an output $\overline{A + B}$; the other input to OR gate 4 is from an AND gate with output A . B.
Therefore for OR gate 4 we can say:
$$Q = \overline{(A + B)} + A.B = \bar{A}.\bar{B} + A.B$$
which is the Boolean expression for an excl-NOR gate.

68 PROGRESS QUESTIONS

2. AND

3. (a)

Line	A	B	C	D	E	F	G
1	0	0	0	1	1	0	1
2	0	0	1	1	1	0	1
3	0	1	0	1	1	0	1
4	1	0	0	1	1	0	1
5	0	1	1	1	0	1	0
6	1	1	0	0	1	1	0
7	1	0	1	1	1	0	1
8	1	1	1	0	0	1	0

(b) Logic level 0
(c) (i) +5 V, i.e. 1 (ii) 0 V, i.e. 0
(d) S_1 only closed (line 5 in truth table)
S_3 only closed (line 6)
none closed (line 8)

4. (a) Common anode because anode of LED is connected directly to +9 V
(b) Common cathode because cathode of LED is connected directly to 0 V. When 4026 output goes 'high', the transistor conducts, i.e. has near zero resistance
In **(a)** the transistor is then in effect at 0 V and in **(b)** at +9 V. In both cases the LED becomes forward biased and conducts. The transistor acts as a current 'sink' in **(a)** and as a current 'source' in **(b)**

6.

A	B	C	D	E
0	0	0	0	0
0	1	0	1	1
1	0	1	0	1
1	1	0	0	0

7.

A	B	C	D	E
0	0	1	0	1
0	1	0	0	0
1	0	0	0	0
1	1	0	1	1

9. (b) $F = A.\bar{B}.C + \bar{A}.B.C + \bar{A}.\bar{B}.\bar{C}$

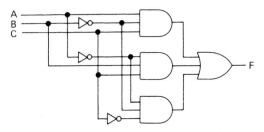

10. (i) $F = A.\bar{B}.C.\bar{D}.\bar{E} + \bar{A}.B.C.\bar{D}.\bar{E}. + \bar{A}.\bar{B}.\bar{C}.D.E$
That is, F is 1 when
A is 1 and B is 0 and C is 1 and D is 0 and E is 0
or
A is 0 and B is 1 and C is 1 and D is 0 and E is 0
or
A is 0 and B is 0 and C is 0 and D is 1 and E is 1
(ii)

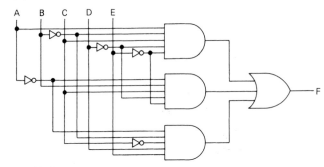

11. $F = \bar{A}.\bar{B}.C$

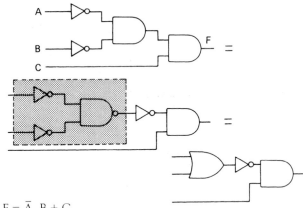

12. $F = \bar{A}.B + C$

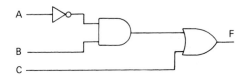

13. **(a)** A, 5.97 V; B, 6.03 V
 (b) $V_A < V_{in} < V_B$, Y high, X low, $V_{out} = 0$ V
 $V_A < V_{in} > V_B$, X and Y high, $V_{out} = 12$ V
 $V_A > V_{in} < V_B$, X and Y low, $V_{out} = 12$ V
 (c) (i) 1010 Ω (ii) 990 Ω
 (d) (i) 24 °C to 28 °C
 (e) See answer to question 2, Chapter 56

69 BISTABLES (I)

2. (i) 1 (ii) 0
3. **(b)**

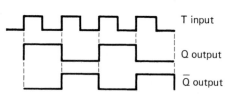

70 BISTABLES (II)

1.

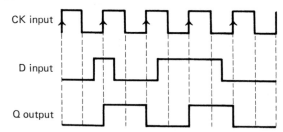

2. 1 Hz
3. **(a)**

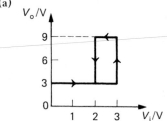

(b)

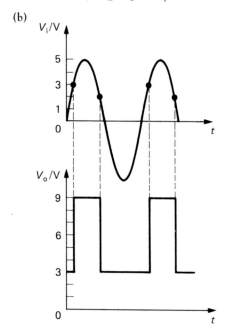

4.

D	CK	S	R	Q
0	1	1	0	0
0	0	1	1	0
1	1	0	1	1
1	0	1	1	1

Q follows D when CK is a 1, therefore it is suitable

71 ASTABLES AND MONOSTABLES

1. (i)

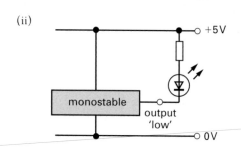

(ii)

2. **(a)** 0.2 s **(b)** 1/0.2 = 5 Hz **(c)** 0.1/0.1 = 1
3. **(a)** Flashing on and off once per second (on for 0.5 s, off for 0.5 s)
 (b) Flashes twice per second
 (c) Flashes once every 2 seconds

72 BINARY COUNTERS

1. **(a)** (i) $2^1 = 2$ (ii) $2^2 = 4$ (iii) $2^3 = 8$ (iv) $2^4 = 16$
 (v) $2^5 = 32$
 (b) 1, 3, 7, 15, 31
2. **(b)** (i) 1 (ii) 3 (iii) 3 (iv) 4 (v) 5
 (c) $f/16$

73 REGISTERS AND MEMORIES

1. **(b)** 8
2. **(c)** (i) 512 (ii) 64

75 PROGRESS QUESTIONS

2. **(a)** (i)

Pulse number	LED A	LED B	LED C
1	1	0	0
2	0	1	0
3	1	1	0
4	0	0	1
5	1	0	1
6	0	1	1
7	1	1	1

(b) (i) Tr A (ii) S_A (iii) rises to about 0.6 V for a silicon transistor (iv) lights (v) close S_B

4. **(b)**

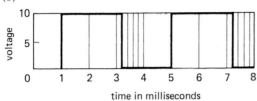

Duration of output pulse = 2.2 ms

6. (i) rises (ii) falls

7. **(a)**

R_1 (kΩ)	R_2 (kΩ)	t_1 (s)	t_2 (s)	T (s)	t_1/t_2
0	100	0.7	0.7	1.4	1
50	100	1.05	0.7	1.75	1.5/1
100	100	1.4	0.7	2.1	2/1

(b) Period of waves
(c) Duty cycle or mark–space ratio
(d) $t_1/t_2 = 1$
(e)

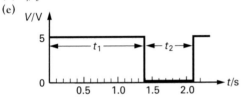

10. **(a)** (i) 35.7 ms (ii) 0.35 ms (iii) ≈ 28 Hz
    (b) Square wave, frequency halved

77 COMMUNICATION SYSTEMS

1. 7000 Hz
2. Lower sideband 790–799 kHz: upper sideband 801–810 kHz

80 AUDIO AMPLIFIERS

3. 1.3 W
4. **(b)** 7 dB

82 PROGRESS QUESTIONS

4. At the two frequencies (i) the power is half and (ii) the voltage is 0.7 of its maximum value at frequencies mid-way between 10 Hz and 50 kHz
5. **(a)** 30 dB **(b)** 40 dB
6. **(c)** (i) 3.2 V (ii) 160 (iii) 300 Hz to 3 kHz
7. (ii) C–D : 0.9955 MHz to 0.99995 MHz; E–F : 1.00005 MHz to 1.0045 MHz
   (iii) 9 kHz

88 PROGRESS QUESTIONS

2. **(c)** 1.4 m
5. Taking f = 50 kHz gives:
   $T = 1/(50 \times 10^3)$ s $= CR = C \times 10^3$
   $\therefore C = \frac{1}{50}$ μF = 20 nF
6. **(c)** 40 MHz or 20 MHz
8. **(c)** (i) $312.5 \times 50 = 15\,625$ Hz (ii) 50 Hz
9. **(d)** 23 pF

92 PROGRESS QUESTIONS

3. $25 \times 8 \times 350\,000 = 70\,000\,000 = 70$ M bit/s
5. **(a)** 1200 baud, since there must be at least one complete time cycle in the tone and the low tone requires more line 'space' than the high tone
   (b) 4 Kbytes = 4096 bytes each of 11 bits gives 4096×11 bits and at a bit-rate of 1200 baud, time = $4096 \times 11/1200 \approx 37.5$ s

96 PROGRAMS AND FLOWCHARTS

1. **(a)**   **(b)**

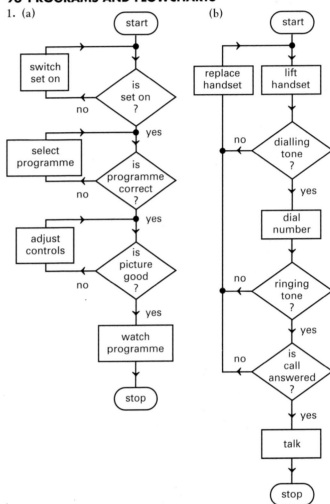

2. Similar to Fig. 96.2

98 PROGRESS QUESTIONS

5. (a)

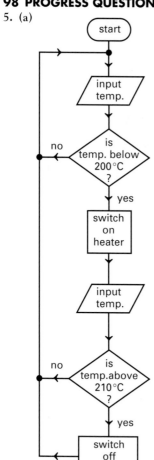

(b)

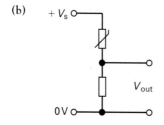

(c) A/D converter

7. (a) (i) address and control (ii) data
   (b) CPU
   (c) RAM is a 'read' and 'write' memory, ROM only 'reads'
   (d) When data has to be changed occasionally; in small production runs; when cost of ROM is not justified
   (e) Through use of tri-state gates which can be disabled so that they can neither send nor receive signals on the bus and are in a high-impedance state
   (f) Parallel port has 8 data lines so that a byte at a time is output; handshaking signals ensure the peripheral knows when data is ready at the port

8. 640 million

INDEX